Linear Algebra

Textbooks in Mathematics

Series editors: Al Boggess and Ken Rosen

Mathematical Modeling with Excel
Brian Albright and William P. Fox

Chromatic Graph Theory, Second Edition
Gary Chartrand and Ping Zhang

Partial Differential Equations: Analytical Methods and Applications
Victor Henner, Tatyana Belozerova, and Alexander Nepomnyashchy

Ordinary Differential Equations: An Introduction to the Fundamentals
Kenneth B. Howell

Algebra: Groups, Rings, and Fields
Louis Rowen

Differential Geometry of Manifolds, Second Edition
Stephen T. Lovett

The Shape of Space, Third Edition
Jeffrey R. Weeks

Differential Equations: A Modern Approach With Wavelets
Steven G. Krantz

Advanced Calculus: Theory And Practice
John Srdjan Petrovic

Advanced Problem Solving Using Maple: Applied Mathematics, Operation Research, Business Analytics, And Decision Analysis
William P. Fox and William C. Bauldry

Nonlinear Optimization: Models And Applications
William Fox

https://www.routledge.com/Textbooks-in-Mathematics/book-series/CANDHTEX-BOOMTH

Linear Algebra

James R. Kirkwood and Bessie H. Kirkwood

CRC Press
Taylor & Francis Group
Boca Raton London New York

CRC Press is an imprint of the
Taylor & Francis Group, an **informa** business

A CHAPMAN & HALL BOOK

First edition published 2021
by CRC Press
6000 Broken Sound Parkway NW, Suite 300, Boca Raton, FL 33487-2742
and by CRC Press
2 Park Square, Milton Park, Abingdon, Oxon, OX14 4RN

ISBN: 978-1-4987-7685-1 (hbk)
ISBN: 978-1-315-15224-0 (ebk)

Typeset in Garamond
by Deanta Global Publishing Services, Chennai, India

Contents

Preface

A linear algebra course has two major audiences that it must satisfy. It provides an important theoretical and computational tool for nearly every discipline that uses mathematics. This includes (but is definitely not limited to) statistics, economics, environmental science, engineering, and computer science. It also provides an introduction to abstract mathematics. That is, it is a transition course. Definitions assume the importance that formulas hold in earlier mathematics. Understanding theorems and the ideas of their proofs is required at a quantum level above courses that are primarily computational.

The study is concerned with vector spaces and a certain class of functions between vector spaces called linear transformations. Matrices – depending on how they are used – are an example of vector spaces and also the main way in which linear transformations are expressed. Accordingly, our study will begin with a study of the algebra of matrices. Elements of this study will be repeated in a more general setting, if one takes an abstract algebra course.

This book has two parts. Chapters 1–7 are written in a style for someone for whom this is an introduction to linear algebra. This part of the book was written with the idea that a typical reader would be one who has completed two semesters of calculus but who has not taken courses that emphasize abstract mathematics. We want to ensure that the computational methods become deeply ingrained and to make the intuition of the theory as transparent as possible.

In some cases, the intuition of a proof is more transparent if an example is presented before the theorem is articulated or if the proof of a theorem is given using concrete cases rather than an abstract argument–for example, by using three vectors instead of n vectors.

Two primary goals of these chapters are to enable students to become adept at computations and to develop an understanding of the theory of basic topics including linear transformations, the role of bases in the representation of linear transformation, and the importance of eigenvectors, to name a few. A great deal of effort is taken in explaining examples. The reason is that examples demonstrate computational methods and, if chosen appropriately, can make the ideas of a proof intuitive.

There are places in Chapters 4–7 where there are results that are important because of their applications, but the theory behind the result is time-consuming and is more advanced than a typical student in a first exposure would be expected to digest. In such cases, the reader is alerted that the result is proven later in the section and omitting the derivation will not compromise the usefulness of the results. Two specific examples of this are the projection matrix and the Gram–Schmidt process.

The exercises span a range from simple computations to fairly direct abstract exercises.

A note about computations. While we expect that many users will want to make use of technology for many of the computations, topics such as matrix multiplication, Gaussian elimination, computation of the inverse and determinant of a matrix are explained in the text in a way that will enable students to do these computations by hand if that is desired.

For users who want to make extensive use of a computer algebra system for computations, there are several systems available. MATLAB is the choice of many, and we have included a tutorial for MATLAB in the Appendix. Because of the extensive use of the program R by statisticians, a tutorial for that program is also included.

Part two, which consists of Chapters 8–14, is at a higher level. It includes topics not usually taught in a first course, such as a detailed justification of the Jordan canonical form, properties of the determinant derived from axioms, the Perron–Frobenius theorem, and bilinear and quadratic forms. These chapters are written in a style that is accessible to someone who has had a first course in linear algebra and are a source for someone with an interest in some of the more important upper-level topics.

Historical notes are included throughout the text. They give a time frame for the discovery of the important ideas of linear algebra and points of interest in the lives of the pioneers.

A Note about Mathematical Proofs

As a text for a first course in linear algebra, this book has a major focus on demonstrating facts and techniques of linear systems that will be invaluable in higher mathematics and fields that use higher mathematics. This entails developing many computational tools. A first course in linear algebra also serves as a bridge to mathematics courses that are primarily theoretical in nature and, as such, necessitates understanding and in some cases developing mathematical proofs. This is a learning process that can be frustrating.

We offer some suggestions that may ease the transition.

First, the most important tool for abstract mathematics is knowing the important definitions well. These are analogous to the formulas of computational mathematics. A casual perusal will not be sufficient.

You will often encounter the question "Is it true that …?" Here you must show that a result holds in all cases allowed by the hypotheses or find one example where the claim does not hold. Such an example is called a counterexample. If you are going to prove that a result always holds, it is helpful (perhaps necessary) to believe that it holds. One way to establish an opinion for its validity is to work through some examples. This may highlight the essential properties that are crucial to a proof. If the result is not valid, this will often yield a counter example. Simple examples are usually the best. In linear algebra this usually means working with small dimensional objects, such as 2×2 matrices.

In the text, we will sometimes give the ideas of a proof rather than a complete proof when we feel that "what is really going on" is obscured by the abstraction; for example, if the idea of a proof seems clear by examining 2×2 matrices but would be difficult to follow for $n \times n$ matrices.

In constructing a mathematical proof, there are two processes involved. One is to develop the ideas of the proof, and the second is to present those ideas in appropriate language.

There are different methods of proof that can be broadly divided into direct methods and indirect methods. An example of an indirect method is proof by contradiction. The contrapositive of "if statement A is true, then statement B is true" is "if statement B is false, then statement A is false". Proving the contrapositive of an "if-then" statement is logically equivalent to proving the original statement. The contrapositive of

If $x = 4$, then x is an integer

is

If x is not an integer, then $x \neq 4$.

The converse of an "if-then" statement reverses the hypothesis and the conclusion. The converse of "if statement A is true, then statement B is true" is "if statement B is true, then statement A is true".

Unlike the contrapositive, the converse is not logically equivalent to the original statement. The converse of

If $x = 4$, then x is an integer

is

If x is an integer, then $x = 4$.

Here the original statement is true, but the converse is false.

You will sometimes have a problem of the type "show 'A if and only if B'". In this case you must prove

If A then B

and

If B then A

are both true.

In constructing a proof it is necessary to completely understand what is given and what is to be shown. Sometimes, what is given can be phrased in different ways, and one way may be more helpful than others. Often in trying to construct a proof it is worthwhile writing explicitly "this is what I know" and "this is what I want to show" to get started.

In the proofs presented in the text we make a concerted effort to present the intuition behind the proof and why something is "the reasonable thing to do".

MATLAB® is a registered trademark of The MathWorks, Inc. For product information, please contact:

The MathWorks, Inc.
3 Apple Hill Drive
Natick, MA 01760-2098 USA
Tel: 508 647 7000
Fax: 508-647-7001
E-mail: info@mathworks.com
Web: www.mathworks.com

Chapter 1

Matrices

Linear algebra is the branch of mathematics that deals with vector spaces and linear transformations. For someone just beginning their study of linear algebra, that is probably a meaningless statement. It does, however, convey the idea that there are two concepts in our study that will be of utmost importance; namely, vector spaces and linear transformations. A primary tool in our study of these topics will be matrices. In this chapter we give the rules that govern matrix algebra.

Section 1.1 Matrix Arithmetic

The term "matrix" was first used in a mathematical sense by James Sylvester in 1850 to denote a rectangular array of numbers from which determinants could be formed. The idea of using arrays to solve systems of linear equations goes as far back as the Chinese circa 300 BC.

A matrix is a rectangular array of numbers. Matrices are the most common way of representing vectors and linear transformations and play a central role in nearly every computation in linear algebra. The size of a matrix is described by its dimensions; i.e., the number of rows and columns in the matrix, with the number of rows given first. A 3×2 (three by two) matrix has three rows and two columns. Matrices are usually represented by upper-case letters.

The following matrix A is an example of a 3×2 matrix.

$$A = \begin{pmatrix} 0 & -5 \\ 6 & 3 \\ -1 & 3 \end{pmatrix}.$$

We will often want to represent a matrix in an abstract form. The following example is typical.

$$A = \begin{pmatrix} a_{11} & a_{12} & \cdots & a_{1n} \\ a_{21} & a_{22} & \cdots & a_{2n} \\ \vdots & \vdots & \cdots & \vdots \\ a_{m1} & a_{m2} & \cdots & a_{mn} \end{pmatrix}$$

2 ■ *Linear Algebra*

It is customary to denote the numbers that make up the matrix, called the entries of the matrix, as the lower-case version of the letter that names the matrix, with two subscripts. The subscripts denote the position of the entry. The entry a_{ij} occupies the ith row and jth column of the matrix A. The notation $(A)_{ij}$ is also common, depending on the setting.

Two matrices A and B are equal if they have the same dimensions and $a_{ij} = b_{ij}$ for every i and j.

Definition:

A square matrix is a matrix that has the same number of rows as columns; that is, an $n \times n$ matrix. If A is a square matrix, the entries $a_{11}, a_{22}, \ldots, a_{nn}$ make up the main diagonal of A. The trace of a square matrix is the sum of the entries on the main diagonal.

A square matrix is a diagonal matrix if the only non-zero entries of A are on the main diagonal. A square matrix is upper (lower) triangular if the only non-zero entries are above (below) or on the main diagonal. A matrix that consists of either a single row or a single column is called a vector. The matrix

$$\begin{pmatrix} 1 \\ 5 \\ 6 \end{pmatrix}$$

is an example of a column vector, and

$$\left(4, -1, 0, 8\right)$$

is an example of a row vector.

In Figure 1.1, matrix (a) is a diagonal matrix, but the others are not.

Matrix Arithmetic

We now define three arithmetic operations on matrices: matrix addition, scalar multiplication, and matrix multiplication.

Matrix Addition

In order to add two matrices, the matrices must have the same dimensions, and then one simply adds the corresponding entries. For example

$$\begin{pmatrix} 2 & 0 & 0 & 0 \\ 0 & 0 & 0 & 0 \\ 0 & 0 & -1 & 0 \\ 0 & 0 & 0 & 5 \end{pmatrix} \qquad \begin{pmatrix} 0 & 0 & 0 & 0 \\ 0 & 0 & 5 & 0 \\ 0 & 3 & 0 & 0 \\ 2 & 0 & 0 & 0 \end{pmatrix}$$

(a)   (b)

$$\begin{pmatrix} 0 & 1 & 0 & 0 \\ 0 & 3 & 0 & 0 \\ 0 & 0 & 7 & 0 \\ 0 & 0 & 0 & 2 \end{pmatrix} \qquad \begin{pmatrix} 1 & 0 & 0 \\ 0 & 2 & 0 \end{pmatrix}$$

(c)   (d)

Figure 1.1 Only matrix (a) is a diagonal matrix.

$$\begin{pmatrix} 7 & -1 & 0 \\ -2 & 6 & 6 \end{pmatrix} + \begin{pmatrix} -2 & -3 & 4 \\ 3 & 5 & 0 \end{pmatrix} = \begin{pmatrix} 7-2 & -1-3 & 0+4 \\ -2+3 & 6+5 & 6+0 \end{pmatrix} = \begin{pmatrix} 5 & -4 & 4 \\ 1 & 11 & 6 \end{pmatrix}$$

but

$$\begin{pmatrix} 7 & -1 & 0 \\ -2 & 6 & 6 \end{pmatrix} + \begin{pmatrix} 1 & 0 \\ 2 & -1 \\ 0 & 5 \end{pmatrix}$$

is not defined.

To be more formal – and to begin to get used to the abstract notation – we could express this idea as

$$\left(A+B\right)_{ij} = A_{ij} + B_{ij}.$$

Scalar Multiplication

A scalar is a number. Scalar multiplication means multiplying a matrix by a number and is accomplished by multiplying every entry in the matrix by the scalar. For example

$$3\begin{pmatrix} 1 & -4 \\ 2 & 0 \end{pmatrix} = \begin{pmatrix} 3 & -12 \\ 6 & 0 \end{pmatrix}.$$

Matrix Multiplication

Matrix multiplication is not as intuitive as the two prior operations. A fact that we will demonstrate later is that every linear transformation can be expressed as multiplication by a matrix. The definition of matrix multiplication is based in part on the idea that composition of two linear transformations should be expressed as the product of the matrices that represent the linear transformations.

We describe matrix multiplication by stages.

Case 1

Multiplying a matrix consisting of one row by a matrix consisting of one column, where the row matrix is on the left and the column matrix is on the right. The process is

$$\left(a_1, a_2, \ldots, a_n\right)\begin{pmatrix} b_1 \\ b_2 \\ \vdots \\ b_n \end{pmatrix} = a_1 b_1 + a_2 b_2 + \cdots + a_n b_n.$$

This can be accomplished only if there are the same number of entries in the row matrix as in the column matrix. Notice that the product of a $(1 \times n)$ matrix and a $(n \times 1)$ matrix is a number that, in this case, we will describe as a (1×1) matrix.

Case 2

Multiplying a matrix consisting of one row by a matrix consisting of more than one column, where the row matrix is on the left. Two examples are

$$\left(a_1,\ldots,a_n\right)\begin{pmatrix} b_1 & c_1 \\ \vdots & \vdots \\ b_n & c_n \end{pmatrix} = \left(a_1b_1 + \cdots + a_nb_n, \quad a_1c_1 + \cdots + a_nc_n\right)$$

$$\left(a_1,\ldots,a_n\right)\begin{pmatrix} b_1 & c_1 & d_1 \\ \vdots & \vdots & \vdots \\ b_n & c_n & d_n \end{pmatrix}$$
$$= \left(a_1b_1 + \cdots + a_nb_n, \quad a_1c_1 + \cdots + a_nc_n, \quad a_1d_1 + \cdots + a_nd_n\right).$$

Thus, the product of a $(1 \times n)$ matrix and a $(n \times k)$ matrix is a $(1 \times k)$ matrix.

It may help to visualize this as applying Case 1 multiple times.

Case 3

Multiplying a matrix consisting of more than one row by a matrix consisting of one column, where the column matrix is on the right. Two examples of this are

$$\left(2 \times n\right) \times \left(n \times 1\right) \quad \begin{pmatrix} a_1,\ldots,a_n \\ b_1,\ldots,b_n \end{pmatrix}\begin{pmatrix} c_1 \\ \vdots \\ c_n \end{pmatrix} = \begin{pmatrix} a_1c_1 + \cdots + a_nc_n \\ b_1c_1 + \cdots + b_nc_n \end{pmatrix} \quad 2 \times 1$$

$$\left(3 \times n\right) \times \left(n \times 1\right) \quad \begin{pmatrix} a_1,\ldots,a_n \\ b_1,\ldots,b_n \\ c_1,\ldots,c_n \end{pmatrix}\begin{pmatrix} d_1 \\ \vdots \\ d_n \end{pmatrix} = \begin{pmatrix} a_1d_1 + \cdots + a_nd_n \\ b_1d_1 + \cdots + b_nd_n \\ c_1d_1 + \cdots + c_nd_n \end{pmatrix} \quad 3 \times 1$$

Case 4

Multiplying any two "compatible" matrices. In order to be compatible for multiplication, the matrix on the left must have the same number of entries in a row as there are entries in a column of the matrix on the right. One example is

$$\left(2 \times 3\right) \times \left(3 \times 2\right) \quad \begin{pmatrix} a_1 & a_2 & a_3 \\ b_1 & b_2 & b_3 \end{pmatrix}\begin{pmatrix} c_1 & c_2 \\ d_1 & d_2 \\ e_1 & e_2 \end{pmatrix}$$
$$= \begin{pmatrix} a_1c_1 + a_2d_1 + a_3e_1 & a_1c_2 + a_2d_2 + a_3e_2 \\ b_1c_1 + b_2d_1 + b_3e_1 & b_1c_2 + b_2d_2 + b_3e_2 \end{pmatrix}$$

which is a (2×2) matrix.

We describe a formula for computing the product of two compatible matrices.

Suppose that A is an $m \times k$ matrix and B is a $k \times n$ matrix, say

$$A = \begin{pmatrix} a_{11} & a_{12} & \cdots & a_{1k} \\ a_{21} & a_{22} & \cdots & a_{2k} \\ \vdots & \vdots & \vdots & \vdots \\ a_{m1} & a_{m2} & \cdots & a_{mk} \end{pmatrix} \quad B = \begin{pmatrix} b_{11} & b_{12} & \cdots & b_{1n} \\ b_{21} & b_{22} & \cdots & b_{2n} \\ \vdots & \vdots & \vdots & \vdots \\ b_{k1} & b_{k2} & \cdots & b_{kn} \end{pmatrix}.$$

The product matrix AB has dimensions $m \times n$ and the formula for the i, j entry of the matrix AB is

$$\left(AB \right)_{ij} = \sum_{l=1}^{k} a_{il} b_{lj}.$$

One can think of this as multiplying the ith row of A by the jth column of B.

A summary of matrix multiplication is:

If A is an $m \times k$ matrix and B is an $l \times n$ matrix, then AB is defined if and only if $k = l$. If $k = l$ then AB is an $m \times n$ matrix whose i, j entry is

$$\left(AB \right)_{ij} = \sum_{s=1}^{k} a_{is} b_{sj}.$$

Another useful way to visualize matrix multiplication is to consider the product AB in the following way: Let $\hat{b}_i$ be the vector that is the ith column of the matrix B, so that if

$$B = \begin{pmatrix} b_{11} & b_{12} & b_{13} \\ b_{21} & b_{22} & b_{23} \\ b_{31} & b_{32} & b_{33} \end{pmatrix}$$

then

$$\hat{b}_1 = \begin{pmatrix} b_{11} \\ b_{21} \\ b_{31} \end{pmatrix}, \hat{b}_2 = \begin{pmatrix} b_{12} \\ b_{22} \\ b_{32} \end{pmatrix}, \hat{b}_3 = \begin{pmatrix} b_{13} \\ b_{23} \\ b_{33} \end{pmatrix}.$$

We can then think of the matrix B as a "vector of vectors"

$$B = \begin{bmatrix} \hat{b}_1 & \hat{b}_2 & \hat{b}_3 \end{bmatrix}$$

and we have

$$AB = \begin{bmatrix} A\hat{b}_1 & A\hat{b}_2 & A\hat{b}_3 \end{bmatrix}.$$

This will often be used in the text.

We verify this in the case

$$A = \begin{pmatrix} a_{11} & a_{12} & a_{13} \\ a_{21} & a_{22} & a_{23} \end{pmatrix}.$$

We have

$$A\hat{b}_1 = \begin{pmatrix} a_{11} & a_{12} & a_{13} \\ a_{21} & a_{22} & a_{23} \end{pmatrix} \begin{pmatrix} b_{11} \\ b_{21} \\ b_{31} \end{pmatrix} = \begin{pmatrix} a_{11}b_{11} + a_{12}b_{21} + a_{13}b_{31} \\ a_{21}b_{11} + a_{22}b_{21} + a_{23}b_{31} \end{pmatrix}$$

$$A\hat{b}_2 = \begin{pmatrix} a_{11} & a_{12} & a_{13} \\ a_{21} & a_{22} & a_{23} \end{pmatrix} \begin{pmatrix} b_{12} \\ b_{22} \\ b_{32} \end{pmatrix} = \begin{pmatrix} a_{11}b_{12} + a_{12}b_{22} + a_{13}b_{32} \\ a_{21}b_{12} + a_{22}b_{22} + a_{23}b_{32} \end{pmatrix}$$

$$A\hat{b}_3 = \begin{pmatrix} a_{11} & a_{12} & a_{13} \\ a_{21} & a_{22} & a_{23} \end{pmatrix} \begin{pmatrix} b_{13} \\ b_{23} \\ b_{33} \end{pmatrix} = \begin{pmatrix} a_{11}b_{13} + a_{12}b_{23} + a_{13}b_{33} \\ a_{21}b_{13} + a_{22}b_{23} + a_{23}b_{33} \end{pmatrix}$$

so

$$\begin{bmatrix} A\hat{b}_1 & A\hat{b}_2 & A\hat{b}_3 \end{bmatrix}$$
$$= \begin{pmatrix} a_{11}b_{11} + a_{12}b_{21} + a_{13}b_{31} & a_{11}b_{12} + a_{12}b_{22} + a_{13}b_{32} & a_{11}b_{13} + a_{12}b_{23} + a_{13}b_{33} \\ a_{21}b_{11} + a_{22}b_{21} + a_{23}b_{31} & a_{21}b_{12} + a_{22}b_{22} + a_{23}b_{32} & a_{21}b_{13} + a_{22}b_{23} + a_{23}b_{33} \end{pmatrix}.$$

Also,

$$AB = \begin{pmatrix} a_{11} & a_{12} & a_{13} \\ a_{21} & a_{22} & a_{23} \end{pmatrix} \begin{pmatrix} b_{11} & b_{12} & b_{13} \\ b_{21} & b_{22} & b_{23} \\ b_{31} & b_{32} & b_{33} \end{pmatrix}$$
$$= \begin{pmatrix} a_{11}b_{11} + a_{12}b_{21} + a_{13}b_{31} & a_{11}b_{12} + a_{12}b_{22} + a_{13}b_{32} & a_{11}b_{13} + a_{12}b_{23} + a_{13}b_{33} \\ a_{21}b_{11} + a_{22}b_{21} + a_{23}b_{31} & a_{21}b_{12} + a_{22}b_{22} + a_{23}b_{32} & a_{21}b_{13} + a_{22}b_{23} + a_{23}b_{33} \end{pmatrix}.$$

As a final cautionary remark, we note that if A is an $m \times k$ matrix and B is a $k \times n$ matrix, then AB is defined and is an $m \times n$ matrix, but BA is not defined unless $m = n = k$.

Historical Note

When tracing the beginning of an idea in mathematics, it is often difficult and controversial to say what came first. The beginning of matrices and determinants arose through the study of systems of linear equations. The Babylonians studied problems that lead to simultaneous linear equations and some of these are preserved on surviving clay tablets from around four thousand years ago. The Chinese, between 200 BC and 100 BC, also studied such problems and came substantially closer to matrices than the Babylonians. The text *Nine Chapters on the Mathematical Art* written during the Han Dynasty gives the first known example of matrix methods. In that work, the Chinese used a method similar to Gaussian elimination to solve systems of linear equations.

Exercises

1. Suppose

$$A = \begin{pmatrix} 2 & 0 \\ 3 & 1 \end{pmatrix}, B = \begin{pmatrix} -1 & 2 \\ -3 & 2 \\ 1 & 4 \end{pmatrix}, C = \begin{pmatrix} 1 & 5 & 2 \\ -3 & 0 & 6 \end{pmatrix}, D = \begin{pmatrix} 4 & -2 & 2 \\ 0 & 1 & 5 \\ 2 & 0 & -3 \end{pmatrix}, E = \begin{pmatrix} 1 & 7 \\ 4 & -3 \end{pmatrix}.$$

Where the computations are possible, find
   (a.) $3A - 2E$
   (b.) BC
   (c.) $CB + 4D$
   (d.) $C - 4B$
   (e.) BA
   (f.) EC

2. Compute $A\hat{b}_1, A\hat{b}_2, \left[A\hat{b}_1, A\hat{b}_2 \right]$ and AB for

$$A = \begin{pmatrix} 2 & -1 \\ 1 & 3 \end{pmatrix}, \quad B = \begin{pmatrix} 1 & 0 & -1 \\ 2 & 1 & 4 \end{pmatrix}.$$

3. Find a 3×3 matrix A for which
   (a.) $A\begin{pmatrix} x \\ y \\ z \end{pmatrix} = \begin{pmatrix} 4x - 2y \\ 3z \\ 0 \end{pmatrix}$

   (b.) $A\begin{pmatrix} x \\ y \\ z \end{pmatrix} = \begin{pmatrix} z \\ y \\ x \end{pmatrix}$

   (c.) $A\begin{pmatrix} x \\ y \\ z \end{pmatrix} = \begin{pmatrix} x + y + z \\ 2x + 2y + 2z \\ -x - y - z \end{pmatrix}$

4. Show that
   (a.) $\begin{pmatrix} a_1 & b_1 \\ a_2 & b_2 \end{pmatrix}\begin{pmatrix} x \\ y \end{pmatrix} = x\begin{pmatrix} a_1 \\ a_2 \end{pmatrix} + y\begin{pmatrix} b_1 \\ b_2 \end{pmatrix}$

   (b.) $\begin{pmatrix} a_1 & b_1 & c_1 \\ a_2 & b_2 & c_2 \\ a_3 & b_3 & c_3 \end{pmatrix}\begin{pmatrix} x \\ y \\ z \end{pmatrix} = x\begin{pmatrix} a_1 \\ a_2 \\ a_3 \end{pmatrix} + y\begin{pmatrix} b_1 \\ b_2 \\ b_3 \end{pmatrix} + z\begin{pmatrix} c_1 \\ c_2 \\ c_3 \end{pmatrix}$

5. If AB is a 5×7 matrix, how many columns does B have?

6. If A and B are 2×2 matrices, show that

$$(AB)_{11} + (AB)_{22} = (BA)_{11} + (BA)_{22}.$$

7. If A and B are matrices so that AB is defined, then jth column of $AB = A\left[j\text{th column of } B \right]$ and

$$i\text{th row of } AB = \left[i\text{th row of } A \right]B.$$

Demonstrate this principle by finding the second row and third column of AB for

$$A = \begin{pmatrix} 3 & -1 & 0 \\ 1 & 5 & -2 \\ 4 & 3 & 1 \end{pmatrix}, \quad B = \begin{pmatrix} 2 & 5 & -3 \\ -1 & 0 & 4 \\ 4 & 2 & -5 \end{pmatrix}.$$

8. Construct a 5×5 matrix, not all of whose entries are zero, that satisfies the given property (three different matrices).

   (a.) $A_{ij} = 0$ if $i < j$

   (b.) $A_{ij} = 0$ if $i \geq j$

   (c.) $A_{ij} = 0$ if $|i - j| = 1$.

9. Compute

$$(a,b)\begin{pmatrix} e & f \\ g & h \end{pmatrix} \text{ and } (c,d)\begin{pmatrix} e & f \\ g & h \end{pmatrix}$$

   and compare with

$$\begin{pmatrix} a & b \\ c & d \end{pmatrix}\begin{pmatrix} e & f \\ g & h \end{pmatrix}.$$

   Can you make any conjectures based on this example?

10. If AB is a 3×3 matrix, show that BA is defined.

11. If

$$A = \begin{pmatrix} 3 & 8 & 0 \\ 4 & -1 & -2 \end{pmatrix}, \quad B = \begin{pmatrix} 2 & 1 & 1 \\ 3 & -1 & 7 \end{pmatrix}$$

    find the matrix C for which

$$3A - 5B + 2C = \begin{pmatrix} 0 & 1 & 2 \\ 2 & -3 & 4 \end{pmatrix}.$$

12. A graph is a set of vertexes and a set of edges between some of the vertices. If the graph is simple, then there is no edge from a vertex to itself. If the graph is not directed, then an edge from vertex i to vertex j is also an edge from j to vertex i. Figure 1.2 shows a simple directed graph.

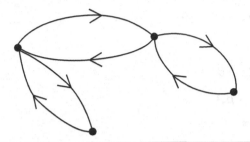

Figure 1.2   A simple directed graph.

Associated with each graph is a matrix A defined by

$$A_{ij} = \begin{cases} 1 & \text{if there is an edge from vertex } i \text{ to vertex } j \\ 0 & \text{otherwise} \end{cases}.$$

(a.) If there are n vertices, what are the dimensions of the associated matrix A? What else can you conclude about A?

We say there is a path of length k from vertex i to vertex j if it is possible to traverse from vertex i to vertex j by traversing across exactly k edges. (Traversing the same edge more than once is allowed.)

(b.) How could the matrix A be used to determine whether there is a path from i to vertex j in exactly 2 steps? In exactly k steps?

(c.) How could you tell how many paths there are from i to vertex j in exactly k steps?

(d.) For the graph in Figure 1.2, number the vertices from 1 to 4
   (i)   Construct the associated matrix.
   (ii) Use the associated matrix to determine whether it is possible to go from vertex 1 to vertex 4 by traversing across exactly 3 edges using the matrix you constructed.

Section 1.2 The Algebra of Matrices

Properties of Matrix Addition, Scalar Multiplication, and Matrix Multiplication

Matrix addition, scalar multiplication, and matrix multiplication obey the rules of the next theorem.

Theorem 1.1:

If the sizes of the matrices A, B and C are so that the operations can be performed, and α and β are scalars, then

(a.) $A + B = B + A$

(b.) $(A + B) + C = A + (B + C)$

(c.) $(AB)C = A(BC)$

(d.) $A(B+C) = AB + AC$

(e.) $(B+C)A = BA + CA$

(f.) $\alpha(A+B) = \alpha A + \alpha B$

(g.) $(\alpha + \beta)A = \alpha A + \beta A$

(h.) $\alpha(\beta A) = (\alpha\beta)A$

(i.) $\alpha(AB) = (\alpha A)B = A(\alpha B)$

This theorem is simply saying that these rules of combining matrices obey the usual laws of arithmetic. When we more closely examine rules that govern matrix multiplication, we will see that some of these rules are different from what we might hope.

The proofs of these results are not deep, but can be tedious. We demonstrate a proof of part (c).

Suppose that A is a $k \times l$ matrix, B is an $l \times m$ matrix and C is an $m \times n$ matrix. Then AB is a $k \times m$ matrix, $(AB)C$ is a $k \times n$ matrix, BC is an $l \times n$ matrix and $A(BC)$ is a $k \times n$ matrix. Thus, $(AB)C$ and $A(BC)$ have the same dimensions. To complete the proof, we must show that the entries of the two matrices are the same. That is, we must show

$$\left[(AB)C\right]_{rs} = \left[A(BC)\right]_{rs} \text{ for } r = 1,\dots,k; s = 1,\dots,n.$$

We have

$$\left[(AB)C\right]_{rs} = \sum_{j=1}^{m}(AB)_{rj}C_{js}$$

$$= \sum_{j=1}^{m}\left[\sum_{i=1}^{l}A_{ri}B_{ij}\right]C_{js} = \sum_{j=1}^{m}\left[\sum_{i=1}^{l}(A_{ri}B_{ij})C_{js}\right] = \sum_{j=1}^{m}\left[\sum_{i=1}^{l}A_{ri}\left(B_{ij}C_{js}\right)\right]$$

$$= \sum_{i=1}^{l}\left[\sum_{j=1}^{m}A_{ri}\left(B_{ij}C_{js}\right)\right] = \sum_{i=1}^{l}A_{ri}\left[\sum_{j=1}^{m}\left(B_{ij}C_{js}\right)\right]$$

$$= \sum_{i=1}^{l}A_{ri}(BC)_{is} = \left[A(BC)\right]_{rs}$$

We have defined the arithmetic operations on matrices. Matrices, together with these operations, form a mathematical structure. To expand this structure, we need to introduce some additional matrices.

In arithmetic, the numbers 0 and 1 have special significance. The number 0 is important because it is the additive identity; i.e.,

$$0 + a = a$$

for every number a. The number 1 is important because it is the multiplicative identity; i.e.,

$$1 \times a = a$$

for every number a.

There are matrices that fulfill these same roles, but we must be sure the dimensions of these matrices are compatible with what we want to accomplish.

A zero matrix is a matrix for which all of the entries are 0. We denote the zero matrix of size $m \times n$ by

$$\tilde{0}_{m \times n}.$$

If the dimensions are obvious, we will simply use $\tilde{0}$. We have the desired property that if A is an $m \times n$ matrix, then

$$A + \tilde{0}_{m \times n} = A.$$

Matrix multiplication presents some challenges that do not occur with matrix addition. The most prominent of these is that matrix multiplication is normally not commutative even if the dimensions allow for multiplication. That is, it is most often the case that for matrices A and B

$$AB \neq BA.$$

We expand on this. If A is an $m \times k$ matrix and B is a $k \times n$ matrix then AB is an $m \times n$ matrix, but BA is not defined unless $m = n$. If $m = n$ then AB is an $m \times m$ matrix, but BA is a $k \times k$ matrix.

Even if A and B are both $m \times m$ matrices it is most often the case that $AB \neq BA$.

Example

Consider

$$A = \begin{pmatrix} 1 & 0 \\ 0 & 0 \end{pmatrix}, \quad B = \begin{pmatrix} 0 & 0 \\ 1 & 1 \end{pmatrix},$$

We have

$$AB = \begin{pmatrix} 1 & 0 \\ 0 & 0 \end{pmatrix}\begin{pmatrix} 0 & 0 \\ 1 & 1 \end{pmatrix} = \begin{pmatrix} 0 & 0 \\ 0 & 0 \end{pmatrix}$$

and

$$BA = \begin{pmatrix} 0 & 0 \\ 1 & 1 \end{pmatrix}\begin{pmatrix} 1 & 0 \\ 0 & 0 \end{pmatrix} = \begin{pmatrix} 0 & 0 \\ 1 & 0 \end{pmatrix}.$$

Not only do we have $AB \neq BA$, but in the case of AB, we see another point of caution. If x and y are numbers, and if $xy = 0$, then it must be that $x = 0$ or $y = 0$.

In the case of BA in the example above, we have the product of two matrices, neither of which is equal to $\tilde{0}$, but the product is $\tilde{0}$. Among the implications of this is that if

$$AB = AC \text{ and } A \neq \tilde{0}$$

we cannot cancel A. That is, we cannot be sure that $B = C$.

We group these results.

Facts:

If A and B are matrices for which AB is defined then

(i) it is not necessarily the case that BA is defined.
(ii) Even if AB and BA are both defined it is not necessarily the case that $AB = BA$
(iii) It is possible that neither A nor B is the zero matrix, but AB is the zero matrix.
(iv) It is possible that there are matrices A, B and C for which $AB = AC$ and $B \neq C$.

The Identity Matrix

The matrix that acts (as closely as possible) like the number 1 as the multiplicative identity is a square matrix that has 1s on the main diagonal and 0s for every other entry. The $n \times n$ identity matrix is denoted by I_n. For example,

$$I_3 = \begin{pmatrix} 1 & 0 & 0 \\ 0 & 1 & 0 \\ 0 & 0 & 1 \end{pmatrix}.$$

If A is an $n \times n$ matrix, then

$$AI_n = I_n A = A.$$

As an example, it is easy to check that

$$I_3 \begin{pmatrix} a_{11} & a_{12} & a_{13} \\ a_{21} & a_{22} & a_{23} \\ a_{31} & a_{32} & a_{33} \end{pmatrix} = \begin{pmatrix} 1 & 0 & 0 \\ 0 & 1 & 0 \\ 0 & 0 & 1 \end{pmatrix} \begin{pmatrix} a_{11} & a_{12} & a_{13} \\ a_{21} & a_{22} & a_{23} \\ a_{31} & a_{32} & a_{33} \end{pmatrix} = \begin{pmatrix} a_{11} & a_{12} & a_{13} \\ a_{21} & a_{22} & a_{23} \\ a_{31} & a_{32} & a_{33} \end{pmatrix}$$

and

$$\begin{pmatrix} a_{11} & a_{12} & a_{13} \\ a_{21} & a_{22} & a_{23} \\ a_{31} & a_{32} & a_{33} \end{pmatrix} I_3 = \begin{pmatrix} a_{11} & a_{12} & a_{13} \\ a_{21} & a_{22} & a_{23} \\ a_{31} & a_{32} & a_{33} \end{pmatrix} \begin{pmatrix} 1 & 0 & 0 \\ 0 & 1 & 0 \\ 0 & 0 & 1 \end{pmatrix} = \begin{pmatrix} a_{11} & a_{12} & a_{13} \\ a_{21} & a_{22} & a_{23} \\ a_{31} & a_{32} & a_{33} \end{pmatrix}.$$

If A is an $n \times m$ matrix, then $I_n A = A$ and $AI_m = A$.

The Inverse of a Square Matrix

In the real numbers, every non-zero number has a multiplicative inverse. For example, the multiplicative inverse of 2 is ½ because

$$2 \times ½ = 1,$$

and 1 is the multiplicative identity.

We want to construct the analogous element for matrix multiplication. That is, if A is a non-zero $n \times n$ matrix, we would like to find an $n \times n$ matrix B for which

$$AB = BA = I_n.$$

This is not always possible, as we see in the next example.

Example

Let

$$A = \begin{pmatrix} 1 & 0 \\ 0 & 0 \end{pmatrix}.$$

We seek a 2×2 matrix B for which

$$AB = BA = I_2.$$

Let

$$B = \begin{pmatrix} a & b \\ c & d \end{pmatrix}.$$

Then

$$AB = \begin{pmatrix} 1 & 0 \\ 0 & 0 \end{pmatrix} \begin{pmatrix} a & b \\ c & d \end{pmatrix} = \begin{pmatrix} a & b \\ 0 & 0 \end{pmatrix}$$

and it is impossible to find a, b, c, and d for which

$$AB = \begin{pmatrix} 1 & 0 \\ 0 & 1 \end{pmatrix}.$$

Definition:

If A is an $n \times n$ matrix, and B is an $n \times n$ matrix for which

$$AB = I_n = BA$$

then B is said to be the inverse of A.

If the inverse of A exists, then it is unique and it is denoted A^{-1}. A matrix that has an inverse is said to be nonsingular. A square matrix that does not have an inverse is singular.

Theorem 1.2:

If A and B are nonsingular matrices, then AB is a nonsingular matrix and

$$\left(AB\right)^{-1} = B^{-1}A^{-1}.$$

Proof:

We have

$$\left(B^{-1}A^{-1}\right)\left(AB\right)=\left[\left(B^{-1}A^{-1}\right)A\right]B=\left[B^{-1}\left(A^{-1}A\right)\right]B=\left(B^{-1}I\right)B=B^{-1}B=I$$

and

$$\left(AB\right)\left(B^{-1}A^{-1}\right)=A\left[B\left(B^{-1}A^{-1}\right)\right]=A\left[\left(BB^{-1}\right)A^{-1}\right]$$
$$=A\left(IA^{-1}\right)=AA^{-1}=I.$$

Finding the inverse of a matrix is often a computationally complex task. In this section we develop an algorithm to determine the inverse of a matrix. In doing so, we introduce several topics that will be important in our later work. These topics include determinants, elementary row operations, and elementary matrices.

Instructions for using MATLAB and R, which perform the computations of linear algebra, are given in the appendices.

Determinants

In Chapter 9 we give an in-depth development of determinants. Here we give an introduction to determinants and basic properties of determinants that are needed for our immediate purposes.

A determinant is a number that is associated with a square matrix.

The most basic way of computing a determinant is expanding by cofactors. We explain this method by way of examples.

For a 2×2 matrix

$$A=\begin{pmatrix} a & b \\ c & d \end{pmatrix}$$

the determinant of A, denoted det (A) or $\begin{vmatrix} a & b \\ c & d \end{vmatrix}$, is

$$\det\left(A\right)\equiv ad-bc.$$

Suppose we have a 3×3 matrix. We assign a + sign to the upper left-hand position, and alternate signs of the positions as we proceed either to the right or down. We then have

$$\begin{pmatrix} + & - & + \\ - & + & - \\ + & - & + \end{pmatrix}.$$

In expanding by cofactors, one chooses any row or any column, and breaks the determinant into pieces as in the next example.

Example

Consider

$$A = \begin{pmatrix} -2 & 3 & 1 \\ 0 & -1 & 2 \\ 4 & 5 & 3 \end{pmatrix}.$$

Suppose we elect to expand along the second column. We compute the determinate by expanding the determinate into three smaller pieces. To get the first piece, we eliminate the entries in the first row and second column to get

$$\begin{pmatrix} 0 & 2 \\ 4 & 3 \end{pmatrix}.$$

According to the array of signs, the sign assigned to the first row and second column is minus and the entry in the first row second column is 3. Thus, the first piece in our computation is

$$(-)(3) \begin{vmatrix} 0 & 2 \\ 4 & 3 \end{vmatrix} = -3(0-8) = 24.$$

Since we are expanding along the second column, we move to the A_{22} entry and repeat the procedure above. This gives

$$(+)(-1) \begin{vmatrix} -2 & 1 \\ 4 & 3 \end{vmatrix} = -(-6-4) = 10.$$

Moving to the A_{32} element gives

$$(-)(5) \begin{vmatrix} -2 & 1 \\ 0 & 2 \end{vmatrix} = 20.$$

The determinant of A is the sum of these three numbers; that is,

$$\det(A) = 24 + 10 + 20 = 54.$$

For larger matrices, we follow a procedure analogous to the one above.

It is a somewhat deep fact that one gets the same value regardless of which row or column one chooses for the expansion.

Example

The determinant for upper and lower triangular matrices is easy to compute. Let

$$A = \begin{pmatrix} a_{11} & a_{12} & a_{13} & a_{14} \\ 0 & a_{22} & a_{23} & a_{24} \\ 0 & 0 & a_{33} & a_{34} \\ 0 & 0 & 0 & a_{44} \end{pmatrix}.$$

Then, expanding along the first column,

$$\det(A) = a_{11} \begin{vmatrix} a_{22} & a_{23} & a_{24} \\ 0 & a_{33} & a_{34} \\ 0 & 0 & a_{44} \end{vmatrix} - 0 \begin{vmatrix} a_{12} & a_{13} & a_{14} \\ 0 & a_{33} & a_{34} \\ 0 & 0 & a_{44} \end{vmatrix} + 0 \begin{vmatrix} a_{12} & a_{13} & a_{14} \\ a_{22} & a_{23} & a_{24} \\ 0 & 0 & a_{44} \end{vmatrix} - 0 \begin{vmatrix} a_{12} & a_{13} & a_{14} \\ a_{22} & a_{23} & a_{24} \\ 0 & a_{33} & a_{34} \end{vmatrix}$$

$$= a_{11} \begin{vmatrix} a_{22} & a_{23} & a_{24} \\ 0 & a_{33} & a_{34} \\ 0 & 0 & a_{44} \end{vmatrix} = a_{11} \left(a_{22} \begin{vmatrix} a_{33} & a_{34} \\ 0 & a_{44} \end{vmatrix} - 0 \begin{vmatrix} a_{23} & a_{24} \\ 0 & a_{44} \end{vmatrix} + 0 \begin{vmatrix} a_{23} & a_{24} \\ a_{33} & a_{34} \end{vmatrix} \right)$$

$$= a_{11} a_{22} \begin{vmatrix} a_{33} & a_{34} \\ 0 & a_{44} \end{vmatrix} = a_{11} a_{22} a_{33} a_{44}.$$

This is an example of why the next theorem holds.

Theorem 1.3:

The determinant of an upper or lower triangular matrix is the product of the main diagonal entries.

Corollary:

The determinant of a diagonal matrix is the product of the main diagonal entries.

The computation of a determinant can be simplified by performing elementary row operations, which are defined as follows.

Definition:

If A is a matrix, the following operations are called elementary row operations on A.

1. Interchanging two rows.
2. Adding a multiple of one row to another.
3. Multiplying a row by a scalar.

The next theorem is proven in Chapter 9.

Theorem 1.4:

Let A be a square matrix and B be the matrix that results when a row operation is applied to A. We have

1. If B is the result of interchanging two rows of A, then $\det(B) = -\det(A)$.
2. If B is the result of adding a multiple of one row of A to another row of A, then $\det(B) = \det(A)$.
3. If B is the result of multiplying a row by a constant k, then $\det(B) = k \det(A)$.

The following corollary is very important.

Corollary:

If A is a square matrix, interchanging two rows or adding a multiple of one row to another will not change the value of the determinant of A. If $k \neq 0$, multiplying a row by k will not change the determinant from a non-zero number to zero.

We will see later that the following is true.

If A is a square matrix, then it is possible to apply elementary row operations (with $k \neq 0$) to A until the result is either the identity matrix or a matrix with a row of all zeroes.

In the next chapter, we demonstrate that applying elementary row operations (if $k \neq 0$) does not change whether a matrix is invertible.

The next example demonstrates how applying elementary row operations can simplify calculating a determinant.

Example

We evaluate

$$\begin{vmatrix} 2 & 4 & -6 & 0 \\ 1 & 5 & 3 & 6 \\ 3 & 1 & 4 & 5 \\ 2 & 6 & -7 & -3 \end{vmatrix}$$

Factoring 2 from the first row gives $2\begin{vmatrix} 1 & 2 & -3 & 0 \\ 1 & 5 & 3 & 6 \\ 3 & 1 & 4 & 5 \\ 2 & 6 & -7 & -3 \end{vmatrix}$

Subtracting the first row from the second row, subtracting 3 times the first row from the third row and subtracting 2 times the first row from the fourth row gives

$$2\begin{vmatrix} 1 & 2 & -3 & 0 \\ 0 & 3 & 6 & 6 \\ 0 & -5 & 13 & 5 \\ 0 & 2 & -1 & -3 \end{vmatrix}$$

Factoring 3 from the second row gives

$$2(3)\begin{vmatrix} 1 & 2 & -3 & 0 \\ 0 & 1 & 2 & 2 \\ 0 & -5 & 13 & 5 \\ 0 & 2 & -1 & -3 \end{vmatrix}$$

Adding 5 times the second row to the third row and adding (-2) times the second row to the fourth row gives

$$2(3)\begin{vmatrix} 1 & 2 & -3 & 0 \\ 0 & 1 & 2 & 2 \\ 0 & 0 & 23 & 15 \\ 0 & 0 & -5 & -7 \end{vmatrix}$$

Factoring 23 from the third row gives

$$2(3)(23)\begin{vmatrix} 1 & 2 & -3 & 0 \\ 0 & 1 & 2 & 2 \\ 0 & 0 & 1 & 15/23 \\ 0 & 0 & -5 & -7 \end{vmatrix}$$

Finally, adding 5 times the third row to the fourth row gives

$$2(3)(23)\begin{vmatrix} 1 & 2 & -3 & 0 \\ 0 & 1 & 2 & 2 \\ 0 & 0 & 1 & 15/23 \\ 0 & 0 & 0 & -86/23 \end{vmatrix}$$

Therefore, the value of the determinant is $2(3)(23)(-86/23) = -516$.

Theorem 1.5:

Properties of the determinant of $n \times n$ matrices include

1. $\det(I_n) = 1$

2. $\det(AB) = \det(A)\det(B)$

3. $\det(A^{-1}) = \dfrac{1}{\det(A)}$ provided $\det(A)$ is not 0

4. $\det(cA) = c^n \det(A)$

5. The determinant of a triangular or diagonal matrix is the product of its diagonal elements.
6. $\det(A) \neq 0$ if and only if A is invertible.
7. If two rows or two columns of a matrix are equal, the determinant is 0.
8. If two rows of a matrix are interchanged, the value of the determinant is multiplied by -1.
9. If a multiple of a row of a matrix is added to another row, the determinant is unchanged.
10. If a row of an $n \times n$ matrix is multiplied by k, the determinant is multiplied by k.

Elementary matrices

In the next chapter we study how to use matrices to solve certain types of equations. In the process of solving the equations, we will manipulate the rows of the matrix with the elementary row operations. These manipulations are:

1. Interchanging two rows of a matrix.

2. Multiplying a row of a matrix by a number.
3. Adding a multiple of one row to another row.

These operations can be executed by multiplying a matrix on the left by matrices called elementary matrices. In this section we study elementary matrices.

We use the following notation:

E_{ij} results from interchanging the ith and jth rows of the identity matrix I.
$E_i(c)$ results from multiplying the ith row of the identity matrix I by the scalar c.
$E_{ij}(c)$ results from adding c times the row i to the row j of the identity matrix I.

Matrices that Interchange Two Rows of a Matrix

Suppose that A is a $3 \times n$ matrix. Let E_{ij} denote the matrix that is obtained when the ith and jth rows of the identity matrix I_3 are interchanged. Then $E_{ij}A$ is the matrix that results when the ith and jth rows of A are interchanged.

Example

Let

$$E_{23} = \begin{pmatrix} 1 & 0 & 0 \\ 0 & 0 & 1 \\ 0 & 1 & 0 \end{pmatrix} \quad \text{and} \quad A = \begin{pmatrix} a_{11} & a_{12} & a_{13} & a_{14} \\ a_{21} & a_{22} & a_{23} & a_{24} \\ a_{31} & a_{32} & a_{33} & a_{34} \end{pmatrix}.$$

Then

$$E_{23}A = \begin{pmatrix} 1 & 0 & 0 \\ 0 & 0 & 1 \\ 0 & 1 & 0 \end{pmatrix}\begin{pmatrix} a_{11} & a_{12} & a_{13} & a_{14} \\ a_{21} & a_{22} & a_{23} & a_{24} \\ a_{31} & a_{32} & a_{33} & a_{34} \end{pmatrix} = \begin{pmatrix} a_{11} & a_{12} & a_{13} & a_{14} \\ a_{31} & a_{32} & a_{33} & a_{34} \\ a_{21} & a_{22} & a_{23} & a_{24} \end{pmatrix}.$$

In general, if A is a $m \times n$ matrix and E_{ij} is the matrix that is obtained when the ith and jth rows of the identity matrix I_m are interchanged, then $E_{ij}A$ is the matrix that results when the ith and jth rows of A are interchanged.

We also have $E_{ij}^{-1} = E_{ij}$ since if the same two rows of a matrix are interchanged twice, the matrix is unchanged.

Multiplying a Row of a Matrix by a Constant

If we want to multiply the second row of a $3 \times n$ matrix by 5, we would multiply the matrix by

$$E_2(5) = \begin{pmatrix} 1 & 0 & 0 \\ 0 & 5 & 0 \\ 0 & 0 & 1 \end{pmatrix}$$

which is the matrix that results when the second row of I_3 is multiplied by 5.

We have

$$E_2(5)\begin{pmatrix} a_{11} & a_{12} & a_{13} & a_{14} \\ a_{21} & a_{22} & a_{23} & a_{24} \\ a_{31} & a_{32} & a_{33} & a_{34} \end{pmatrix} = \begin{pmatrix} 1 & 0 & 0 \\ 0 & 5 & 0 \\ 0 & 0 & 1 \end{pmatrix}\begin{pmatrix} a_{11} & a_{12} & a_{13} & a_{14} \\ a_{21} & a_{22} & a_{23} & a_{24} \\ a_{31} & a_{32} & a_{33} & a_{34} \end{pmatrix}$$

$$= \begin{pmatrix} a_{11} & a_{12} & a_{13} & a_{14} \\ 5a_{21} & 5a_{22} & 5a_{23} & 5a_{24} \\ a_{31} & a_{32} & a_{33} & a_{34} \end{pmatrix}$$

Note that

$$E_2(5)^{-1} = \begin{pmatrix} 1 & 0 & 0 \\ 0 & 1/5 & 0 \\ 0 & 0 & 1 \end{pmatrix} = E_2(1/5).$$

In general, if A is a $m \times n$ matrix and $E_j(k)$ is the matrix that is obtained when the jth row of the identity matrix I_m is multiplied by k, then $E_j(k)\,A$ is the matrix that results when the jth row of A is multiplied by k.

Also, if $k \neq 0$, then

$$E_j(k)^{-1} = E_j\left(\frac{1}{k}\right).$$

Adding a Multiple of One Row to Another Row

If we want to multiply the second row of a $3 \times n$ matrix by 5 and add it to the third row, we would multiply the matrix by

$$E_{23}(5) = \begin{pmatrix} 1 & 0 & 0 \\ 0 & 1 & 0 \\ 0 & 5 & 1 \end{pmatrix}.$$

If we add k times the ith row of I_m to the jth row of I_m we obtain the matrix $E_{ij}(k)$. If A is an $m \times n$ matrix, then the product $E_{ij}(k)A$ is the matrix obtained when k times the ith row of A is added to the jth row of A.

For example, we have

$$E_{23}(5)\begin{pmatrix} a_{11} & a_{12} & a_{13} & a_{14} \\ a_{21} & a_{22} & a_{23} & a_{24} \\ a_{31} & a_{32} & a_{33} & a_{34} \end{pmatrix} = \begin{pmatrix} 1 & 0 & 0 \\ 0 & 1 & 0 \\ 0 & 5 & 1 \end{pmatrix}\begin{pmatrix} a_{11} & a_{12} & a_{13} & a_{14} \\ a_{21} & a_{22} & a_{23} & a_{24} \\ a_{31} & a_{32} & a_{33} & a_{34} \end{pmatrix}$$

$$= \begin{pmatrix} a_{11} & a_{12} & a_{13} & a_{14} \\ a_{21} & a_{22} & a_{23} & a_{24} \\ a_{31}+5a_{21} & a_{32}+5a_{22} & a_{33}+5a_{23} & a_{34}+5a_{24} \end{pmatrix}$$

Also

$$E_{23}(5)^{-1} = \begin{pmatrix} 1 & 0 & 0 \\ 0 & 1 & 0 \\ 0 & -5 & 1 \end{pmatrix} = E_{23}(-5).$$

Computing the Inverse of a Matrix

Normally, we will calculate the inverse of a matrix using computer software. In this subsection, we describe one method of finding the inverse of a matrix by using elementary matrices. These are the ideas involved.

Suppose that A is an $n \times n$ invertible matrix. We concatenate A with I_n as shown in Figure 1.3(a.)

We multiply the concatenated matrix by elementary matrices to get the matrix shown in Figure 1.3(b.). The matrix B in Figure 1.3(b.) is the inverse of A.

Example

Let

$$A = \begin{pmatrix} 2 & 4 & 0 \\ 3 & 9 & -6 \\ -4 & 8 & 12 \end{pmatrix}$$

so that

$$(I_3|A) = \left(\begin{array}{ccc|ccc} 1 & 0 & 0 & 2 & 4 & 0 \\ 0 & 1 & 0 & 3 & 9 & -16 \\ 0 & 0 & 1 & -4 & 8 & 12 \end{array} \right).$$

We first multiply the first row by ½. This gives

$$E_1\left(\frac{1}{2}\right)(I_3|A) = \left(\begin{array}{ccc|ccc} \frac{1}{2} & 0 & 0 & 1 & 2 & 0 \\ 0 & 1 & 0 & 3 & 9 & -6 \\ 0 & 0 & 1 & -4 & 8 & 12 \end{array} \right).$$

Add -3 times the first row to the second row to get

$$E_{12}(-3)E_1\left(\frac{1}{2}\right)(I_3|A) = \left(\begin{array}{ccc|ccc} \frac{1}{2} & 0 & 0 & 1 & 2 & 0 \\ -\frac{3}{2} & 1 & 0 & 0 & 3 & -6 \\ 0 & 0 & 1 & -4 & 8 & 12 \end{array} \right).$$

$$(I_n|A) \qquad\qquad (B|I_n)$$

$$\text{(a)} \qquad\qquad \text{(b)}$$

Figure 1.3   Concatenated matrices before and after row reduction.

Add 4 times the first row to the third row to get

$$E_{13}(4)E_{12}(-3)E_1\left(\frac{1}{2}\right)(I_3|A) = \left(\begin{array}{ccc|ccc} \frac{1}{2} & 0 & 0 & 1 & 2 & 0 \\ -\frac{3}{2} & 1 & 0 & 0 & 3 & -6 \\ 2 & 0 & 1 & 0 & 16 & 12 \end{array}\right)$$

Multiply the second row by 1/3

$$E_2\left(\frac{1}{3}\right)E_{13}(4)E_{12}(-3)E_1\left(\frac{1}{2}\right)(I_3|A) = \left(\begin{array}{ccc|ccc} \frac{1}{2} & 0 & 0 & 1 & 2 & 0 \\ -\frac{1}{2} & \frac{1}{3} & 0 & 0 & 1 & -2 \\ 2 & 0 & 1 & 0 & 16 & 12 \end{array}\right)$$

Add –16 times the second row to the third row to get

$$E_{23}(-16)E_2\left(\frac{1}{3}\right)E_{13}(4)E_{12}(-3)E_1\left(\frac{1}{2}\right)(I_3|A) = \left(\begin{array}{ccc|ccc} \frac{1}{2} & 0 & 0 & 1 & 2 & 0 \\ -\frac{1}{2} & \frac{1}{3} & 0 & 0 & 1 & -2 \\ 10 & -\frac{16}{3} & 1 & 0 & 0 & 44 \end{array}\right)$$

Multiply the third row by 1/44 to get

$$E_3\left(\frac{1}{44}\right)E_{23}(-16)E_2\left(\frac{1}{3}\right)E_{13}(4)E_{12}(-3)E_1\left(\frac{1}{2}\right)(I_3|A)$$

$$= \left(\begin{array}{ccc|ccc} \frac{1}{2} & 0 & 0 & 1 & 2 & 0 \\ -\frac{1}{2} & \frac{1}{3} & 0 & 0 & 1 & -2 \\ \frac{10}{44} & \frac{-16}{132} & \frac{1}{44} & 0 & 0 & 1 \end{array}\right)$$

Add 2 times the third row to the second row to get

$$E_{23}(2)E_3\left(\frac{1}{44}\right)E_{23}(-16)E_2\left(\frac{1}{3}\right)E_{13}(4)E_{12}(-3)E_1\left(\frac{1}{2}\right)(I_3|A)$$

$$= \left(\begin{array}{ccc|ccc} \frac{1}{2} & 0 & 0 & 1 & 2 & 0 \\ \frac{-1}{22} & \frac{1}{11} & \frac{1}{22} & 0 & 1 & 0 \\ \frac{10}{44} & \frac{-16}{132} & \frac{1}{44} & 0 & 0 & 1 \end{array}\right)$$

Add –2 times the second row to the first row to get

$$E_{21}(-2)E_{23}(2)E_3\left(\frac{1}{44}\right)E_{23}(-16)E_2\left(\frac{1}{3}\right)E_{13}(4)E_{12}(-3)E_1\left(\frac{1}{2}\right)(I_3|A)$$

$$=\begin{pmatrix} \dfrac{13}{22} & \dfrac{-2}{11} & \dfrac{-1}{11} & 1 & 0 & 0 \\[2mm] \dfrac{-1}{22} & \dfrac{1}{11} & \dfrac{1}{22} & 0 & 1 & 0 \\[2mm] \dfrac{10}{44} & \dfrac{-16}{132} & \dfrac{1}{44} & 0 & 0 & 1 \end{pmatrix}$$

Thus,

$$A^{-1}=\begin{pmatrix} \dfrac{13}{22} & \dfrac{-2}{11} & \dfrac{-1}{11} \\[2mm] \dfrac{-1}{22} & \dfrac{1}{11} & \dfrac{1}{22} \\[2mm] \dfrac{10}{44} & \dfrac{-16}{132} & \dfrac{1}{44} \end{pmatrix}.$$

One can check that

$$\begin{pmatrix} \dfrac{13}{22} & \dfrac{-2}{11} & \dfrac{-1}{11} \\[2mm] \dfrac{-1}{22} & \dfrac{1}{11} & \dfrac{1}{22} \\[2mm] \dfrac{5}{22} & \dfrac{-4}{33} & \dfrac{1}{44} \end{pmatrix}\begin{pmatrix} 2 & 4 & 0 \\ 3 & 9 & -6 \\ -4 & 8 & 12 \end{pmatrix}=\begin{pmatrix} 2 & 4 & 0 \\ 3 & 9 & -6 \\ -4 & 8 & 12 \end{pmatrix}\begin{pmatrix} \dfrac{13}{22} & \dfrac{-2}{11} & \dfrac{-1}{11} \\[2mm] \dfrac{-1}{22} & \dfrac{1}{11} & \dfrac{1}{22} \\[2mm] \dfrac{5}{22} & \dfrac{-4}{33} & \dfrac{1}{44} \end{pmatrix}$$

$$=\begin{pmatrix} 1 & 0 & 0 \\ 0 & 1 & 0 \\ 0 & 0 & 1 \end{pmatrix}$$

If $E_1,\dots,E_k$ are elementary row operations for which

$$E_k\cdots E_1A=I$$

then

$$A^{-1}=\left(E_k\cdots E_1A\right)A^{-1}=\left(E_k\cdots E_1\right)\left(AA^{-1}\right)=\left(E_k\cdots E_1\right)I=\left(E_k\cdots E_1\right).$$

Also, since

$$A^{-1}=\left(E_k\cdots E_1\right)$$

then

$$A=\left(E_k\cdots E_1\right)^{-1}=E_1^{-1}\cdots E_k^{-1}.$$

We thus have the following results.

Theorem 1.6:

If A is an $n \times n$ invertible matrix, then A can be expressed as the product of elementary matrices.

Theorem 1.7:

A square matrix has an inverse if and only if the determinant of the matrix is nonzero.

The Transpose of a Matrix

If A is an $m \times n$ matrix, the transpose of the matrix A, denoted A^T, is the $n \times m$ matrix given by

$$\left(A^T\right)_{ij} = A_{ji}.$$

One way to view the transpose of the matrix A is as the matrix that is created when the rows of A become the columns of A^T.

Example

If

$$A = \begin{pmatrix} -1 & 4 & 0 \\ 2 & -3 & 6 \end{pmatrix}$$

then

$$A^T = \begin{pmatrix} -1 & 2 \\ 4 & -3 \\ 0 & 6 \end{pmatrix}.$$

Theorem 1.8:

If the sizes of the matrices A and B are so that the operations can be performed, then

(a.) $\left(A^T\right)^T = A$

(b.) $\left(A + B\right)^T = A^T + B^T$

(c.) $\left(\alpha A\right)^T = \alpha\left(A^T\right)$ for any scalar α

(d.) $\left(AB\right)^T = \left(B^T\right)\left(A^T\right).$

(e.) If A is a square matrix, then $\det\left(A\right) = \det\left(A^T\right)$

Proof of part (d.):

Let A be an $m \times k$ matrix and B be an $k \times n$ matrix. Then AB is an $m \times n$ matrix and $(AB)^T$ is an $n \times m$ matrix and

$$\left((AB)^T\right)_{ji} = (AB)_{ij} = \sum_{l=1}^{k} A_{il} B_{lj}.$$

Also, B^T is an $n \times k$ matrix and A^T is an $k \times m$ matrix so $B^T A^T$ is an $n \times m$ matrix with

$$\left[\left(B^T\right)\left(A^T\right)\right]_{ji} = \sum_{l=1}^{k} \left(B^T\right)_{jl} \left(A^T\right)_{li} = \sum_{l=1}^{k} B_{lj} A_{il} = \sum_{l=1}^{k} A_{il} B_{lj} = (AB)_{ij}$$

$$= \left((AB)^T\right)_{ji}.$$

Definition:

A matrix A is symmetric if $A^T = A$.

We will see later that symmetric matrices are particularly important in some applications.

Historical Note

Historically, determinants were used before matrices. Originally, a determinant was defined as a property of a system of linear equations. There, the determinant "determines" whether the system has a unique solution. The Japanese mathematician Seki Takakazu is credited with the discovery of the determinant (at first in 1683, the complete version by 1710). In Europe, Gabriel Cramer (1704–1752) added to the theory, treating the subject in relation to sets of equations. Laplace, in 1772, described computing a determinant by expanding in terms of its minors. In 1801 Gauss introduced the word determinant (previous vocabulary used resultant).

Arthur Cayley gave an abstract definition of a matrix (1858) and used two vertical lines to denote the determinant (1841).

Exercises

1. Show that if A, B and C are $n \times n$ matrices with $AB = I_n$ and $BC = I_n$ then $A = C = B^{-1}$.
2. Find a non-zero 2×2 matrix A for which $A^2 = 0$.
3. Find all the matrices that commute with

$$\begin{pmatrix} 1 & 0 \\ 1 & 1 \end{pmatrix}.$$

4. Show that matrices of the form

$$\begin{pmatrix} a & 0 \\ 0 & a \end{pmatrix}$$

are the only matrices that commute with all 2×2 matrices.
5. For

$$A = \begin{pmatrix} 1 & -3 \\ 4 & 0 \end{pmatrix}, \ B = \begin{pmatrix} 2 & 0 \\ -1 & 5 \end{pmatrix}$$

find $A - 3B$, BA^T, AB^T, $\left(AB^T \right)^T$.

6. A stochastic matrix is a square matrix whose entries are nonnegative and each row sums to 1. Show that if

$$A = \begin{pmatrix} a_{11} & a_{12} \\ a_{21} & a_{22} \end{pmatrix} \quad \text{and} \quad B = \begin{pmatrix} b_{11} & b_{12} \\ b_{21} & b_{22} \end{pmatrix}$$

are stochastic matrices, then AB is a stochastic matrix. Can you generalize this to larger stochastic matrices?

7. What can you say about the matrix A if A^2 is defined?

8. (a.) Suppose that

$$A = \begin{pmatrix} a_{11} & a_{12} & a_{13} \\ 0 & 0 & 0 \\ a_{31} & a_{32} & a_{33} \end{pmatrix} \quad B = \begin{pmatrix} b_{11} & b_{12} & b_{13} \\ b_{21} & b_{22} & b_{23} \\ b_{31} & b_{32} & b_{33} \end{pmatrix}.$$

Compute AB and BA.

(b.) Suppose that A and B are square matrices of the same dimension and A has a row of all zeroes. Is there anything you can say about AB or BA?

(c.) Without using determinants show that any matrix with a row of all zeroes or a column of all zeroes cannot be invertible.

9. (a.) Suppose that

$$A = \begin{pmatrix} a_{11} & a_{12} & a_{13} \\ a_{11} & a_{11} & a_{11} \\ a_{31} & a_{32} & a_{33} \end{pmatrix} \quad B = \begin{pmatrix} b_{11} & 0 & b_{13} \\ b_{21} & 0 & b_{23} \\ b_{31} & 0 & b_{33} \end{pmatrix}.$$

Compute AB and BA.

(b.) Suppose that A and B are square matrices of the same dimension and B has a column of all zeroes. Is there anything you can say about AB or BA?

10. (a.) Show that if A and B are matrices with $AB = \tilde{0}$, and A is invertible, then $B = \tilde{0}$.

(b.) Show that if A, B and C are matrices with $AB = AC$, and A is invertible, then $B = C$.

(c.) Give an example of non-zero matrices A, B and C for which $AB = AC$ but $B \neq C$.

11. Let

$$A = \begin{pmatrix} 1 & 2 \\ 3 & 4 \end{pmatrix} \quad B = \begin{pmatrix} 5 & 6 \\ 7 & 8 \end{pmatrix}.$$

(a.) Compute $(A + B)^2$ and $A^2 + 2AB + B^2$.

(b.) Give a necessary and sufficient condition that will ensure

$$(A+B)^2 = A^2 + 2AB + B^2.$$

12. Let

$$A = \begin{pmatrix} 0 & a & b & c \\ 0 & 0 & d & e \\ 0 & 0 & 0 & f \\ 0 & 0 & 0 & 0 \end{pmatrix}.$$

Find A^2, A^3, and A^4. If A is an upper triangular $n \times n$ matrix with all zeroes on the main diagonal, describe the structure of A^2, A^3,..., A^n.

13. If A is an $n \times n$ matrix, C is an $m \times m$ matrix and B is a matrix so that $A \times B \times C$ is defined, what are the dimensions of B?

14. Let

$$A = \begin{pmatrix} a & b \\ c & d \end{pmatrix}$$

and suppose that A commutes with the matrix

$$C = \begin{pmatrix} 0 & -1 \\ 1 & 0 \end{pmatrix}.$$

(a.) What can you say about the relationships of the elements of A?
(b.) If A and B each commute with C, show that A commutes with B.

15. Let

$$A = \begin{pmatrix} a_{11} & 0 & 0 \\ a_{21} & a_{22} & 0 \\ a_{31} & a_{32} & a_{33} \end{pmatrix} \qquad B = \begin{pmatrix} b_{11} & 0 & 0 \\ b_{21} & b_{22} & 0 \\ b_{31} & b_{32} & b_{33} \end{pmatrix}$$

(a.) Find AB.
(b.) Why is it true that $(AB)_{23} = 0$?
(c.) Show that the product of two lower triangular matrices is lower triangular.

16. Find a matrix B so that $3A - 2B = C$ if

$$A = \begin{pmatrix} 1 & -2 \\ 7 & 3 \\ -5 & 0 \end{pmatrix} \quad \text{and} \quad C = \begin{pmatrix} 6 & 1 \\ 0 & 3 \\ 2 & -4 \end{pmatrix}.$$

17. Show that if A is an $n \times m$ matrix then $I_n A = A$ and $A I_m = A$.
18. Which of the matrices has an inverse?

$$\begin{pmatrix} 1 & 0 & 7 & 8 \\ 0 & 3 & 5 & 9 \\ 0 & 0 & -4 & 0 \\ 0 & 0 & 0 & 2 \end{pmatrix} \begin{pmatrix} 1 & 9 & 2 & 1 \\ 2 & 4 & 0 & 2 \\ 8 & -1 & -6 & 8 \\ 9 & 5 & 5 & 9 \end{pmatrix}$$

$$\begin{pmatrix} 1 & 1 & 5 & 7 \\ 0 & 4 & 8 & 0 \\ 0 & 0 & 9 & 3 \\ 0 & 0 & 0 & 0 \end{pmatrix}.$$

19. (a.) Find the determinant of

$$A = \begin{pmatrix} a & b & c \\ a & b & c \\ d & e & f \end{pmatrix}$$

by expanding along the first column.

   (b.) Find the determinant of

$$A = \begin{pmatrix} a & b & c \\ d & e & f \\ a & b & c \end{pmatrix}$$

by expanding along the first column and using the answer to part (a.).

   (c.) Make a conjecture about the determinant of a square matrix that has two identical rows.

   (d.) Make a conjecture about the determinant of a square matrix that has two identical columns.

20. Show that if A and B are $n \times n$ matrices, then AB is invertible if and only if A and B are each invertible.

21. Suppose that A and B are 2×2 symmetric matrices. Show that AB is symmetric if and only if $AB = BA$.

22. (a.) Show that if A is an $n \times n$ matrix and if

$$A^3 + 4A^2 + 5A - 2I_n = 0_{n \times n}$$

then A^{-1} exists.

23. (a.) Let

$$A = \begin{pmatrix} 1 & 1/2 \\ a & b \end{pmatrix}.$$

Find a and b for which $A^2 = 0$.

   (b.) Let

$$A = \begin{pmatrix} 1 & 1/n \\ a & b \end{pmatrix}.$$

Find a and b for which $A^2 = 0$.

24. (a.) Find the values of a for which

$$\begin{pmatrix} a & a \\ a & a \end{pmatrix}^2 = \begin{pmatrix} a & a \\ a & a \end{pmatrix}.$$

(b.) Find the values of a for which

$$\begin{pmatrix} a & a & a \\ a & a & a \\ a & a & a \end{pmatrix}^2 = \begin{pmatrix} a & a & a \\ a & a & a \\ a & a & a \end{pmatrix}.$$

(c.) Make a conjecture about the values of a for which the $n \times n$ matrix A, all of whose entries are A satisfies $A^2 = A$.

Can you prove your conjecture is true?

25. If A and B are $n \times n$ matrices, the commutator of A and B, denoted $[A, B]$ is $AB - BA$.
    (a.) Show that if A and B are 2×2 matrices then the trace of $[A, B]$ is 0.
    (b.) Show that if A is a 2×2 matrix whose trace is 0 then $A^2 = cI_2$ for some number c.

26. Which matrices below are elementary matrices?

(a.) $\begin{pmatrix} 1 & 0 & 2 \\ 0 & 1 & 0 \\ 0 & 0 & 1 \end{pmatrix}$

(b.) $\begin{pmatrix} 1 & 0 & -\sqrt{3} \\ 0 & 2 & 0 \\ 0 & 0 & 1 \end{pmatrix}$

(c.) $\begin{pmatrix} 2 & 0 & 2 \\ 0 & 1 & 0 \\ 0 & 0 & 1 \end{pmatrix}$

(d.) $\begin{pmatrix} -2 & 1 \\ 0 & 1 \end{pmatrix}$

27. Multiply each matrix below by the appropriate elementary matrix to return it to the identity matrix.

(a.) $\begin{pmatrix} 1 & 0 & 0 \\ 0 & 3 & 0 \\ 0 & 0 & 1 \end{pmatrix}$

(b.) $\begin{pmatrix} 1 & -2 & 0 \\ 0 & 1 & 0 \\ 0 & 0 & 1 \end{pmatrix}$

(c.) $\begin{pmatrix} 0 & 1 & 0 \\ 1 & 0 & 0 \\ 0 & 0 & 1 \end{pmatrix}$

28. Give the elementary matrix for which multiplying by that matrix gives the result shown.

(a.) $E \begin{pmatrix} 1 & -2 & 0 \\ 3 & -1 & 4 \\ 1 & 2 & 1 \end{pmatrix} = \begin{pmatrix} 1 & 2 & 1 \\ 3 & -1 & 4 \\ 1 & -2 & 0 \end{pmatrix}$

(b.) $E \begin{pmatrix} 1 & -2 & 0 \\ 3 & -1 & 4 \\ 1 & 2 & 1 \end{pmatrix} = \begin{pmatrix} 1 & -2 & 0 \\ 0 & -7 & 1 \\ 1 & 2 & 1 \end{pmatrix}$

(c.) $E \begin{pmatrix} 1 & -2 & 0 \\ 3 & -1 & 4 \\ 1 & 2 & 1 \end{pmatrix} = \begin{pmatrix} 3 & -6 & 0 \\ 3 & -1 & 4 \\ 1 & 2 & 1 \end{pmatrix}$

29. (a.) Find

$$\begin{pmatrix} 1 & -2 & 0 \\ 3 & -1 & 4 \\ 1 & 2 & 1 \end{pmatrix} \begin{pmatrix} 1 & 0 & 0 \\ 0 & 0 & 1 \\ 0 & 1 & 0 \end{pmatrix}.$$

(b.) Describe the result when a matrix is multiplied on the right by an elementary matrix.

30. Find the inverse of the matrices below by multiplying the matrix by elementary matrices until the identity is achieved.

(a.) $\begin{pmatrix} 2 & 4 \\ 1 & 3 \end{pmatrix}$

(b.) $\begin{pmatrix} 0 & -2 \\ 2 & 5 \end{pmatrix}$

(c.) $\begin{pmatrix} 1 & 4 \\ 5 & 5 \end{pmatrix}$

(d.) $\begin{pmatrix} 1 & -1 & 4 \\ 0 & 3 & 6 \\ -2 & 1 & 0 \end{pmatrix}$

(e.) $\begin{pmatrix} 1 & 0 & 4 \\ -2 & 1 & 3 \\ 1 & 2 & 5 \end{pmatrix}$

(f.) $\begin{pmatrix} 0 & 2 & 1 \\ 3 & 3 & 0 \\ 1 & 0 & 4 \end{pmatrix}$

$$(g.) \begin{pmatrix} a & 1 & 0 \\ 0 & b & 1 \\ 0 & 0 & c \end{pmatrix} \quad abc \neq 0$$

31. If A is a matrix and c is a scalar with $cA = 0$, show that $A = 0$ or $c = 0$.

Historical Note

A partial timeline of the evolution of the ideas of linear algebra is given below.

200 BC: Han dynasty, coefficients are written on a counting board.
1545 Girolamo Cardano gives for 2×2 matrices what later becomes known as the Cramer rule.
1683 Seki and Leibnitz independently give the first appearance of determinants.
1812 Louis Cauchy gives the multiplication formula of determinant; i.e., det(AB) = det(A) det(B).
1844 Hermann Grassmann publishes his "Theory of Extension" which includes foundational new topics of what is today called linear algebra.
1848 J.J. Sylvester gives the first use of term "matrix".
1858 Arthur Cayley pioneers the development of matrix algebra. In this, he defined matrix multiplication to describe the composition of linear functions. 1870 The Jordan canonical form is presented.
1888 Giuseppe Peano defines the axioms of abstract vector space.

Section 1.3 The *LU* Decomposition of a Square Matrix (Optional)

The material in this section will not be used in later sections and is optional.

In some applications it may reduce the complexity of the computations to express a square matrix as the product of a lower triangular matrix L and an upper triangular matrix U. This is called the *LU* decomposition of the matrix. In this section we demonstrate how to accomplish this factorization.

Suppose that A is a $n \times n$ matrix. Our strategy is to apply elementary row operations to A until the result is an upper triangular matrix. The elementary row operations are accomplished by multiplication by elementary matrices, each of which is lower triangular. Suppose these matrices are $E_1, \ldots, E_k$ where the subscript denotes the order in which they are applied. We then have

$$E_k \cdots E_1 A = U$$

$$A = E_1^{-1} \cdots E_k^{-1} U.$$

We have shown in the exercises that the product of lower triangular matrices is a lower triangular matrix. Each E_i is lower triangular so each E_i^{-1} is lower triangular, and thus

$$E_1^{-1} \cdots E_k^{-1} = L$$

is lower triangular.

Example

Let

$$A = \begin{pmatrix} 1 & 2 & -2 \\ 2 & -4 & 6 \\ 2 & 16 & 8 \end{pmatrix}.$$

1. Add (–2) times the first row of A to the second row of A. This is done by multiplying A on the left by

$$E_1 = \begin{pmatrix} 1 & 0 & 0 \\ -2 & 1 & 0 \\ 0 & 0 & 1 \end{pmatrix}.$$

Now

$$E_1 A = \begin{pmatrix} 1 & 2 & -2 \\ 0 & -8 & 10 \\ 2 & 16 & 8 \end{pmatrix}.$$

2. For the matrix found in Step (1.), add (–2) times the first row of $E_1 A$ to the third row of $E_1 A$. This is done by multiplying of $E_1 A$ on the left by

$$E_2 = \begin{pmatrix} 1 & 0 & 0 \\ 0 & 1 & 0 \\ -2 & 0 & 1 \end{pmatrix}.$$

Now

$$E_2 E_1 A = \begin{pmatrix} 1 & 2 & -2 \\ 0 & -8 & 10 \\ 0 & 12 & 12 \end{pmatrix}.$$

3. For the matrix found in Step (2.), add (1.5) times the second row of $E_2 E_1 A$ to the third row of $E_2 E_1 A$. This is done by multiplying of $E_2 E_1 A$ on the left by

$$E_3 = \begin{pmatrix} 1 & 0 & 0 \\ 0 & 1 & 0 \\ 0 & 1.5 & 1 \end{pmatrix}.$$

Now

$$E_3 E_2 E_1 A = \begin{pmatrix} 1 & 2 & -2 \\ 0 & -8 & 10 \\ 0 & 0 & 27 \end{pmatrix}$$

which is an upper triagular matrix.

We thus have

$$A = E_1^{-1} E_2^{-1} E_3^{-1} \begin{pmatrix} 1 & 2 & -2 \\ 0 & -8 & 10 \\ 0 & 0 & 27 \end{pmatrix}.$$

4. Let

$$L = E_1^{-1} E_2^{-1} E_3^{-1} = \begin{pmatrix} 1 & 0 & 0 \\ 2 & 1 & 0 \\ 2 & -3/2 & 1 \end{pmatrix}; U = \begin{pmatrix} 1 & 2 & -2 \\ 0 & -8 & 10 \\ 0 & 0 & 27 \end{pmatrix}.$$

Then

$$LU = \begin{pmatrix} 1 & 0 & 0 \\ 2 & 1 & 0 \\ 2 & -3/2 & 1 \end{pmatrix} \begin{pmatrix} 1 & 2 & -2 \\ 0 & -8 & 10 \\ 0 & 0 & 27 \end{pmatrix} = \begin{pmatrix} 1 & 2 & -2 \\ 2 & -4 & 6 \\ 2 & 16 & 8 \end{pmatrix} = A.$$

We note that this method does not always work in the original form of the matrix. An example where it does not work is a matrix whose first row is all zeroes. In this case, one must permute the rows to arrive at a form where the method will work. It is common to denote the matrix that permutes the rows as P and one then finds the LU decomposition of PA. Computer programs will do the permutation automatically (sometimes, even when it is unnecessary).

In the next chapter we study how to solve certain systems of equations using a process called Gaussian elimination. The LU decomposition can provide a more efficient method of solving systems of linear equations than Gaussian elimination in that fewer steps are involved.

The website https://en.wikipedia.org/wiki/LU_decomposition has many enhancements to this process.

Exercises

In Exercises 1–8 find the LU decomposition of the given matrices:

1. $\begin{pmatrix} 1 & 3 & 6 \\ 2 & 9 & -2 \\ 3 & 6 & 5 \end{pmatrix}$

2. $\begin{pmatrix} 2 & 4 & 6 \\ 1 & 3 & 5 \\ -1 & 1 & 4 \end{pmatrix}$

3. $\begin{pmatrix} 3 & 1 & 4 \\ 6 & 1 & 8 \\ 0 & 8 & 5 \end{pmatrix}$

4. $\begin{pmatrix} 1 & 2 & 2 \\ 3 & 4 & 8 \\ 2 & 12 & 16 \end{pmatrix}$

5. $\begin{pmatrix} 1 & 2 & -4 \\ 2 & -5 & 8 \\ 4 & 6 & 9 \end{pmatrix}$

6. $\begin{pmatrix} 1 & 7 & -5 \\ 3 & -2 & 1 \\ -4 & 1 & 0 \end{pmatrix}$

7. $\begin{pmatrix} 2 & 2 & 0 & 1 \\ 4 & 0 & -3 & 5 \\ 1 & 3 & 3 & 6 \\ -3 & 5 & 7 & 0 \end{pmatrix}$

8. $\begin{pmatrix} 4 & 4 & 6 & 3 \\ 3 & 1 & -2 & -5 \\ 1 & 2 & -4 & 0 \\ 2 & -1 & 0 & 7 \end{pmatrix}$

Chapter 2

Systems of Linear Equations

In this chapter we discuss finding the solution of a system of linear equations. While this is an important topic in itself, the techniques we learn here will carry over to many other types of problems.

Historical Note

Seki Takakazu, also called **Seki Kōwa**, (born c. 1640, died 1708), was the most important figure of the *wasan* ("Japanese calculation") tradition that
flourished from the early seventeenth century until the opening of Japan to the West in the mid-nineteenth century. He was instrumental in recovering mathematical knowledge from ancient Chinese sources and then extending and generalizing the main results.

Seki anticipated several of the discoveries of Western mathematics and was the first person to study determinants in 1683. Ten years later, Leibniz independently used determinants to solve simultaneous equations, although Seki's version was the more general.

He developed a theory of determinants that predated work by the eighteenth-century German mathematician, Gottfried Leibniz.

Seki's discovery, around 1680, of a general theory of elimination – a method of solving simultaneous equations by reducing the number of variables one by one – wasn't matched until more than a century later, by Étienne Bézout (1730–1783).

Section 2.1 Basic Definitions

Definition:

A linear equation in the variables $x_1, \ldots, x_n$ is an expression of the form

$$a_1 x_1 + \cdots + a_n x_n = b$$

where $a_1,\ldots,a_n$ and b are constants.

A solution to the equation

$$a_1 x_1 + \cdots + a_n x_n = b$$

is an ordered n-tuple of numbers $\left(s_1,\ldots,s_n\right)$ for which

$$a_1 s_1 + \cdots + a_n s_n = b.$$

The solution set to the equation

$$a_1 x_1 + \cdots + a_n x_n = b$$

is the set of all solutions to the equation.

A system of linear equations in the variables $x_1,\ldots,x_n$ is a finite set of linear equations of the form

$$a_{11} x_1 + \cdots + a_{1n} x_n = b_1$$
$$a_{21} x_1 + \cdots + a_{2n} x_n = b_2$$
$$\vdots$$
$$a_{m1} x_1 + \cdots + a_{mn} x_n = b_m$$

A solution to the system of equations above is an ordered n-tuple of numbers $\left(s_1,\ldots,s_n\right)$ that is a solution to each of the equations in the system. The set of all possible solutions is called the solution set of the system.

Exercises

In Problems 1–3 determine whether the given values form a solution to the system of equations

1. $2x + y - z = 5$
   $x + y = 2$
   $2y - z = 1$
   $x = 2,\ y = 0,\ z = -1.$

2. $x - y = 4$
   $2x + 3y = 5$
   $x = \dfrac{17}{5},\quad y = -\dfrac{3}{5}$

3. $x - y + 2z = 0$
   $3x - 2y = 4$
   $y + z = 3$
   $x = 1,\ y = -1,\ z = -1.$

4. Which of the following are linear equations? If an equation is not a linear equation, tell why.

(a.) $3x - xy + 2z = 7$

(b.) $3a - 4b = \sqrt{7}$

(c.) $\sin(30°)x - 4y - 7z = 0$

(d.) $w - \sqrt{5}x + \dfrac{4}{z} = 0$

(e.) $4\sin x + y^2 + 3^z = -5$

(f.) $3w - 2y = 6z = -8$

Section 2.2 Solving Systems of Linear Equations (Gaussian Elimination)

The solution set of a system of linear equations has an important property that is not necessarily shared by equations that are not linear; namely, there is either

(i) No solution.
(ii) Exactly one solution.
(iii) Infinitely many solutions.

We will see why this is the case later in this chapter.

Definition:

A system of linear equations that has either exactly one solution or infinitely many solutions is said to be consistent. A system of linear equations that has no solutions is inconsistent.

We give an example of each possibility.

Example

For the system

$$2x - 3y = 5$$

$$x + 3y = -2$$

there is exactly one solution, namely $x = 1$, $y = -1$. The graph of the system of equations

$$2x - 3y = 5$$

$$3x - y = 4$$

$$x + y = 1$$

is shown in Figure 2.1(a.). Since there is no point where the graphs of all three equations meet, there is no solution to the system of equations. Figure 2.1(b.) shows a system of three equations with exactly one solution.

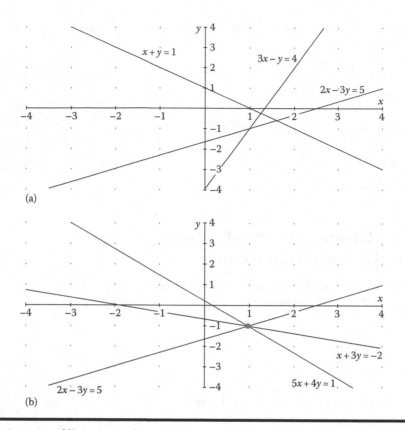

Figure 2.1 Systems of linear equations with (a) no solution and (b) one solution.

The system of equations

$$x + y = 3$$

$$x + y = 4$$

obviously has no solution. The graph is shown in Figure 2.2(a.). The system of equations

$$x + y = 3$$

$$2x + 2y = 6$$

has infinitely many solutions. The graph is shown in Figure 2.2(b.).

A special and important class of systems of linear equations that always have at least one solution is the class of homogeneous equations. These are systems where every $b_i = 0$. One solution to such a system will always be when each variable is 0. This means for a system of two or three variables, the graph of each equation in a homogeneous system passes through the origin.

Solving Systems of Linear Equations

The most common way of solving a system of linear equations is to convert the given system to an equivalent system where the solution to the equivalent system is obvious. By "an equivalent system" we mean that the two systems have the same variables and the same solution set.

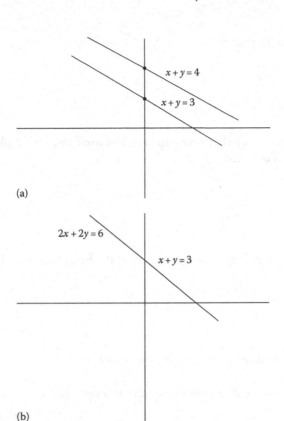

Figure 2.2 (a) A system of equations with no solution. (b) A system of equations with infinitely many solutions.

Most often, the solution to a system of linear equations is found by a computer. We demonstrate below how to find the solution by hand. A computer program uses a very similar technique but, of course, is much more rapid.

The conversion is accomplished by applying the following operations that do not affect the solution set of a system of equations:

(i) Multiplying an equation by a non-zero number.
    The equation

$$3x + 4y = 7$$

is equivalent to

$$6x + 8y = 14.$$

(ii) Interchanging the order of equations.
    The system of equations

$$2x - 4y + 3z = 5$$

$$9x + 12y + z = -10$$

is equivalent to the system

$$9x + 12y + z = -10$$

$$2x - 4y + 3z = 5.$$

(iii) Replacing one equation by the sum or difference of itself and a multiple of another equation. In the system of equations

$$3x + 2y = 10$$

$$x - y = 4$$

we replace the second equation by the sum of the first plus twice the second to get

$$3x + 2y = 10$$

$$5x = 18$$

which is an equivalent system that is easier to solve.

Notice the parallel of these conditions to elementary row operations.

Example

We apply these operations to a system of equations to yield an equivalent system whose solution is obvious. Consider

$$3x + 4y - 2z = 5$$
$$x - 2y + z = 0$$
$$7x + 4y - 3z = 3$$

Interchange the first and second equations to get

$$x - 2y + z = 0$$
$$3x + 4y - 2z = 5$$
$$7x + 4y - 3z = 3$$

Replace the second equation by the second equation minus three times the first equation to get

$$x - 2y + z = 0$$
$$0x + 10y - 5z = 5$$
$$7x + 4y - 3z = 3$$

Replace the third equation by the sum of the third equation and minus seven times the first equation to get

$$x - 2y + z = 0$$
$$0x + 10y - 5z = 5$$
$$0x + 18y - 10z = 3$$

Note that now the variable x appears in only the first equation.
Multiply the second equation by 1/5 to get

$$x-2y+z=0$$
$$0x+2y-z=1$$
$$0x+18y-10z=3$$

Replace the first equation by the sum of the first equation plus the second equation to get

$$x+0y+0z=1$$
$$0x+2y-z=1$$
$$0x+18y-10z=3$$

We now know that $x=1$.
Replace the third equation by the sum of the third equation minus nine times the second equation to get

$$x+0y+0z=1$$
$$0x+2y-z=1$$
$$0x+0y-1z=-6$$

We now know that $z=6$.
Multiply the third equation by –1 to get

$$x+0y+0z=1$$
$$0x+2y-z=1$$
$$0x+0y+z=6$$

Replace the second equation by the sum of the second equation and the third equation to get

$$x+0y+0z=1$$
$$0x+2y+0z=7$$
$$0x+0y+z=6$$

Multiply the second equation by 1/2 to get

$$x+0y+0z=1$$
$$0x+y+0z=7/2$$
$$0x+0y+z=6$$

One typically does not list the terms whose coefficients are 0, so we have

$$x=1$$
$$y=7/2.$$
$$z=6$$

The procedure that we have just executed is an example of Gaussian elimination, which is one of the most commonly used methods of solving systems of linear equations. A modified version of Gaussian elimination is what most computer systems and calculators use to solve systems of linear equations.

Using Technology to Accomplish Gaussian Elimination

The operations we did above can be done by a computer, which will be our approach in most cases. In order to give the information in a form that is easy to communicate with a computer, it is convenient to use matrices. This also simplifies the notation in solving the system by hand.

Instructions to use MATLAB or R to perform the computations of linear algebra, are given in the appendices.

Consider the system of equations in the example above. We associate the system of linear equations with the matrix that consists of the coefficients of the variables on the left columns of the matrix and the numbers on the right column for each equation. Thus we have

$$3x + 4y - 2z = 5 \qquad \begin{pmatrix} 3 & 4 & -2 & | & 5 \\ 1 & -2 & 1 & | & 0 \\ 7 & 4 & -3 & | & 3 \end{pmatrix}. \tag{1}$$
$$x - 2y + z = 0 \quad \leftrightarrow$$
$$7x + 4y - 3z = 3$$

The matrix on the right in Equation (1) is called the augmented matrix for the system of equations. The vertical line is simply to help separate the coefficients of the variables from the numbers on the right side of the equations, and will not normally appear.

There is also the matrix of coefficients associated with a system of linear equations that is occasionally used. In Equation (1) this is

$$\begin{pmatrix} 3 & 4 & -2 \\ 1 & -2 & 1 \\ 7 & 4 & -3 \end{pmatrix}.$$

Representing systems of linear equations by matrices is advantageous for computer solutions of the systems.

Exercises

In Exercises 1–3 find the augmented matrix for the systems of equations

1. $3x + 5y = 9$
   $x - 2y = -7$
   $y = 5$

2. $x - y + 3z = 6$
   $2x + 4y - 3z = 2$
   $x + y + z = 0$
   $3y - 2z = -8$

3. $x + y + z = 0$
   $x + y + z = 3$
   $x + 2y + z = 5$

In Exercises 4–6 find the systems of equations associated with the augmented matrix. Denote the variables as $x_1, x_2, \ldots$

4. $\begin{pmatrix} 3 & -1 & 2 & | & 0 \\ 0 & 4 & 1 & | & 6 \\ -2 & 5 & 1 & | & -8 \end{pmatrix}$

5. $\begin{pmatrix} 8 & 3 & 0 & | & 6 \\ -4 & 2 & 1 & | & 2 \end{pmatrix}$

6. $\begin{pmatrix} 1 & 0 & 4 & 2 & | & 6 \\ 3 & 2 & 1 & -7 & | & -3 \\ 0 & 0 & 2 & 0 & | & 2 \\ -2 & 4 & -1 & 5 & | & 0 \end{pmatrix}$

Section 2.3 Equivalent Systems of Linear Equations

The manipulations of a system of equations that yield an equivalent system of equations have corresponding manipulations with the associated matrix. These manipulations correspond to multiplying the augmented matrix by particular elementary matrices. In particular,

(i) Multiplying an equation by a non-zero number corresponds to multiplying the corresponding row of the matrix by the same non-zero number.
(ii) Interchanging the order of equations corresponds to interchanging the corresponding rows of the matrix, and
(iii) Replacing one equation by the sum of itself and a multiple of another equation corresponds to replacing the corresponding row by the sum of itself and the multiple of another row.

In Chapter 1, we saw that these operations can be done by multiplying the given matrix on the left by an elementary matrix.

Definition:
   The operations on a matrix given above are called elementary row operations.

Definition:
   If a matrix B can be obtained from the matrix A by applying elementary row operations to matrix A, then A and B are said to be row equivalent.
   From what has been noted above about elementary row operations and systems of linear equations, we have the following result:

Theorem 2.1:
   Two systems of linear equations that are expressed by matrices that are row equivalent, have the same solution set.

   Example
   We redo the previous example, showing how the associated matrices are changed. We have

$$
\begin{aligned}
3x + 4y - 2z &= 5 \\
x - 2y + z &= 0 \\
7x + 4y - 3z &= 3
\end{aligned}
\quad \leftrightarrow \quad
\begin{pmatrix}
3 & 4 & -2 & 5 \\
1 & -2 & 1 & 0 \\
7 & 4 & -3 & 3
\end{pmatrix}
$$

$$
\begin{aligned}
x - 2y + z &= 0 \\
3x + 4y - 2z &= 5 \\
7x + 4y - 3z &= 3
\end{aligned}
\quad \leftrightarrow \quad
\begin{pmatrix}
1 & -2 & 1 & 0 \\
3 & 4 & -2 & 5 \\
7 & 4 & -3 & 3
\end{pmatrix}
$$

$$
\begin{aligned}
x - 2y + z &= 0 \\
0x + 10y - 5z &= 5 \\
7x + 4y - 3z &= 3
\end{aligned}
\quad \leftrightarrow \quad
\begin{pmatrix}
1 & -2 & 1 & 0 \\
0 & 10 & -5 & 5 \\
7 & 4 & -3 & 3
\end{pmatrix}
$$

$$
\begin{aligned}
x - 2y + z &= 0 \\
0x + 10y - 5z &= 5 \\
0x + 18y - 10z &= 3
\end{aligned}
\quad \leftrightarrow \quad
\begin{pmatrix}
1 & -2 & 1 & 0 \\
0 & 10 & -5 & 5 \\
0 & 18 & -10 & 3
\end{pmatrix}
$$

$$
\begin{aligned}
x - 2y + z &= 0 \\
0x + 2y - z &= 1 \\
0x + 18y - 10z &= 3
\end{aligned}
\quad \leftrightarrow \quad
\begin{pmatrix}
1 & -2 & 1 & 0 \\
0 & 2 & -1 & 1 \\
0 & 18 & -10 & 3
\end{pmatrix}
$$

$$
\begin{aligned}
x + 0y + 0z &= 1 \\
0x + 2y - z &= 1 \\
0x + 18y - 10z &= 3
\end{aligned}
\quad \leftrightarrow \quad
\begin{pmatrix}
1 & 0 & 0 & 1 \\
0 & 2 & -1 & 1 \\
0 & 18 & -10 & 3
\end{pmatrix}
$$

$$
\begin{aligned}
x + 0y + 0z &= 1 \\
0x + 2y - z &= 1 \\
0x + 0y - 1z &= -6
\end{aligned}
\quad \leftrightarrow \quad
\begin{pmatrix}
1 & 0 & 0 & 1 \\
0 & 2 & -1 & 1 \\
0 & 0 & -1 & -6
\end{pmatrix}
$$

$$
\begin{aligned}
x + 0y + 0z &= 1 \\
0x + 2y - z &= 1 \\
0x + 0y + z &= 6
\end{aligned}
\quad \leftrightarrow \quad
\begin{pmatrix}
1 & 0 & 0 & 0 \\
0 & 2 & -1 & 1 \\
0 & 0 & 1 & 6
\end{pmatrix}
$$

$$
\begin{aligned}
x + 0y + 0z &= 1 \\
0x + 2y + 0z &= 7 \\
0x + 0y + z &= 6
\end{aligned}
\quad \leftrightarrow \quad
\begin{pmatrix}
1 & 0 & 0 & 1 \\
0 & 2 & 0 & 7 \\
0 & 0 & 1 & 6
\end{pmatrix}
$$

$$
\begin{aligned}
x + 0y + 0z &= 1 \\
0x + y + 0z &= 7/2 \\
0x + 0y + z &= 6
\end{aligned}
\quad \leftrightarrow \quad
\begin{pmatrix}
1 & 0 & 0 & 1 \\
0 & 1 & 0 & 7/2 \\
0 & 0 & 1 & 6
\end{pmatrix}.
$$

For a procedure like the one above, we will almost always use computer technology. The typical way that we will solve systems of linear equations is to enter the augmented matrix and command the computer to put the matrix into row reduced form. We must then interpret the output.

Interpreting the output to get a solution is simple when there is exactly one solution as there was in the example above. Our next task is to understand the output when there is no solution or there are infinitely many solutions.

Row Reduced Form of a Matrix

The row reduced form of a system of linear equations is the simplest form from which to interpret the solution for a system of linear equations.

A matrix in row reduced form has the following characteristics:

Reading from left to right, the first non-zero entry in each row will be a 1. The first 1 in each row is called a leading 1. The corresponding variable is called a leading variable.
A leading 1 is the only non-zero entry in its column.
A leading 1 in a row is always strictly to the right of a leading 1 of a row above it.
Any rows that are all 0 will appear as the bottom rows of the row reduced form.

The matrix below is in row reduced form.

$$\begin{pmatrix} 0 & 1 & 3 & 0 & 2 & 0 & 6 \\ 0 & 0 & 0 & 1 & 0 & 5 & 3 \\ 0 & 0 & 0 & 0 & 0 & 0 & 0 \end{pmatrix}$$

If the variables for the matrix above are x_1, x_2, x_3, x_4, x_5, and x_6, then the leading variables are x_2 and x_4.

$$\begin{pmatrix} 1 & 0 & 0 & 5 & 0 \\ 0 & 1 & 0 & 4 & 0 \\ 0 & 0 & 1 & 6 & 0 \\ 0 & 0 & 0 & 0 & 0 \end{pmatrix}$$

If the variables for the matrix above are x_1, x_2, x_3, and x_4, then the leading variables are x_1, x_2, and x_3.

The following matrices are not in row reduced form.

$$\begin{pmatrix} 0 & 1 & 2 & 0 & 3 \\ 1 & 0 & 0 & 0 & 9 \\ 0 & 0 & 0 & 1 & 5 \end{pmatrix}$$

$$\begin{pmatrix} 1 & 1 & 0 & 0 & 3 \\ 0 & 0 & 2 & 0 & 9 \\ 0 & 0 & 0 & 1 & 5 \end{pmatrix}$$

$$\begin{pmatrix} 1 & 0 & 0 & 5 & 0 \\ 0 & 1 & 0 & 1 & 0 \\ 0 & 0 & 0 & 0 & 0 \\ 0 & 0 & 0 & 0 & 1 \end{pmatrix}$$

Exercises

1. Argue that an $n \times n$ matrix in row reduced form either has a row of all 0s or is the identity matrix.

2. Which of the following matrices are in row reduced form?

(a.) $\begin{pmatrix} 1 & 0 & 0 & 2 & 3 \\ 0 & 0 & 1 & 0 & 6 \\ 0 & 1 & 0 & 3 & 5 \\ 0 & 0 & 0 & 0 & 0 \end{pmatrix}$

(b.) $\begin{pmatrix} 1 & 3 & 0 & 2 & 0 \\ 0 & 0 & 1 & 7 & 0 \\ 0 & 0 & 0 & 0 & 1 \\ 0 & 0 & 0 & 0 & 0 \end{pmatrix}$

(c.) $\begin{pmatrix} 0 & 1 & 0 & 4 & 0 \\ 0 & 0 & 0 & 0 & 1 \\ 0 & 0 & 0 & 0 & 0 \\ 0 & 0 & 0 & 0 & 0 \end{pmatrix}$

(d.) $\begin{pmatrix} 0 & 0 & 0 & 1 & 0 \\ 0 & 1 & 2 & 0 & 0 \\ 0 & 0 & 0 & 0 & 0 \\ 1 & 0 & 0 & 0 & 0 \end{pmatrix}$

(e.) $\begin{pmatrix} 1 & 0 & 0 & 0 & 0 \\ 0 & 1 & 0 & 0 & 0 \\ 0 & 0 & 0 & 1 & 0 \\ 1 & 0 & 0 & 0 & 0 \end{pmatrix}$

Section 2.4 Expressing the Solution of a System of Linear Equations

Systems of Linear Equations That Have No Solutions

The system of equations

$$3x + 4y + 5z = 5$$

$$3x + 4y + 5z = 9$$

has no solution. The augmented matrix for the system is

$$\begin{pmatrix} 3 & 4 & 5 & 5 \\ 3 & 4 & 5 & 9 \end{pmatrix}$$

which, when row reduced,

$$\begin{pmatrix} 1 & 4/3 & 5/3 & 0 \\ 0 & 0 & 0 & 1 \end{pmatrix}.$$

The row $\begin{pmatrix} 0 & 0 & 0 & 1 \end{pmatrix}$ corresponds to the equation

$$0x + 0y + 0z = 1$$

which is impossible.

A system of linear equations has no solution if and only if one of the rows in the row reduced matrix of the augmented matrix for system is $\begin{pmatrix} 0 & 0 & \cdots & 0 & 1 \end{pmatrix}.$

Systems of Linear Equations That Have Exactly One Solution

A system of linear equations will have exactly one solution if and only if each variable is a leading variable and there is no row of the form $\begin{pmatrix} 0 & 0 & \cdots & 0 & 1 \end{pmatrix}$ when the system is in row reduced form.

Systems of Linear Equations That Have Infinitely Many Solutions

Example

Consider the system of equations

$$3x_1 - 2x_3 + x_4 + 6x_5 - x_6 = 3$$

$$x_1 + 5x_2 - 2x_3 - 3x_4 + x_5 + 2x_6 = 10.$$

The augmented matrix and the row reduced matrix for the system are

$$\begin{pmatrix} 3 & 0 & -2 & 1 & 6 & -1 & 3 \\ 1 & 5 & -2 & -3 & 1 & 2 & 10 \end{pmatrix} \rightarrow \begin{pmatrix} 1 & 0 & \frac{-2}{3} & \frac{1}{3} & 2 & \frac{-1}{3} & 1 \\ 0 & 1 & \frac{-4}{15} & \frac{-2}{3} & \frac{-1}{5} & \frac{7}{15} & \frac{9}{5} \end{pmatrix}$$

There is a solution because there is no row $\begin{pmatrix} 0 & 0 & 0 & 0 & 0 & 0 & 1 \end{pmatrix}.$

Because not every variable is a leading variable there are infinitely many solutions as we show in more detail below. This, together with the previous cases we have seen, is why a linear system has no solutions, exactly one solution, or infinitely many solutions. Note that when there are more variables than equations, there will be at least one variable that is not a leading variable, and so such a system that is consistent will have infinitely many solutions.

In the case where there are infinitely many solutions, it is sometimes helpful to convert from the row reduced form of the matrix back to the form where the variables are explicitly listed. In this case, we have

$$x_1 - \frac{2}{3}x_3 + \frac{1}{3}x_4 + 2x_5 - \frac{1}{3}x_6 = 1$$

$$x_2 - \frac{4}{15}x_3 - \frac{2}{3}x_4 - \frac{1}{5}x_5 + \frac{7}{15}x_6 = \frac{9}{5}.$$

When there are infinitely many solutions, there is a protocol for expressing the solution in standard form, which we now describe.

The variables that are not leading variables are called free variables. In this example, x_1 and x_2 are leading variables and x_3, x_4, x_5, and x_6 are free variables. Any time there are free variables in a consistent system, there will be infinitely many solutions.

The first step to convert to the standard form of the solution is to express each leading variable in terms of the free variables. In this case, we have

$$x_1 = \frac{2}{3}x_3 - \frac{1}{3}x_4 - 2x_5 + \frac{1}{3}x_6 + 1$$

$$x_2 = \frac{4}{15}x_3 + \frac{2}{3}x_4 + \frac{1}{5}x_5 - \frac{7}{15}x_6 + \frac{9}{5}$$

We generate a solution by choosing any values for the free variables, and those values will determine the values of the leading variables. For example, if we take

$$x_3 = 15, x_4 = 3, x_5 = 5 \text{ and } x_6 = 15$$

then

$$x_1 = \frac{2}{3}(15) - \frac{1}{3}(3) - 2(5) + \frac{1}{3}(15) + 1 = 10 - 1 - 10 + 5 + 1 = 5$$

$$x_2 = \frac{4}{15}(15) + \frac{2}{3}(3) + \frac{1}{5}(5) - \frac{7}{15}(15) + \frac{9}{5} = 4 + 2 + 1 - 7 + \frac{9}{5} = \frac{9}{5}$$

and one can check that

$$x_1 = 5, x_2 = \frac{9}{5}, x_3 = 15, x_4 = 3, x_5 = 5 \text{ and } x_6 = 15$$

is a solution to

$$3x_1 - 2x_3 + x_4 + 6x_5 - x_6 = 3$$

$$x_1 + 5x_2 - 2x_3 - 3x_4 + x_5 + 2x_6 = 10.$$

The preferred format for the infinite solutions case is to assign a parameter to each free variable. In the present example, we let

$$x_3 = r, \ x_4 = s, \ x_5 = t, \ x_6 = u$$

and then express each leading variable in terms of the free variables. Here, we have

$$x_1 = \frac{2}{3}r - \frac{1}{3}s - 2t + \frac{1}{3}u + 1$$

$$x_2 = \frac{4}{15}r + \frac{2}{3}s + \frac{1}{5}t - \frac{7}{15}u + \frac{9}{5}.$$

Example

The augmented matrix for the system of equations

$$3w - 2x + 4y - z = 5$$

$$w + y = 7$$

$$x + y + z = 0$$

is

$$\begin{pmatrix} -3 & -2 & 4 & -1 & 5 \\ 1 & 0 & 1 & 0 & 7 \\ 0 & 1 & 1 & 1 & 0 \end{pmatrix}.$$

The row reduced form of the matrix is

$$\begin{pmatrix} 1 & 0 & 0 & -1/9 & 37/9 \\ 0 & 1 & 0 & 8/9 & -26/9 \\ 0 & 0 & 1 & 1/9 & 26/9 \end{pmatrix}$$

So w, x and y are leading variables and z is a free variable. Let $z = t$.
Then

$$w = \frac{1}{9}z + \frac{37}{9} = \frac{1}{9}t + \frac{37}{9}$$

$$x = \frac{-8}{9}z - \frac{26}{9} = \frac{-8}{9}t - \frac{26}{9}$$

$$y = \frac{-1}{9}z + \frac{26}{9} = \frac{-1}{9}t + \frac{26}{9}$$

In the appendix we give the commands that MATLAB uses to convert a matrix to row reduced form.

Application of Linear Systems to Curve Fitting

Example

We find the equation of the circle that passes through the points (1,1), (3,1) and (2,4). The equation of a circle whose radius is r and center (h,k) is

$$(x-h)^2 + (y-k)^2 = r^2$$

or

$$x^2 - 2xh + h^2 + y^2 - 2yk + k^2 = r^2$$

or

$$x^2 + y^2 - 2xh - 2yk + k^2 + h^2 - r^2 = 0$$

which can be written

$$x(-2h) + y(-2k) + (k^2 + h^2 - r^2) = -x^2 - y^2. \qquad (2)$$

If we let $a = -2h, b = -2k, c = k^2 + h^2 - r^2$ then Equation (2) becomes

$$xa + yb + c = -x^2 - y^2$$

and each point (x,y) on the circle yields a linear equation in a, b, and c.
   We have

$$(1,1) \leftrightarrow a + b + c = -2$$

$$(3,1) \leftrightarrow 3a + b + c = -10$$

$$(2,4) \leftrightarrow 2a + 4b + c = -20$$

When the augmented matrix

$$\begin{pmatrix} 1 & 1 & 1 & -2 \\ 3 & 1 & 1 & -10 \\ 2 & 4 & 1 & -20 \end{pmatrix}$$

is row reduced, the result is

$$\begin{pmatrix} 1 & 0 & 0 & -4 \\ 0 & 1 & 0 & -14/3 \\ 0 & 0 & 1 & 20/3 \end{pmatrix}$$

so

$$a = -4 \text{ and } h = \frac{a}{-2} = 2, \; b = -\frac{14}{3} \text{ and } k = \frac{b}{-2} = \frac{7}{3},$$

$$c = \frac{20}{3} \text{ and } c = k^2 + h^2 - r^2$$

so

$$r^2 = k^2 + h^2 - c = 4 + \frac{49}{9} - \frac{20}{3} = \frac{36 + 49 - 60}{9} = \frac{25}{9}$$

and

$$r = \frac{5}{3}.$$

Thus, the equation of the circle is

$$(x-2)^2 + \left(y - \frac{7}{3}\right)^2 = \frac{25}{9}$$

In the next example we find the equation of a plane that passes through three given points. There will be such a plane if and only if the points are not collinear. If there is a plane, it will have an equation of the form

$$ax + by + cz = d.$$

Our task is, knowing three points on the plane, find *a*, *b*, *c*, and *d*.

Our method is simple, but there are two cases. Case 1 is where $d \neq 0$ and case 2 is where $d = 0$.

Example

This is for the case where $d \neq 0$.

Find the equation of the plane that passes through the points (8,2,6), (4,4,4), (2,6,1).

The equation of a plane is of the form

$$ax + by + cz = d \text{ or } \frac{a}{d}x + \frac{b}{d}y + \frac{c}{d}z = 1.$$

The division by *d* requires $d \neq 0$.

We will use the second form. We want to find the coefficients of *x*, *y*, and *z*. We substitute the given points, which yields the three equations

$$a'8 + b'2 + c'6 = 1$$

$$a'4 + b'4 + c'4 = 1$$

$$a'2 + b'6 + c'1 = 1$$

where

$$a' = \frac{a}{d}, b' = \frac{b}{d}, c' = \frac{c}{d}.$$

The augmented matrix for the system of equations is

$$\begin{pmatrix} 8 & 2 & 6 & 1 \\ 4 & 4 & 4 & 1 \\ 2 & 6 & 1 & 1 \end{pmatrix}.$$

When row reduced, this becomes

$$\begin{pmatrix} 1 & 0 & 0 & \dfrac{1}{28} \\ 0 & 1 & 0 & \dfrac{1}{7} \\ 0 & 0 & 1 & \dfrac{1}{14} \end{pmatrix}$$

so

$$a' = \frac{1}{28}, b' = \frac{1}{7}, c' = \frac{1}{14}$$

and the equation of the plane is

$$\frac{1}{28}x + \frac{1}{7}y + \frac{1}{14}z = 1$$

or

$$x + 4y + 2z = 28.$$

If we had thought $d = 0$, our equations would have been

$$a8 + b2 + c6 = 0$$

$$a4 + b4 + c4 = 0$$

$$a2 + b6 + c1 = 0$$

The augmented matrix would be

$$\begin{pmatrix} 8 & 2 & 6 & 0 \\ 6 & 4 & 4 & 0 \\ 2 & 6 & 1 & 0 \end{pmatrix}$$

which, when row reduced is

$$\begin{pmatrix} 1 & 0 & 0 & 0 \\ 0 & 1 & 0 & 0 \\ 0 & 0 & 1 & 0 \end{pmatrix}$$

This implies that the equation for the plane is

$$0x + 0y + 0z = 0$$

which is not a plane at all.

Case 2. $d = 0$.

Example:

Find the equation of the plane that passes through the points

$$(1,1,-1),(3,2,0),\left(1,0,\frac{-2}{5}\right).$$

In this case, if we did not know $d=0$ (and when we begin the problem, we typically do not know that) and set up the equations as before, the augmented matrix would be

$$\begin{pmatrix} 1 & 1 & -1 & 1 \\ 3 & -2 & 0 & 1 \\ 1 & 0 & -2/5 & 1 \end{pmatrix}$$

which, when row reduced, is

$$\begin{pmatrix} 1 & 0 & -2/5 & 0 \\ 0 & 1 & -3/5 & 0 \\ 0 & 0 & 0 & 1 \end{pmatrix}.$$

This indicates there is no solution and our assumption that $d \neq 0$ was erroneous.

We modify the matrix with $d=0$ to get

$$\begin{pmatrix} 1 & 1 & -1 & 0 \\ 3 & -2 & 0 & 0 \\ 1 & 0 & -2/5 & 0 \end{pmatrix}$$

which, when row reduced is

$$\begin{pmatrix} 1 & 0 & -2/5 & 0 \\ 0 & 1 & -3/5 & 0 \\ 0 & 0 & 0 & 0 \end{pmatrix}.$$

Thus c is the free variable,

$$a = \frac{2}{5}c, \; b = \frac{3}{5}c$$

and the equation of the plane is

$$\frac{2}{5}cx + \frac{3}{5}cy + cz = 0$$

or

$$c(2x + 3y + z) = 0.$$

Exercises

In Exercises 1–4, we give the row reduced form of the augmented matrix for a system of linear equations. Tell whether the system of equations is consistent, and if it is consistent give the solution in standard form.

1. $\begin{pmatrix} 1 & 2 & 0 & -1 & 3 & 4 \\ 0 & 0 & 1 & 2 & 0 & -2 \end{pmatrix}$

2. $\begin{pmatrix} 1 & 0 & 6 & 0 & 1 \\ 0 & 1 & 2 & 0 & 3 \\ 0 & 0 & 0 & 1 & 0 \\ 0 & 0 & 0 & 0 & 0 \\ 0 & 0 & 0 & 0 & 0 \end{pmatrix}$

3. $\begin{pmatrix} 1 & 0 & 0 \\ 0 & 1 & 0 \\ 0 & 0 & 1 \end{pmatrix}$

4. $\begin{pmatrix} 1 & 4 & 0 & 0 & 6 \\ 0 & 0 & 1 & 0 & 3 \\ 0 & 0 & 0 & 1 & 2 \\ 0 & 0 & 0 & 0 & 0 \end{pmatrix}$

In Exercises 5–10, we give the augmented matrix for a system. Find the row reduced form of the matrix and if a solution exists, give the solution in standard form.

5. $2x - 3y = 6$
   $x + 5y = 9$

6. $x - 2y + 3z = 5$
   $5x - y + 2z = 4$

7. $3x_1 - 5x_2 + x_3 - 2x_4 = 0$
   $6x_1 - 10x_2 + 2x_3 - 4x_4 = -6$

8. $x + 2y - 3z = -1$
   $6x - 2y + 4z = 14$
   $10x + 6y - 8z = 4$

9. $4x + 3y - 2z = 14$
   $-2x + y - 4z = -2$
   $2x + 4y - 6z = 12$

10. $4x - y = 8$
    $2x + 7y = 6$
    $-3x - 5y = 0$

11. (a.) Is it possible for a system of three equations and two unknowns to have a unique solution?

    (b.) Is it possible for a system of three equations and four unknowns to have a unique solution?

12. A consistent system of linear equations has four equations and six variables. How many solutions does it have?

13. For what value(s) of h does the system

$$3x - 4y = 7$$

$$2x + hy = 9$$

have exactly one solution?
14. For what value(s) of h does the system

$$x - 2y = 7$$

$$hx + 2y = 9$$

have no solution?
15. Find the line of intersection of the two planes

$$3x - 2y + 5z = 12$$

$$2x + 6y - z = 7$$

16. Find the polynomial $p(x) = ax^2 + bx + c$ for which
    (a.) $p(1) = 8, \ p(2) = -6, \ p(4) = 16$
    (b.) $p(0) = 4, \ p(-3) = 7, \ p(2) = 1$
    (c.) $p(-2) = 5, \ p(-4) = 9, \ p(1) = 0$
17. Find the equation of the plane that passes through the points
    (a.) $(3, -9, 2), (-4, 0, 6), (1, 1, 1)$
    (b.) $(9, 3, 7), (1, 6, -2), (5, 0, 0)$
    (c.) $(4, 9, 6), (2, 7, -8), (4, 4, 12)$
    (d.) $(4, 1, 0), (2, 1, 2), (3, 2, 5)$
18. Is there a plane that passes through the points $(3, -4, 5), (5, -3, 4), (7, -2, 3)$?
    Is there a line that passes through these points?
19. Center of mass problems. The center of mass of a group of particles is the unique point that is the average position of all parts of the system weighted according to their masses. This is useful because often systems of many points can be treated as a single point located at the center of mass.

In a system of k particles the center of mass is computed according to the formula

center of mass $= \dfrac{1}{m} \left(m_1 \hat{v}_1 + \cdots + m_k \hat{v}_k \right)$

where m_i is the mass of the ith particle and $\hat{v}_i$ is the location of the ith particle and $m = m_1 + \cdots + m_k$.

Note that the center of mass is a linear combination of the vectors $\{\hat{v}_1,\ldots,\hat{v}_n\}$
Find the center of mass for the following systems:
(a.)

Location	Mass
(2, –3, 7)	5 g
(5, 8, 0)	8 g
(2, –8, –9)	10 g

(b.)

Location	mass
(4, 9, 1)	1 g
(7, –6, –10)	6 g
(–8, 0, 0)	3 g

(c.)In parts (a.) and (b.) find the mass and location of a particle so that if that particle were added to the system the center of mass would be at (0, 0, 0).

Section 2.5 Expressing Systems of Linear Equations in Other Forms

Representing a System of Linear Equations as a Vector Equation

Matrices are equal when they have the same dimensions and the corresponding entries coincide. In this section, we will use the fact that

$$\begin{pmatrix} a_1 \\ \vdots \\ a_n \end{pmatrix} = \begin{pmatrix} b_1 \\ \vdots \\ b_n \end{pmatrix}$$

if and only if $a_1 = b_1$, ..., $a_n = b_n$.
   Consider

$$x \begin{pmatrix} 4 \\ 2 \\ -7 \end{pmatrix} + y \begin{pmatrix} 1 \\ 0 \\ 5 \end{pmatrix} + z \begin{pmatrix} 3 \\ -2 \\ 6 \end{pmatrix} = \begin{pmatrix} 8 \\ 12 \\ -5 \end{pmatrix} \tag{3}$$

This could, for example, be a problem in physics or engineering, where there are forces available in the directions

$$\begin{pmatrix} 4 \\ 2 \\ -7 \end{pmatrix}, \begin{pmatrix} 1 \\ 0 \\ 5 \end{pmatrix}, \begin{pmatrix} 3 \\ -2 \\ 6 \end{pmatrix}$$

and we want to know how much of each force should be applied to achieve the resultant force

$$\begin{pmatrix} 8 \\ 12 \\ -5 \end{pmatrix}$$

The expression in Equation (3) is a vector equation.

If we expand the left side of Equation (3), we get

$$x\begin{pmatrix} 4 \\ 2 \\ -7 \end{pmatrix} + y\begin{pmatrix} 1 \\ 0 \\ 5 \end{pmatrix} + z\begin{pmatrix} 3 \\ -2 \\ 6 \end{pmatrix} = \begin{pmatrix} 4x \\ 2x \\ -7x \end{pmatrix} + \begin{pmatrix} 1y \\ 0y \\ 5y \end{pmatrix} + \begin{pmatrix} 3z \\ -2z \\ 6z \end{pmatrix} = \begin{pmatrix} 4x+1y+3z \\ 2x+0y-2z \\ -7x+5y+6z \end{pmatrix}$$

so we must have

$$\begin{pmatrix} 4x+1y+3z \\ 2x+0y-2z \\ -7x+5y+6z \end{pmatrix} = \begin{pmatrix} 8 \\ 12 \\ -5 \end{pmatrix}$$

which is true if and only if each of the equations

$$4x+1y+3z = 8$$

$$2x+0y-2z = 12$$

$$-7x+5y+6z = -5$$

is satisfied. Thus, we have converted a vector equation into a system of linear equations. This is worthwhile knowing because there are instances in which it might be advantageous to formulate a problem in one setting, but easier to solve the problem in another setting.

Example

Express the vector equation

$$x_1\begin{pmatrix} 2 \\ 0 \\ -1 \\ 1 \end{pmatrix} + x_2\begin{pmatrix} -6 \\ 4 \\ 3 \\ 7 \end{pmatrix} + x_3\begin{pmatrix} 4 \\ 9 \\ 0 \\ -4 \end{pmatrix} = \begin{pmatrix} 5 \\ 8 \\ 0 \\ -2 \end{pmatrix}$$

as a system of linear equations.

Following the example above, we have

$$2x_1 - 6x_2 + 4x_3 = 5$$

$$4x_2 + 9x_3 = 8$$

$$-x_1 + 3x_2 = 0$$

$$x_1 + 7x_2 - 4x_3 = -2$$

Example

Express the system of linear equations

$$8w - 3x + 4y + 2z = 1$$

$$w + x + 6y = 9$$

as a vector equation.

Reversing the ideas above we get

$$w\begin{pmatrix} 8 \\ 1 \end{pmatrix} + x\begin{pmatrix} -3 \\ 1 \end{pmatrix} + y\begin{pmatrix} 4 \\ 6 \end{pmatrix} + z\begin{pmatrix} 2 \\ 0 \end{pmatrix} = \begin{pmatrix} 1 \\ 9 \end{pmatrix}.$$

Equivalence of a System of Linear Equations and a Matrix Equation

The equation

$$\begin{pmatrix} 2 & -1 & 5 \\ 6 & 0 & 3 \end{pmatrix}\begin{pmatrix} x \\ y \\ z \end{pmatrix} = \begin{pmatrix} 3 \\ -4 \end{pmatrix}$$

is an example of a matrix equation.

We want to find x, y, and z. The expression on the left is the product of a 2×3 matrix with a 3×1 matrix which is a 2×1 matrix, so the dimensions are correct, but the answer may not be not obvious. If we expand the expression on the left, we get

$$\begin{pmatrix} 2x - y + 5z \\ 6x + 3z \end{pmatrix} = \begin{pmatrix} 3 \\ -4 \end{pmatrix}$$

which gives the system of linear equations

$$2x - y + 5z = 3$$

$$6x + 3z = -4.$$

Similarly, the system of linear equations

$$2w - 3x + y - 5z = 7$$

$$w + 5x - 7y = 0$$

is equivalent to the matrix equation

$$\begin{pmatrix} 2 & -3 & 1 & -5 \\ 1 & 5 & -7 & 0 \end{pmatrix} \begin{pmatrix} w \\ x \\ y \\ z \end{pmatrix} = \begin{pmatrix} 7 \\ 0 \end{pmatrix}.$$

Example

If there are the same number of equations as unknowns, and if the matrix is invertible, then there is another method of solution to a matrix equation. Suppose

$$\begin{pmatrix} 3 & 1 & 9 \\ 0 & -2 & 5 \\ -1 & -3 & 6 \end{pmatrix} \begin{pmatrix} x \\ y \\ z \end{pmatrix} = \begin{pmatrix} -4 \\ 1 \\ 2 \end{pmatrix}.$$

If

$$\begin{pmatrix} 3 & 1 & 9 \\ 0 & -2 & 5 \\ -1 & -3 & 6 \end{pmatrix}$$

is invertible then

$$\begin{pmatrix} 3 & 1 & 9 \\ 0 & -2 & 5 \\ -1 & -3 & 6 \end{pmatrix}^{-1} \begin{pmatrix} 3 & 1 & 9 \\ 0 & -2 & 5 \\ -1 & -3 & 6 \end{pmatrix} \begin{pmatrix} x \\ y \\ z \end{pmatrix} = \begin{pmatrix} 3 & 1 & 9 \\ 0 & -2 & 5 \\ -1 & -3 & 6 \end{pmatrix}^{-1} \begin{pmatrix} -4 \\ 1 \\ 2 \end{pmatrix}$$

so

$$\begin{pmatrix} x \\ y \\ z \end{pmatrix} = \begin{pmatrix} 3 & 1 & 9 \\ 0 & -2 & 5 \\ -1 & -3 & 6 \end{pmatrix}^{-1} \begin{pmatrix} -4 \\ 1 \\ 2 \end{pmatrix} = \begin{pmatrix} -1/14 \\ -17/14 \\ -2/7 \end{pmatrix}.$$

This demonstrates that the matrix equation $A\hat{x} = \hat{b}$, where A is a $n \times n$ matrix, will have a unique solution if A^{-1} exists.

Thus, we have the equivalency of the three forms: a system of linear equations, a vector equation, and a matrix equation.

Example

Express the vector equation

$$x \begin{pmatrix} 0 \\ 2 \\ -3 \end{pmatrix} + y \begin{pmatrix} 9 \\ 4 \\ 1 \end{pmatrix} + z \begin{pmatrix} -2 \\ 6 \\ -1 \end{pmatrix} = \begin{pmatrix} 3 \\ 8 \\ 5 \end{pmatrix}$$

as a matrix equation and a system of linear equations.

An equivalent matrix equation is

$$\begin{pmatrix} 0 & 9 & -2 \\ 2 & 4 & 6 \\ -3 & 1 & -1 \end{pmatrix} \begin{pmatrix} x \\ y \\ z \end{pmatrix} = \begin{pmatrix} 3 \\ 8 \\ 5 \end{pmatrix}$$

and an equivalent system of linear equations is

$$9y - 2z = 3$$

$$2x + 4y + 6z = 8$$

$$-3x + y - z = 5$$

We repeat an idea that we have previously seen for emphasis. For A a matrix, the expression $A\hat{x}$ is a linear combination of the columns of A. Similarly, $\hat{y}A$ is a linear combination of the rows of A.

Definition:

If $\hat{v}_1, \hat{v}_2, \ldots, \hat{v}_n$ and $\hat{b}$ are vectors, we say that $\hat{b}$ is a linear combination of $\hat{v}_1, \hat{v}_2, \ldots, \hat{v}_n$ if there are scalars $a_1, a_2, \ldots, a_n$ for which

$$a_1 \hat{v}_1 + a_2 \hat{v}_2 + \cdots a_n \hat{v}_n = \hat{b}.$$

In later chapters, an important question will be whether a given vector can be expressed as a linear combination of a given set of vectors. We now have the tools to answer this type of question.

Example

Determine if $\hat{b}$ is a linear combination of $\hat{v}_1$, $\hat{v}_2$ and $\hat{v}_3$ for

$$\hat{v}_1 = \begin{pmatrix} 1 \\ 0 \\ -2 \end{pmatrix}, \hat{v}_2 = \begin{pmatrix} 3 \\ 5 \\ 1 \end{pmatrix}, \hat{v}_3 = \begin{pmatrix} 8 \\ 2 \\ -1 \end{pmatrix}, \hat{b} = \begin{pmatrix} 4 \\ 7 \\ -6 \end{pmatrix}.$$

Another way to phrase this question is does the vector equation

$$a_1 \begin{pmatrix} 1 \\ 0 \\ -2 \end{pmatrix} + a_2 \begin{pmatrix} 3 \\ 5 \\ 1 \end{pmatrix} + a_3 \begin{pmatrix} 8 \\ 2 \\ -1 \end{pmatrix} = \begin{pmatrix} 4 \\ 7 \\ -6 \end{pmatrix} \tag{4}$$

have a solution.

Converting Equation (4) to a system of linear equations gives

$$1a_1 + 3a_2 + 8a_3 = 4$$

$$5a_2 + 2a_3 = 7$$

$$-2a_1 + a_2 - a_3 = -6$$

The augmented matrix for this system of linear equations is

$$\begin{pmatrix} 1 & 3 & 8 & 4 \\ 0 & 5 & 2 & 7 \\ -2 & 1 & -1 & -6 \end{pmatrix}$$

which, when row reduced, is

$$\begin{pmatrix} 1 & 0 & 0 & 233/61 \\ 0 & 1 & 0 & 101/61 \\ 0 & 0 & 1 & -39/61 \end{pmatrix}$$

Thus, there is a unique solution; namely,

$$a_1 = \frac{233}{61}, \; a_1 = \frac{101}{61}, \; a_1 = \frac{-39}{61}$$

Exercises

1. Write the following systems of linear equations as a (i) vector equation (ii) matrix equation.

(a.)
$$\begin{aligned} 2x_1 - 5x_2 + x_3 &= -2 \\ -4x_1 + x_3 &= 7 \\ x_1 + 6x_2 - 4x_3 &= 0 \end{aligned}$$

(b.)
$$\begin{aligned} x_1 - x_2 + x_3 + x_4 &= 0 \\ x_3 &= 9 \end{aligned}$$

2. Write the following vector equations as a (i) system of linear equations (ii) matrix equation.

(a.) $\quad x\begin{pmatrix} 2 \\ 8 \\ 1 \end{pmatrix} + y\begin{pmatrix} -3 \\ 0 \\ 6 \end{pmatrix} + z\begin{pmatrix} 5 \\ -3 \\ -1 \end{pmatrix} = \begin{pmatrix} 4 \\ 9 \\ 0 \end{pmatrix}$

(b.) $\quad x_1\begin{pmatrix} 0 \\ 3 \\ 0 \\ -5 \end{pmatrix} + x_2\begin{pmatrix} 1 \\ 4 \\ 2 \\ 7 \end{pmatrix} + x_3\begin{pmatrix} 0 \\ -2 \\ -3 \\ 6 \end{pmatrix} + x_4\begin{pmatrix} 3 \\ 1 \\ 1 \\ -2 \end{pmatrix} + x_5\begin{pmatrix} 1 \\ 8 \\ 4 \\ 2 \end{pmatrix} = \begin{pmatrix} 3 \\ -1 \\ -6 \\ -5 \end{pmatrix}$

3. Write the following matrix equations as a (*i*) system of linear equations (*ii*) vector equation.

(a.) $\quad \begin{pmatrix} 0 & -3 & 5 & 2 \\ 2 & 4 & -2 & 3 \\ 1 & 1 & 2 & 6 \end{pmatrix}\begin{pmatrix} w \\ x \\ y \\ z \end{pmatrix} = \begin{pmatrix} -5 \\ 7 \\ 0 \end{pmatrix}$

(b.) $\quad \begin{pmatrix} 1 & 5 & 4 \\ -3 & 2 & 1 \\ 6 & 4 & 0 \end{pmatrix}\begin{pmatrix} x \\ y \\ z \end{pmatrix} = \begin{pmatrix} 9 \\ -6 \\ 8 \end{pmatrix}$

In Exercises 4–6 solve the vector equations

4. $\quad a\begin{pmatrix} 1 \\ -2 \\ 0 \end{pmatrix} + b\begin{pmatrix} 5 \\ 3 \\ 4 \end{pmatrix} + c\begin{pmatrix} 1 \\ 0 \\ -3 \end{pmatrix} = \begin{pmatrix} 6 \\ 1 \\ -5 \end{pmatrix}$

5. $a \begin{pmatrix} 3 \\ 2 \\ -1 \end{pmatrix} + b \begin{pmatrix} 6 \\ 0 \\ 3 \end{pmatrix} = \begin{pmatrix} 4 \\ 2 \\ 7 \end{pmatrix}$

6. $a \begin{pmatrix} 0 \\ 3 \\ 0 \end{pmatrix} + b \begin{pmatrix} -2 \\ 6 \\ 1 \end{pmatrix} + c \begin{pmatrix} 4 \\ 1 \\ -3 \end{pmatrix} + d \begin{pmatrix} 7 \\ 0 \\ -4 \end{pmatrix} = \begin{pmatrix} 16 \\ -3 \\ -5 \end{pmatrix}$

7. Suppose that A is an $n \times n$ matrix and $\det(A) \neq 0$. Let $\hat{a}_1, \ldots, \hat{a}_n$ denote the columns of A.
   (a.) Show that for any $\hat{b}$ in $\mathbb{R}^n$ $A\hat{x} = \hat{b}$ has a solution.
   (b.) Show that any vector in $\mathbb{R}^n$ can be written as a linear combination of $\hat{a}_1, \ldots, \hat{a}_n$.
   (c.) Show that the only solution to $A\hat{x} = \hat{0}$ is $\hat{x} = \hat{0}$.
   (d.) Show that $x_1\hat{a}_1 + \cdots + x_n\hat{a}_n = \hat{0}$ if and only if every $x_i = 0$.
   (e.) Show that $A\hat{x} = \hat{b}$ has the same solution set as $x_1\hat{a}_1 + \cdots + x_n\hat{a}_n = \hat{b}$

8. Determine if $\hat{b}$ is a linear combination of $\hat{v}_1$, $\hat{v}_2$ and $\hat{v}_3$

   (a.) $\hat{v}_1 = \begin{pmatrix} 3 \\ 1 \\ -2 \end{pmatrix}, \hat{v}_2 = \begin{pmatrix} 4 \\ -2 \\ 3 \end{pmatrix}, \hat{v}_3 = \begin{pmatrix} 10 \\ 0 \\ -1 \end{pmatrix}, \hat{b} = \begin{pmatrix} 3 \\ 7 \\ 11 \end{pmatrix}$

   (b.) $\hat{v}_1 = \begin{pmatrix} 1 \\ -1 \\ 2 \end{pmatrix}, \hat{v}_2 = \begin{pmatrix} 7 \\ 3 \\ 6 \end{pmatrix}, \hat{v}_3 = \begin{pmatrix} 1 \\ 0 \\ 5 \end{pmatrix}, \hat{b} = \begin{pmatrix} 6 \\ -14 \\ 5 \end{pmatrix}$

9. Is $\hat{b}$ a linear combination of the columns of A if

   (a.) $A = \begin{pmatrix} 1 & 2 & 9 \\ 5 & 0 & 7 \\ -3 & 2 & -4 \end{pmatrix}$ $\hat{b} = \begin{pmatrix} 12 \\ 0 \\ -5 \end{pmatrix}$

   (b.) $A = \begin{pmatrix} 2 & 0 & 4 \\ 3 & 1 & -6 \\ 8 & 2 & -8 \end{pmatrix}$ $\hat{b} = \begin{pmatrix} 3 \\ 6 \\ -1 \end{pmatrix}$

Section 2.6 Applications

Flow Problems

Linear algebra can be used to model "flow problems". Two examples of flow problems are traffic on a network of streets and flow of water through a water main system.

In such problems we have a lattice where a quantity can enter the system and exit the system. The quantity entering the system is balanced by the quantity leaving the system.

Consider the traffic flow in Figure 2.3. We hypothesize a direction of flow between the intersections. At each intersection, the input traffic must equal the output traffic. In order to be valid solution, all variables must be non-negative.

At intersection A: $x_1 + x_3 = 200$

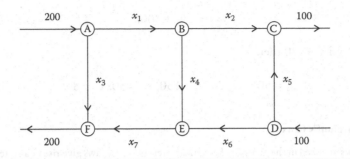

Figure 2.3   A traffic flow diagram.

At intersection B: $x_1 = x_2 + x_4$
At intersection C: $x_2 + x_5 = 100$
At intersection D: $x_5 + x_7 = 300$
At intersection E: $x_4 + x_7 = x_6$
At intersection D: $x_3 + x_6 = 400$

The augmented matrix for this system of equations is

$$\begin{pmatrix} 1 & 0 & 1 & 0 & 0 & 0 & 0 & 200 \\ 1 & -1 & 0 & -1 & 0 & 0 & 0 & 0 \\ 0 & 1 & 0 & 0 & 1 & 0 & 0 & 100 \\ 0 & 0 & 0 & 0 & 1 & 0 & 1 & 300 \\ 0 & 0 & 0 & 0 & 0 & -1 & 1 & 0 \\ 0 & 0 & 1 & 0 & 0 & 1 & 0 & 400 \end{pmatrix}$$

When row reduced, the result is

$$\begin{pmatrix} 1 & 0 & 0 & 0 & 0 & -1 & 0 & -200 \\ 0 & 1 & 0 & 0 & 0 & 0 & -1 & -200 \\ 0 & 0 & 1 & 0 & 0 & 1 & 0 & 400 \\ 0 & 0 & 0 & 1 & 0 & -1 & 1 & 0 \\ 0 & 0 & 0 & 0 & 1 & 0 & 1 & 300 \\ 0 & 0 & 0 & 0 & 0 & 0 & 0 & 0 \end{pmatrix}$$

Thus, x_6 and x_7 are free variables and

$$x_1 = x_6 - 200,$$

$$x_2 = x_7 - 200$$

$$x_3 = -x_6 + 400$$

$$x_4 = x_6 - x_7$$

$$x_5 = -x_7 + 300.$$

If we let $x_6 = 250$ and $x_7 = 220$ then

$$x_1 = 50, x_2 = 20, x_3 = 150, x_4 = 30, x_5 = 80.$$

Example: Kirchoff's Laws

Kirchoff's laws are used in the analysis of electrical circuits. These laws give rise to a system of linear equations.

The circuits that we consider will have power source(s) designated V or E and load(s), designated R. The amount of power of a power source is measured in volts and the amount of load is measured in ohms. When a circuit is activated, current, I, measured in amperes (more often called amps) flows through the circuit and power is dissipated according to $V = IR$. A loop is a path through a circuit that begins and ends at the same point. A junction is a point where at least two circuit paths meet and a branch is a path connecting two junctions.

The junction rule says that the sum of the currents flowing into a node is equal to the sum of currents flowing out of a node. Said another way, the algebraic sum of currents at a node is equal to zero. This says that current is conserved.

The closed loop rule says that the algebraic sum of the potential differences around a closed loop is zero.

Example

An electrical circuit is shown in Figure 2.4.

Assign directions for I_1, I_2, and I_3. If these directions are not consistent with the actual current flow, the calculations will give negative values.

Conservation of current

At node A

$$I_1 \rightarrow A \rightarrow I_2$$

$$I_3 \uparrow$$

so

$$I_1 + I_3 = I_2$$

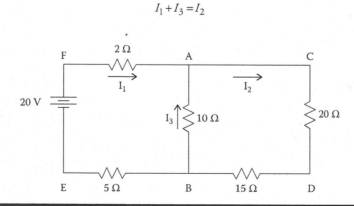

Figure 2.4   An electrical circuit diagram.

or

$$I_1 - I_2 + I_3 = 0$$

Node B gives the same equation.
  Conservation of energy
  There are three loops: *FABEF, ACBDA,* and *FACDBEF*

For loop *ACDBA* $10I_3 + 20I_2 + 15I_2 = 0$ or $35I_2 + 10I_3 = 0$
For loop *FABEF* $2I_1 - 10I_3 + 5I_1 = 20$ or $7I_1 - 10I_3 = 20$.

The three equations

$$I_1 - I_2 + I_3 = 0$$

$$35I_2 + 10I_3 = 0$$

$$7I_1 - 10I_3 = 20$$

are sufficient to solve the system.
  We create the augmented matrix

$$\begin{pmatrix} 1 & -1 & 1 & 0 \\ 0 & 35 & 10 & 0 \\ 7 & -10 & 0 & 20 \end{pmatrix}$$

which, when row reduced, gives

$$\begin{pmatrix} 1 & 0 & 0 & 180/43 \\ 0 & 1 & 0 & 40/43 \\ 0 & 0 & 1 & -140/43 \end{pmatrix}$$

so

$$I_1 = \frac{180}{43} \approx 4.19, \ I_2 = \frac{40}{43} \approx .93, \ I_3 = -\frac{140}{43} \approx -3.26.$$

The fact that I_3 is negative indicates the assumed direction of I_3 in the diagram is incorrect.

Balancing Chemical Equations Using Linear Algebra

Consider the chemical reaction

$$CH_4 + O_2 \rightarrow CO_2 + H_2O.$$

By balancing an equation, we mean assigning integers to each molecule so that there are the same number of atoms of each element on both sides of the equation. As it stands now, for example, there are 2 atoms of oxygen on the left side of the equation and 3 atoms of oxygen on the right side of the equation. We form the reaction

$$wCH_4 + xO_2 \rightarrow yCO_2 + zH_2O.$$

Each element gives rise to an equation as follows:

$$\text{Oxygen} \quad 2x = 2y + z \text{ or } 2x - 2y - z = 0$$

$$\text{Carbon} \quad w = y \text{ or } w - y = 0$$

$$\text{Hydrogen} \quad 4w = 2z \text{ or } 4w - 2z = 0.$$

The augmented matrix for the system of equations is

$$\begin{pmatrix} 0 & 2 & -2 & -1 & 0 \\ 1 & 0 & -1 & 0 & 0 \\ 4 & 0 & 0 & -2 & 0 \end{pmatrix}.$$

When row reduced, this gives

$$\begin{pmatrix} 1 & 0 & 0 & -1/2 & 0 \\ 0 & 1 & 0 & -1 & 0 \\ 0 & 0 & 1 & -1/2 & 0 \end{pmatrix}$$

so z is the free variable and

$$w = \frac{1}{2}z, \ x = z, \ y = \frac{1}{2}z.$$

Thus, the solution is of the form

$$\begin{pmatrix} w \\ x \\ y \\ z \end{pmatrix} = \begin{pmatrix} \frac{1}{2}z \\ z \\ \frac{1}{2}z \\ z \end{pmatrix} = z \begin{pmatrix} \frac{1}{2} \\ 1 \\ \frac{1}{2} \\ 1 \end{pmatrix}.$$

We are not totally free in our choice of z. Each of the variables must be a positive integer. The simplest way to do this is to set $z = 2$, so that $w = 1$, $x = 2$, $y = 1$.

Exercises

1. Balance the following chemical equations.

(a.) $Ca + H_3PO_4 \rightarrow Ca_3P_2O_8 + H_2$

(b.) $C_3H_8 + O_2 \rightarrow CO_2 + H_2O$

(c.) $P_2I_4 + P_4 + H_2O \rightarrow PH_4I + H_3PO_4$

(d.) $KI + KClO_3 + HCl \rightarrow I_2 + H_2O + KCl$

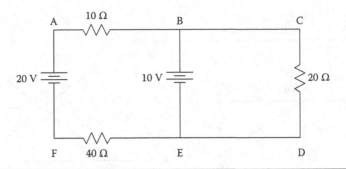

Figure 2.5   Electrical circuit diagram for Exercise 2.

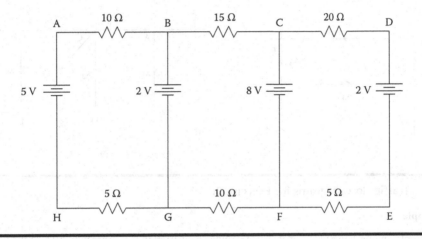

Figure 2.6   Electrical circuit diagram for Exercise 2.

   (e.) $Al + Fe_2O_3 \rightarrow Al_2O_3 + Fe$

2. Determine the currents in the networks in Figures 2.5 and 2.6.
3. Solve the traffic flow problems depicted in Figure 2.7.

Markov Chains

Markov chains is a branch of probability in which matrices and linear algebra play a particularly crucial role. In a Markov chain, we have a collection of states and a process that occupies exactly one of these states at any given time. The model that we consider has a finite number of states and the process can change states only at a discrete unit of time. Thus, the process is described by the ordered pair (s,t) where s is the state and t is time. There is a collection of transition probabilities that describe how the process changes (possibly remaining in the same state) when time changes. The Markov property is that these probabilities depend only on where the process is at the present time; previous history is irrelevant.

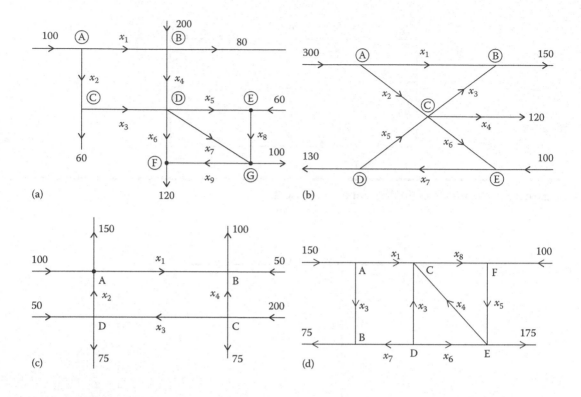

Figure 2.7   Traffic flow diagrams for Exercise 3.

Example

Alice is playing Beth in a game where Alice is somewhat better; in fact, Alice wins 60% of the time. They bet $1 on each play of the game. Suppose that between them they have $5. We will make Alice's fortune the states. The state space would then be $\{0,1,2,3,4,5,\}$.

If Alice's fortune at time t is $2, then at time $t+1$ her fortune will be $3 with probability .6 and will be $1 with probability .4. If her fortune is $0, then it will remain there (she has no money to bet) and if her fortune is $5, then it will remain there (Beth has no money to bet). The states $0 and $5 are called absorbing states, because once the process enters an absorbing state, it never leaves.

From these "transition probabilities" we can construct the "transition matrix" shown in Figure 2.8.

$$P = \begin{pmatrix} 1 & 0 & 0 & 0 & 0 & 0 \\ .4 & 0 & .6 & 0 & 0 & 0 \\ 0 & .4 & 0 & .6 & 0 & 0 \\ 0 & 0 & .4 & 0 & .6 & 0 \\ 0 & 0 & 0 & .4 & 0 & .6 \\ 0 & 0 & 0 & 0 & 0 & 1 \end{pmatrix}$$

The number P_{ij} is the probability of going from state i to state j in one step. Since we must go somewhere in one step from state i, the sum of the entries in each row is 1. This is not necessarily true for the columns.

Thinking of an abstract Markov chain rather than the example above, we develop a formula that describes how to go from state i to step j in two steps. This is one place where our definition

$$P = \begin{pmatrix} 1 & 0 & 0 & 0 & 0 & 0 \\ .4 & 0 & .6 & 0 & 0 & 0 \\ 0 & .4 & 0 & .6 & 0 & 0 \\ 0 & 0 & .4 & 0 & .6 & 0 \\ 0 & 0 & 0 & .4 & 0 & .6 \\ 0 & 0 & 0 & 0 & 0 & 1 \end{pmatrix}$$

Figure 2.8   A transition matrix for a Markov chain.

of matrix multiplication is indispensable. Suppose the process is in state i at time t and we want to compute the probability that the process is in step j at time $t+2$. This occurs if the process goes from state i at time t to some state k at time $t+1$ (which occurs with probability P_{ik}) and then goes to state j at time $t+2$ (which occurs with probability P_{kj}). The probability of both of these transitions occurring is $P_{ik}P_{kj}$. Since this is true for every intermediate step k, the probability of going from state i to step j in two steps is

$$\sum_{k=1}^{n} P_{ik}P_{kj}$$

if there are a total of n states. Note that

$$\sum_{k=1}^{n} P_{ik}P_{kj} = \left(P^2\right)_{ij}.$$

Extending this reasoning, the probability of going from state i to step j in m steps is $\left(P^m\right)_{ij}$.

It is most often the case that our process has a probability of being in different states initially and we would like to determine the probability that it will be in a particular state some time later.

Example

Suppose there are three weather conditions, rain (R), snow (S) and clear (C) and we believe that daily weather conditions have the following properties:

If it rains today, then tomorrow it will rain with probability .3, snow with probability .2 and be clear with probability .5
If it snows today, then tomorrow it will rain with probability .1, snow with probability .3 and be clear with probability .6.
If it is clear today, then tomorrow it will rain with probability .2, snow with probability .5 and be clear with probability .3

If we enumerate our states as 1 = rain, 2 = snow, 3 = clear then the transition matrix is

$$\begin{pmatrix} .3 & .2 & .5 \\ .1 & .3 & .6 \\ .2 & .5 & .3 \end{pmatrix}.$$

Suppose that on October 31, there is a 30% chance of rain, 10% chance it will snow and 60% chance it will be clear. We determine the likelihood of the different weather states on November 1 using

$$(.3,.1,.6)\begin{pmatrix} .3 & .2 & .5 \\ .1 & .3 & .6 \\ .2 & .5 & .3 \end{pmatrix} = (.22,.39,.39).$$

This says that given there was a 30% probability of rain, a 10% probability of snow and a 60% probability of a clear day on October 31, the probability that it will rain on November 1 is 22%, the probability of snow on November 1 is 39% and the probability of a clear day on November 1 is 39%.

To calculate the probabilities of the weather on November 2, given the same data as before for the weather on October 31, we compute

$$(.3,.1,.6)\begin{pmatrix} .3 & .2 & .5 \\ .1 & .3 & .6 \\ .2 & .5 & .3 \end{pmatrix}^2 = (.183,.356,.461).$$

An interesting result occurs in this case if we consider the long-term probabilities. If we project letting the process run for 100 days we get

$$(.3,.1,.6)\begin{pmatrix} .3 & .2 & .5 \\ .1 & .3 & .6 \\ .2 & .5 & .3 \end{pmatrix}^{100} = (.1810,.3714,.4476)\,(\text{to 4 decimal places}).$$

More importantly, a high power of the matrix converges to a matrix that has identical rows. In this case

$$\begin{pmatrix} .3 & .2 & .5 \\ .1 & .3 & .6 \\ .2 & .5 & .3 \end{pmatrix}^{100} = \begin{pmatrix} .1810 & .3714 & .4476 \\ .1810 & .3714 & .4476 \\ .1810 & .3714 & .4476 \end{pmatrix}.$$

When this happens, the initial probabilities have no effect on the long-term behavior. If x is the probability it rains on October 31, y is the probability it snows on October 31 and x is the probability it is clear on October 31 then

$$(x,y,z)\begin{pmatrix} .1810 & .3714 & .4476 \\ .1810 & .3714 & .4476 \\ .1810 & .3714 & .4476 \end{pmatrix}$$

$$= (.1810x + .1810y + .1810z, .3714x + .3714y + .3714z, .4476x + .4476y + .4476z)$$

$$= (.1810(x+y+z), .3714(x+y+z), .4476(x+y+z)) = (.1810, .3714, .4476).$$

since $x + y + z = 1$.

The probability vector $(.1810 \quad .3714 \quad .4476)$ is the equilibrium state for the process because

$$(.1810 \quad .3714 \quad .4476)\begin{pmatrix} .3 & .2 & .5 \\ .1 & .3 & .6 \\ .2 & .5 & .3 \end{pmatrix} = (.1810 \quad .3714 \quad .4476).$$

It predicts that in the long run, 18% of the days will have rain, 37% of the days will have snow, and 45% of the days will be clear.

This type of behavior occurs anytime the entries of the transition matrix (or any power of the transition matrix) has all nonzero entries. This is because when this occurs it is possible to go from any one state to any other state (but not necessarily in one step).

The gambling problem of Alice and Beth has quite a different evolution. The game ends whenever Alice's fortune hits either $0 or $5, so we would expect the game to end eventually. In this case, the most pertinent questions include

1. If we know Alice's initial fortune, what is the probability that Alice will wind up the winner of all the money?
2. If we know Alice's initial fortune, how many plays would we expect before the game ends?

The game ends whenever the process enters the state 5 or the state 0. These are called absorbing states. Absorbing states are identified by having a 1 on the main diagonal.

For purposes of calculation, it is convenient to rearrange the states so that absorbing states are listed first. The modified transition matrix in this example is

$$
P^* = \begin{array}{c} \\ 0 \\ 5 \\ 1 \\ 2 \\ 3 \\ 4 \end{array}
\begin{array}{c} \begin{array}{cccccc} 0 & 5 & 1 & 2 & 3 & 4 \end{array} \\
\left(\begin{array}{cccccc}
1 & 0 & 0 & 0 & 0 & 0 \\
0 & 1 & 0 & 0 & 0 & 0 \\
.4 & 0 & 0 & .6 & 0 & 0 \\
0 & 0 & .4 & 0 & .6 & 0 \\
0 & 0 & 0 & .4 & 0 & .6 \\
0 & .6 & 0 & 0 & .4 & 0
\end{array} \right)
\end{array}
$$

We partition P^* as

$$ P^* = \begin{pmatrix} I & O \\ R & Q \end{pmatrix} $$

where

$$ I = \begin{pmatrix} 1 & 0 \\ 0 & 1 \end{pmatrix} \quad O = \begin{pmatrix} 0 & 0 & 0 & 0 \\ 0 & 0 & 0 & 0 \end{pmatrix} $$

$$ R = \begin{pmatrix} .4 & 0 \\ 0 & 0 \\ 0 & 0 \\ 0 & .6 \end{pmatrix} \quad Q = \begin{pmatrix} 0 & .6 & 0 & 0 \\ .4 & 0 & .6 & 0 \\ 0 & .4 & 0 & .6 \\ 0 & 0 & .4 & 0 \end{pmatrix}. $$

We have

$$ (I - Q)^{-1} = \frac{1}{211} \begin{pmatrix} 325 & 285 & 225 & 135 \\ 190 & 475 & 375 & 225 \\ 100 & 250 & 475 & 285 \\ 40 & 100 & 190 & 325 \end{pmatrix} \equiv N $$

Theorem 2.1:

For a finite state absorbing Markov chain with k non-absorbing states, the expected number of transitions that the non-absorbing state i undergoes before absorption is

$$N_{i1} + N_{i2} + \cdots + N_{ik}$$

where $N = (I - Q)^{-1}$.

This can also be expressed as follows:

Let t_i be the expected time until absorption, beginning in the non-absorbing state i and let

$$\hat{t} = \begin{pmatrix} t_1 \\ \vdots \\ t_k \end{pmatrix}.$$

Then

$$\hat{t} = N \begin{pmatrix} 1 \\ \vdots \\ 1 \end{pmatrix}.$$

In our example,

$$N \begin{pmatrix} 1 \\ \vdots \\ 1 \end{pmatrix} = \frac{1}{211} \begin{pmatrix} 325 & 285 & 225 & 135 \\ 190 & 475 & 375 & 225 \\ 100 & 250 & 475 & 285 \\ 40 & 100 & 190 & 325 \end{pmatrix} \begin{pmatrix} 1 \\ 1 \\ 1 \\ 1 \end{pmatrix} = \frac{1}{211} \begin{pmatrix} 970 \\ 1265 \\ 1110 \\ 655 \end{pmatrix} \approx \begin{pmatrix} 4.60 \\ 6.00 \\ 5.26 \\ 3.10 \end{pmatrix}.$$

So if Alice begins with \$1, the expected number of plays until someone is bankrupt is 4.60; if Alice begins with \$2, the expected number of plays until someone is bankrupt is 6.00, etc.

The previous theorem gives the average number of plays until someone goes bankrupt. The next theorem tells who is likely to win.

Theorem 2.2:

The probability that beginning in the non-absorbing state i the system is absorbed in the absorbing state j is $(NR)_{ij}$.

In our example,

$$NR = \frac{1}{211} \begin{pmatrix} 325 & 285 & 225 & 135 \\ 190 & 475 & 375 & 225 \\ 100 & 250 & 475 & 285 \\ 40 & 100 & 190 & 325 \end{pmatrix} \begin{pmatrix} .4 & 0 \\ 0 & 0 \\ 0 & 0 \\ 0 & .6 \end{pmatrix} = \frac{1}{211} \begin{pmatrix} 130 & 81 \\ 76 & 135 \\ 40 & 171 \\ 16 & 195 \end{pmatrix}$$

$$\approx \begin{pmatrix} .62 & .38 \\ .36 & .64 \\ .19 & .81 \\ .08 & .92 \end{pmatrix}.$$

So if Alice begins with $1 and Beth begins with $4, Alice will be the ultimate winner 38% of the time; if Alice begins with $2 and Beth begins with $3, Alice will be the ultimate winner 64% of the time; if Alice begins with $3 and Beth begins with $2, Alice will be the ultimate winner 81% of the time; and if Alice begins with $4 and Beth begins with $1, Alice will be the ultimate winner 92% of the time.

Exercises

1. Carl and David play a game where Carl wins 55% of the time. They bet $1 at each play of the game and will play until one person goes broke. If Carl starts with $2 and David starts with $3,
    (a.) who is more likely to be the long-term winner?
    (b.) What is the expected number of plays until the game ends?
2. Repeat Problem 1 with the assumption that Carl wins 51% of the time.
3. Ellen and Fred are engaged to be married. The ceremony will be held in a location where it either rains or is sunny. If it rains one day, then it will be sunny the next day with probability .7, and if it is sunny one day then it will be sunny the next day with probability .4.
    (a.) If it is raining one week before the ceremony, what is the probability it will be sunny for the wedding?
    (b.) If it is raining two weeks before the ceremony, what is the probability it will be sunny for the wedding?
    (c.) What is the long-range distribution of the weather in this region?

Chapter 3

Vector Spaces

Historical Note

The ideas of modern linear algebra were developed by Hermann Grassmann beginning in the early 1830s. Grassmann lived from 1809 to 1877. He studied Theology and Philosophy, but had no formal training in mathematics. In *Die Lineale Ausdehnungslehre*, of 1844, he introduced several new ideas that would be basic to linear algebra, but his ideas were not well received by contemporary mathematicians, in part because his style of writing was difficult to understand.

He rewrote *The Ausdehnungslehre* of 1844 and published the second *Ausdehnungslehre* in 1862. The topics include most of what is covered in a first linear algebra text today, including solution of systems of linear equations, the theory of linear independence and dimension, subspaces, projection of elements onto subspaces, change of coordinates, orthogonality, linear transformations, matrix representation, rank-nullity theory, eigenvalues and eigenspaces. (See also the Note about Grassman in Chapter 4.)

The fundamental objects of study in linear algebra are vector spaces and linear transformations. In this chapter, we define what a vector space is, derive additional properties of a vector space implied by the axioms of the definition, and examine several examples. Linear transformations are a particular type of function between vector spaces and are the topic of Chapter 4.

We will see that vector spaces have different special collections of building blocks called bases. We will spend substantial effort examining characteristics of bases.

We will use the language that includes the term "scalar" as we did in Chapter 1. Recall that in linear algebra, the term "scalar" is synonymous with "number", and the numbers will be real or complex numbers.

A central theme of mathematics is to study a familiar structure that has objects and rules for combining the objects, and then from the familiar structure ascertain what principles are truly important. One then abstracts the principles to create a more general structure. This is what we will do in this chapter. In the first section we study $\mathbb{R}^n$ to gain some intuition for the more abstract structure. In later sections, we will define a vector space to be a mathematical structure that obeys a group of axioms. A vector will be an object in a vector space that will not necessarily have any geometric properties attached to it.

The purpose of Section 3.1 is to develop some intuition for vector spaces using the most elementary and intuitive example. This section covers material often discussed in multivariable calculus and to some degree in physics. Readers with a background in those areas can omit this section.

Section 3.1 Vector Spaces in $\mathbb{R}^n$

To develop some intuition for vector spaces we start by giving some examples of vector spaces. We first consider the Cartesian plane in two dimensions, which is the set of ordered pairs of real numbers. We denote this set as $\mathbb{R}^2$. So

$$\mathbb{R}^2 = \{(x, y) \mid x, y \in \mathbb{R}\}$$

An advantage of this particular vector space is that it is easily visualized. An aid in the visualization is to associate with each point (x, y) in the plane the unique arrow from the origin $(0,0)$ to the point (x, y). A description of a vector that is often given in beginning physics is a "vector is a quantity that has magnitude and direction", and arrows provide an excellent visual representation of such quantities. If one takes the definition of equality in this setting as meaning two vectors are equal if they have the same magnitude and direction, then we can move the arrow associated with a vector as long as we don't change the length or orientation of the arrow. It is common to denote a vector in $\mathbb{R}^2$ in one of two ways; either an ordered pair of numbers as originally described, or as a single letter that is usually distinguished by having the letter with a "hat" $(\hat{a})$. Associated with each vector space is a field of scalars. For our purposes, the scalar field will be the real numbers or complex numbers. If the vector space is $\mathbb{R}^n$ then the scalar field is the real numbers. Two arithmetic operations that one defines on this set are scalar multiplication, that is, multiplication by a real number, and vector addition.

Arithmetically, if $\hat{a} = (a_1, a_2)$, $\hat{b} = (b_1, b_2)$ and λ is a scalar, then we define

$$\lambda \hat{a} = (\lambda a_1, \lambda a_2) \text{ and } \hat{a} + \hat{b} = (a_1 + b_1, a_2 + b_2).$$

The effect of scalar multiplication by a positive real number is to change the length but not the direction of the vector and if the scalar is negative, the effect is to point the arrow in the opposite direction and also change the magnitude by the absolute value of the scalar (see Figure 3.1).

Geometrically, vector addition can be described by placing the tail of the second vector on the head of the first. The sum of the vectors is the arrow from the tail of the first vector to the head of the second (see Figure 3.2).

To create an algebra, we need the zero vector, which is defined as $\hat{0} = (0,0)$. We also need the negative of a vector. If

$$\hat{v} = (v_1, v_2)$$

then we define

$$-\hat{v} = (-1)\hat{v} = (-v_1, -v_2).$$

Note that

$$\hat{v} + (-\hat{v}) = \hat{0}.$$

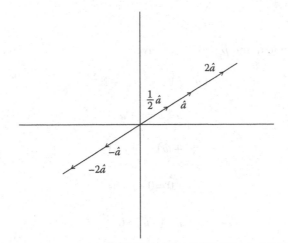

Figure 3.1   Scalar multiples of a vector.

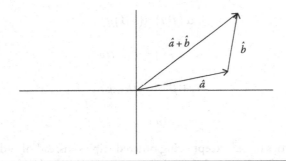

Figure 3.2   Vector addition.

We will normally write

$$\hat{v} + \left(-\hat{w}\right)$$

as

$$\hat{v} - \hat{w}.$$

Examples

Let

$$\hat{u} = \left(2,-3\right), \hat{v} = \left(0,1\right), \hat{w} = \left(-2,-5\right).$$

Then

$$\hat{u} - \hat{v} = \left(2,-4\right)$$
$$3\hat{u} + 4\hat{w} = 3\left(2,-3\right) + 4\left(-2,-5\right) = \left(6,-9\right) + \left(-8,-20\right) = \left(-2,-29\right)$$
$$4\left(\hat{u} + 3\hat{v}\right) = 4\left[\left(2,-3\right) + 3\left(0,1\right)\right] = 4\left[\left(2,-3\right) + \left(0,3\right)\right] = 4\left(2,0\right) = \left(8,0\right)$$

With these definitions, we have the following properties of $\mathbb{R}^2$:

Theorem 3.1:

For $\hat{u}$, $\hat{v}$ and $\hat{w}$ vectors and α and β scalars we have

$$\left(\hat{u}+\hat{v}\right)\in\mathbb{R}^2, \alpha\hat{u}\in\mathbb{R}^2$$

$$\hat{u}+\hat{v}=\hat{v}+\hat{u}$$

$$\hat{u}+\left(\hat{v}+\hat{w}\right)=\left(\hat{u}+\hat{v}\right)+\hat{w}$$

$$\hat{u}+\hat{0}=\hat{0}+\hat{u}=\hat{u}$$

$$\hat{u}+\left(-\hat{u}\right)=\hat{0}$$

$$0\hat{u}=\hat{0}$$

$$\alpha\left(\beta\hat{u}\right)=\left(\alpha\beta\right)\hat{u}$$

$$\alpha\left(\hat{u}+\hat{v}\right)=\alpha\hat{u}+\alpha\hat{v}$$

$$\left(\alpha+\beta\right)\hat{u}=\alpha\hat{u}+\beta\hat{u}$$

$$1\hat{u}=\hat{u}$$

We have similar properties in $\mathbb{R}^3$ except using ordered triples instead of ordered pairs. In $\mathbb{R}^n$ we use ordered n-tuples but do not have the capability of drawing vectors as arrows.

For an abstract vector space, we hypothesize a collection of objects (vectors), a scalar field and two ways of combining the objects that we will call vector addition and scalar multiplication that obey the properties in Theorem 3.1. This will be the definition of a vector space.

While a vector space will only be required to have the structure above, the vector spaces $\mathbb{R}^n$ have an additional property that enables us to determine the length of a vector and the angle between two vectors. This is the (Euclidean) dot product of vectors.

Definition:

If $\hat{u}=\left(u_1,\ldots,u_n\right)$ and $\hat{v}=\left(v_1,\ldots,v_n\right)$ are vectors in $\mathbb{R}^n$ the Euclidean dot product of $\hat{u}$ and $\hat{v}$, denoted $\hat{u}\cdot\hat{v}$, is defined by

$$\hat{u}\cdot\hat{v}=u_1v_1+\cdots+u_nv_n.$$

The Euclidean dot product has the following properties:

Theorem 3.2:

For vectors

$$\hat{u}=\left(u_1,\ldots,u_n\right), \hat{v}=\left(v_1,\ldots,v_n\right), \text{and}$$

$\hat{w} = (w_1, \ldots, w_n)$ and real numbers α and β
we have

$$\hat{u} \cdot \hat{v} = \hat{v} \cdot \hat{u}$$

$$\hat{u} \cdot (\hat{v} + \hat{w}) = \hat{u} \cdot \hat{v} + \hat{u} \cdot \hat{w}$$

$$\alpha (\hat{u} \cdot \hat{v}) = (\alpha \hat{u}) \cdot \hat{v} = \hat{u} \cdot (\alpha \hat{v})$$

$$\hat{u} \cdot \hat{u} \geq 0 \text{ and}$$

$$\hat{u} \cdot \hat{u} = 0 \text{ if and only if } \hat{u} = \hat{0}.$$

The dot product provides a way to define the length or norm of a vector in $\mathbb{R}^n$ that is an extension of Pythagoras' theorem.

Definition:

If $\hat{u} = (u_1, \ldots, u_n)$ is a vector in $\mathbb{R}^n$, the Euclidean norm of $\hat{u}$, denoted $\|\hat{u}\|$, is defined by

$$\|\hat{u}\| = \sqrt{\hat{u} \cdot \hat{u}} = \sqrt{u_1^2 + \cdots + u_n^2}.$$

The Euclidean distance between $\hat{u} = (u_1, \ldots, u_n)$ and $\hat{v} = (v_1, \ldots, v_n)$, denoted $\|\hat{u} - \hat{v}\|$, is defined by

$$\|\hat{u} - \hat{v}\| = \sqrt{(u_1 - v_1)^2 + \cdots + (u_n - v_n)^2}.$$

A vector of norm 1 is called a unit vector.

Theorem 3.3:

The norm has the properties

$$\|\hat{u}\| \geq 0 \text{ and}$$

$$\|\hat{u}\| = 0 \text{ if and only if } \hat{u} = \hat{0}$$

$$\|\alpha \hat{u}\| = |\alpha| \|\hat{u}\|$$

$$\|\hat{u} + \hat{v}\| \leq \|\hat{u}\| + \|\hat{v}\|.$$

We will not prove this result now because we will give proofs in a more general context later.

Theorem 3.4:

In $\mathbb{R}^2$ and $\mathbb{R}^3$ the angle between two non-zero vectors $\hat{a}$ and $\hat{b}$ is determined by

$$\cos \theta = \frac{a \cdot b}{\|a\| \|b\|}.$$

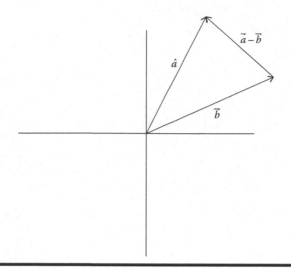

Figure 3.3   Vector subtraction.

Proof:

Consider the diagram in Figure 3.3.

By the law of cosines, we have

$$\left\|\hat{a}-\hat{b}\right\|^2 = \left\|\hat{a}\right\|^2 + \left\|\hat{b}\right\|^2 - 2\left\|\hat{a}\right\|\left\|\hat{b}\right\|\cos\theta.$$

Now

$$\left\|\hat{a}-\hat{b}\right\|^2 = \left(\hat{a}-\hat{b}\right)\cdot\left(\hat{a}-\hat{b}\right) = \hat{a}\cdot\hat{a} - \hat{a}\cdot\hat{b} - \hat{b}\cdot\hat{a} - \hat{b}\cdot\hat{b}$$

$$= \left\|\hat{a}\right\|^2 + \left\|\hat{b}\right\|^2 - 2\hat{a}\cdot\hat{b}$$

so

$$\left\|\hat{a}\right\|^2 + \left\|\hat{b}\right\|^2 - 2\hat{a}\cdot\hat{b} = \left\|\hat{a}\right\|^2 + \left\|\hat{b}\right\|^2 - 2\left\|\hat{a}\right\|\left\|\hat{b}\right\|\cos\theta$$

and

$$\hat{a}\cdot\hat{b} = \left\|\hat{a}\right\|\left\|\hat{b}\right\|\cos\theta.$$

We will see that a particularly important relationship that two non-zero vectors can have with one another is to be orthogonal. In $\mathbb{R}^n$, this mean they are perpendicular, and this occurs exactly when $\hat{u}\cdot\hat{v}=0$.

In abstract settings, there are some vector spaces that have a structure called an inner product, which is similar to the Euclidean dot product. The inner product of vectors $\hat{u}$ and $\hat{v}$ is a number that is denoted $\langle\hat{u},\hat{v}\rangle$ and the square of the norm of a vector $\hat{u}$ is

$$\|\hat{u}\|^2 = \langle \hat{u}, \hat{u} \rangle.$$

One can still form

$$\frac{\langle \hat{u}, \hat{v} \rangle}{\|\hat{u}\|\|\hat{v}\|}$$

as long as $\|\hat{u}\|\|\hat{v}\| \neq 0$, but one must be cautious about a geometrical interpretation.

Example

Let $\hat{u} = (3,-1,6)$ $\hat{v} = (0,2,2)$. Then

$$\hat{u} \cdot \hat{v} = (3)(0) + (-1)(2) + (6)(2) = 10$$

$$\|\hat{u}\| = \sqrt{3^2 + (-1)^2 + 6^2} = \sqrt{46}$$

$$\|\hat{v}\| = \sqrt{0^2 + 2^2 + 2^2} = \sqrt{8}$$

$$\hat{u} + \hat{v} = (3,-1,6) + (0,2,2) = (3,1,8)$$

$$\|\hat{u} + \hat{v}\| = \sqrt{3^2 + 1^2 + 8^2} = \sqrt{74}$$

$$\cos\theta = \frac{u \cdot v}{\|u\|\|v\|} = \frac{10}{\sqrt{8}\sqrt{74}} \approx 0.411, \quad \theta \approx 65.7°.$$

Normalizing a Vector

It is often convenient to have a vector of length 1 that has the same direction as a given vector. If $\hat{v}$ is a non-zero vector, then

$$\frac{1}{\|\hat{v}\|}\hat{v} \quad \text{or} \quad \frac{\hat{v}}{\|\hat{v}\|}$$

is a vector of length 1 that has the same direction as $\hat{v}$, as we show in the exercises.

Example

Let $\hat{v} = (2,-1,-2)$. Then

$$\|\hat{v}\| = \sqrt{2^2 + (-1)^2 + (-2)^2} = 3$$

so

$$\frac{\hat{v}}{\|\hat{v}\|} = \frac{1}{3}(2,-1,-2) = (2/3,-1/3,-2/3).$$

Exercises

1. Let $\hat{u} = (2,3)$, $\hat{v} = (6,-2)$.

   Find $\hat{u} \cdot \hat{v}$, $\|\hat{u}\|$, $\|\hat{v}\|$, $\|\hat{u} + \hat{v}\|$ and the angle between $\hat{u}$ and $\hat{v}$.

2. Let $\hat{u} = (1,-4,2)$, $\hat{v} = (3,6,5)$.

   Find $\hat{u} \cdot \hat{v}$, $\|\hat{u}\|$, $\|\hat{v}\|$, $\|\hat{u} + \hat{v}\|$ and the angle between $\hat{u}$ and $\hat{v}$.

3. $\hat{u} = (1,0,-4)$. Find

   (a.) $\|\hat{u}\|$

   (b.) $\dfrac{\hat{u}}{\|\hat{u}\|}$

   (c.) $\left\|\dfrac{\hat{u}}{\|\hat{u}\|}\right\|$

   (d.) If $\hat{v}$ is any non-zero vector, show that $\left\|\dfrac{\hat{v}}{\|\hat{v}\|}\right\| = 1$.

4. Find a unit vector in the same direction as the given vectors.

   (a.) $(6,-2)$

   (b.) $(4,6)$

   (c.) $(3,1,-3)$

   (d.) $(4,2,0)$

5. The points $(2,1,-4)$ and $\left(-1,-3,\sqrt{11}\right)$ are points on a sphere of radius $\sqrt{21}$. How far is it

   between the two points traveling on the surface of the sphere?

6. Find y so that $(2, -3)$ is orthogonal to $(4, y)$ with the Euclidean dot product.

7. Find two vectors of length 1 that are orthogonal to the non-zero vector (a, b).

8. Solve for x and y in the equation

   $$x(2,3) + y(1,4) = (6,-8).$$

9. Why is it not possible to solve for x and y in the equation

   $$x(2,3) + y(4,6) = (6,6)?$$

10. Find unit two vectors that are perpendicular to both of the vectors

    $$(2,1,-4) \text{ and } (0,2,4).$$

11. Show that if $\hat{u}$ is orthogonal to $\hat{v}$, then $\|\hat{u} + \hat{v}\|^2 = \|\hat{u}\|^2 + \|\hat{v}\|^2$.

12. Show that $\|\hat{u} + \hat{v}\|^2 + \|\hat{u} - \hat{v}\|^2 = 2\|\hat{u}\|^2 + 2\|\hat{v}\|^2$.

13. In subsequent sections, we will define an inner product on the vector space V of continuous functions by

$$\langle f(x), g(x) \rangle = \int_{-1}^{1} f(x) g(x) \, dx.$$

    (a.)   Show that this satisfies the properties of the Euclidean dot product.

    (b.)   Show that

$$f(x) = 1 \text{ and } g(x) = x^3$$

are orthogonal with respect to this inner product.

Section 3.2 Axioms and Examples of Vector Spaces

In the remainder of the chapter, we will study vector spaces in a more abstract sense.

A vector space V over a scalar field $\mathcal{F}$ is a mathematical structure for which two operations can be performed.

In this text, the scalar field will be either the real numbers or complex numbers.

The two operations are vector addition and scalar multiplication.

We hypothesize that if $\hat{u}$, $\hat{v} \in V$ and $\alpha \in \mathcal{F}$ then

$$\hat{u} + \hat{v} \in V \text{ and } \alpha \hat{u} \in V.$$

The first condition says that V is closed under vector addition and the second condition says that V is closed under scalar multiplication.

These operations must satisfy the rules of arithmetic that are given below.

Definition:

A vector space V over a field $\mathcal{F}$ (the scalars) is a nonempty set of objects (the vectors) together with the two operations, addition (+) and scalar multiplication (·) for which

$$+ : V \times V \rightarrow V$$

$$\cdot : \mathcal{F} \times V \rightarrow V$$

that satisfy the following axioms for all vectors $\hat{u}$, $\hat{v}$, $\hat{w} \in V$, and scalars $a, b \in \mathcal{F}$:

1. $\hat{u} + \hat{v} = \hat{v} + \hat{u}$

2. $(\hat{u} + \hat{v}) + \hat{w} = \hat{u} + (\hat{v} + \hat{w})$

3. There is a vector $\hat{0} \in V$ for which $\hat{0} + \hat{u} = \hat{u}$ for all $\hat{u} \in V$.

4. For each vector $\hat{u} \in V$, there is a vector $-\hat{u} \in V$ for which

$$\hat{u} + \left(-\hat{u}\right) = \hat{0}.$$

Note: We normally denote scalar multiplication by juxtaposition; i.e.,

$$a \cdot b = ab \text{ and } a \cdot \hat{u} = a\hat{u}.$$

5. $a\left(\hat{u} + \hat{v}\right) = a\hat{u} + a\hat{v}$

6. $\left(a + b\right)\hat{u} = a\hat{u} + b\hat{u}$

7. $a\left(b\hat{u}\right) = \left(ab\right)\hat{u}$

8. For each vector $\hat{u}$, $1\hat{u} = \hat{u}$.

The vector $\hat{0}$ is called the additive identity on V and the vector $-\hat{u}$ is called the additive inverse of $\hat{u}$.

Notice that in the definition we never say what vectors are, only what they do. This means we can use a single theory to deal with very different objects.

The rules are simply what one would expect of a mathematical structure.

As was stated above, the scalar field for the vector spaces we use will be assumed to be either the real numbers $\mathbb{R}$ or the complex numbers $\mathbb{C}$. In some scenarios there will be differences in results depending on which field is the scalar field. In those cases, the scalar field that is being used will be emphasized.

Examples

Some examples of vector spaces are

$$\mathbb{R}^2 = \left\{ \begin{pmatrix} a \\ b \end{pmatrix} \middle| a, b \in \mathbb{R} \right\} \text{ or } \left\{ (a, b) \middle| a, b \in \mathbb{R} \right\}$$

$$\mathbb{R}^3 = \left\{ \begin{pmatrix} a \\ b \\ c \end{pmatrix} \middle| a, b, c \in \mathbb{R} \right\} \text{ or } \left\{ (a, b, c) \middle| a, b, c \in \mathbb{R} \right\}$$

$$\mathbb{R}^n = \left\{ \begin{pmatrix} x_1 \\ \vdots \\ x_n \end{pmatrix} \middle| x_1, \ldots, x_n \in \mathbb{R} \right\} \text{ or } \left\{ (x_1, \ldots, x_n) \middle| x_1, \ldots, x_n \in \mathbb{R} \right\}$$

$$\mathbb{C}^n = \left\{ \begin{pmatrix} x_1 \\ \vdots \\ x_n \end{pmatrix} \middle| x_1, \ldots, x_n \in \mathbb{C} \right\} \text{ or } \left\{ (x_1, \ldots, x_n) \middle| x_1, \ldots, x_n \in \mathbb{C} \right\}$$

Note that the vectors in these vector spaces are particular types of matrices.

The operations on these vector spaces are the matrix addition and scalar multiplication that were defined in Chapter 1.

Other examples of vector spaces include:

(i)   The set of $m \times n$ matrices with entries from $\mathbb{R}$, denoted $\mathcal{M}_{m \times n}(\mathbb{R})$ with the usual scalar multiplication and matrix addition.

(ii) $F[a,b]$ = real valued functions on $[a, b]$ with the usual addition of functions and multiplication of functions by numbers.

(iii) $C[a,b]$ = functions that are continuous for $a \leq x \leq b$ with the usual addition of functions and multiplication of functions by numbers.

(iv) Polynomials of degree less than or equal to n with coefficients in $\mathbb{R}$, denoted $\mathcal{P}_n(\mathbb{R})$, with the usual addition of polynomials and multiplication of polynomials by numbers.

(v)   The set of polynomials $\{P(x) | P(0) = 0\}$, with the usual addition of polynomials and multiplication of polynomials by numbers.

(vi) Solutions to the differential equation

$$y''(t) + a(t) y'(t) + b(t) y(t) = 0$$

where $a(t)$ and $b(t)$ are continuous functions with the usual addition of functions and multiplication of functions by numbers.

Some Examples of Sets That Are Not Vector Spaces

Polynomials of degree n do not form a vector space because the sum of two such polynomials is not necessarily a polynomial of that type. For example:

$$2x^3 + 5x - 3 \quad \text{and} \quad -2x^3 + 7x^2$$

are both polynomials of degree three, but their sum, $7x^2 + 5x - 3$, is a polynomial of degree two.

The 2×2 matrices whose determinant is 0 do not constitute a vector space because the sum of two such matrices of that type is not necessarily a matrix of that type. For example

$$\begin{pmatrix} 1 & 0 \\ 0 & 0 \end{pmatrix} + \begin{pmatrix} 0 & 0 \\ 0 & 1 \end{pmatrix} = \begin{pmatrix} 1 & 0 \\ 0 & 1 \end{pmatrix}.$$

The set of polynomials

$$\{P(x) | P(0) = 1\}$$

is not a vector space because if

$$P(0) = 1 \quad \text{and} \quad Q(0) = 1, \text{ then } (P + Q)(0) = 1 + 1 = 2.$$

Example

We show that in $M_{2 \times 2}$, the set of matrices of the type

$$\begin{pmatrix} a & a \\ 0 & b \end{pmatrix}$$

is closed under scalar multiplication and addition. Later, we will show that is sufficient to ensure they form a vector space.

This means we are considering matrices A where the A_{11} and A_{12} entries are equal, the A_{24} entry can be any value and the A_{23} entry is 0. We want to know (i) whether the sum of two matrices with those characteristics also has those characteristics and (ii) whether a scalar multiple of a matrix with those characteristics also has those characteristics.

Let

$$A = \begin{pmatrix} a_{11} & a_{11} \\ 0 & \alpha \end{pmatrix}, \ B = \begin{pmatrix} b_{11} & b_{11} \\ 0 & \beta \end{pmatrix}.$$

Then

$$A + B = \begin{pmatrix} a_{11} & a_{11} \\ 0 & \alpha \end{pmatrix} + \begin{pmatrix} b_{11} & b_{11} \\ 0 & \beta \end{pmatrix} = \begin{pmatrix} a_{11} + b_{11} & a_{11} + b_{11} \\ 0 & \alpha + \beta \end{pmatrix}$$

which has the required properties, and

$$cA = c \begin{pmatrix} a_{11} & a_{11} \\ 0 & \alpha \end{pmatrix} = \begin{pmatrix} ca_{11} & ca_{11} \\ 0 & c\alpha \end{pmatrix}$$

which also has the required properties.

However, as another example, consider matrices of the type

$$\begin{pmatrix} a & a \\ 1 & b \end{pmatrix}$$

where the A_{11} and A_{12} entries are equal, the A_{24} entry can be any value and the A_{23} entry is 1. We then have for

$$A = \begin{pmatrix} a_{11} & a_{11} \\ 1 & \alpha \end{pmatrix}, \ B = \begin{pmatrix} b_{11} & b_{11} \\ 1 & \beta \end{pmatrix}$$

that

$$A + B = \begin{pmatrix} a_{11} & a_{11} \\ 1 & \alpha \end{pmatrix} + \begin{pmatrix} b_{11} & b_{11} \\ 1 & \beta \end{pmatrix} = \begin{pmatrix} a_{11} + b_{11} & a_{11} + b_{11} \\ 2 & \alpha + \beta \end{pmatrix}$$

which is not a matrix of that type since the $(A + B)_{23}$ entry is 2.

Additional Properties of Vector Spaces

The next two theorems note additional properties of vector spaces.

Theorem 3.5:

(a.) The additive identity of a vector space is unique.
(b.) The additive inverse of a vector is unique.

Proof:

(a.) The typical way to show that something is unique is to suppose that there are two objects that satisfy the defining property, and then show the two objects are equal.

Suppose that $\hat{0}_1$ and $\hat{0}_2$ are both additive identities.

Since $\hat{0}_1$ is an additive identity, we have

$$\hat{0}_1 + \hat{0}_2 = \hat{0}_2$$

and since $\hat{0}_2$ is an additive identity, we have

$$\hat{0}_1 + \hat{0}_2 = \hat{0}_1.$$

Thus,

$$\hat{0}_1 = \hat{0}_2.$$

(b.) Suppose that $\hat{u}$ and $\hat{v}$ are additive inverses of $\hat{w}$. Then
$$\hat{v} = \hat{v} + \hat{0} = \hat{v} + (\hat{w} + \hat{u}) = (\hat{v} + \hat{w}) + \hat{u} = \hat{0} + \hat{u} = \hat{u}.$$

Theorem 3.6:

If V is a vector space over $\mathcal{F}$ and $\hat{u} \in V$, $a \in \mathcal{F}$, then

(a). $0\hat{u} = \hat{0}$

(b). $a\hat{0} = \hat{0}$

(c). $(-1)\hat{u} = -\hat{u}.$

Proof:

(a.) We have

$$0\hat{u} = (0 + 0)\hat{u} = 0\hat{u} + 0\hat{u}$$

so

$$\hat{0} = (-0\hat{u}) + 0\hat{u} = (-0\hat{u}) + (0\hat{u} + 0\hat{u})$$
$$= ((-0\hat{u}) + 0\hat{u}) + 0\hat{u} = \hat{0} + 0\hat{u} = 0\hat{u}.$$

(b.) We have

$$a\hat{0} = a(\hat{0} + \hat{0}) = a\hat{0} + a\hat{0}$$

so

$$\hat{0} = (-a\hat{0}) + a\hat{0} = (-a\hat{0}) + a\hat{0} + a\hat{0} = ((-a\hat{0}) + a\hat{0}) + a\hat{0} = a\hat{0}.$$

(c.) We have

$$\hat{0} = 0\hat{u} = \big[(-1)+1\big]\hat{u} = (-1)\hat{u} + 1\hat{u} = (-1)\hat{u} + \hat{u}.$$

Thus, $(-1)\hat{u}$ is the additive inverse of $\hat{u}$, but by definition $-\hat{u}$ is the additive inverse of $\hat{u}$, and additive inverses are unique, so

$$(-1)\hat{u} = -\hat{u}.$$

Example

In a vector space V over a field $\mathcal{F}$, if $a\hat{u} = \hat{0}$, then $a = 0$ or $\hat{u} = \hat{0}$.

Suppose that $a \neq 0$ and $a\hat{u} = \hat{0}$. then

$$\hat{u} = \frac{1}{a}a\hat{u} = \frac{1}{a}\hat{0} = \hat{0}.$$

Exercises

In Exercises 1–12 determine whether the set together with the operations defines a vector space. If it is not a vector space give the axioms that are not satisfied.

1. Polynomials of the form

$$p(t) = \alpha t$$

with the usual addition and scalar multiplication.

2. Vectors in $\mathbb{R}^2$ with vector addition and scalar multiplication defined as

$$\begin{pmatrix} u_1 \\ u_2 \end{pmatrix} + \begin{pmatrix} v_1 \\ v_2 \end{pmatrix} = \begin{pmatrix} u_1^2 + v_1^2 \\ u_2^2 + v_2^2 \end{pmatrix}, \; \alpha \begin{pmatrix} u_1 \\ u_2 \end{pmatrix} = \begin{pmatrix} \alpha u_1 \\ \alpha u_2 \end{pmatrix}$$

3. Polynomials of the form

$$p(t) = \alpha + t$$

with the usual addition and scalar multiplication.

4. $M_{2\times2}$ matrices of the form

$$\begin{pmatrix} a & a+b \\ a+b & b \end{pmatrix}$$

with the usual addition and scalar multiplication.

5. $M_{2\times2}$ matrices of the form

$$\begin{pmatrix} a & 1 \\ 0 & b \end{pmatrix}$$

with the usual addition and scalar multiplication.

6. Continuous functions that satisfy $f(2) = 2$, with the usual addition and scalar multiplication.
7. Continuous functions that satisfy $f(2) = 0$, with the usual addition and scalar multiplication.
8. Vectors in $\mathbb{R}^2$ with vector addition and scalar multiplication defined as

$$\begin{pmatrix} u_1 \\ u_2 \end{pmatrix} + \begin{pmatrix} v_1 \\ v_2 \end{pmatrix} = \begin{pmatrix} u_1 + v_1 \\ u_2 + v_2 \end{pmatrix}, \; \alpha \begin{pmatrix} u_1 \\ u_2 \end{pmatrix} = \begin{pmatrix} \alpha u_1 \\ 0 \end{pmatrix}$$

9. Vectors in $\mathbb{R}^2 \begin{pmatrix} x \\ y \end{pmatrix}$ with $xy \geq 0$, with the usual addition and scalar multiplication.

10. Ordered pairs of the form $\begin{pmatrix} 1 \\ y \end{pmatrix}$ with

$$\begin{pmatrix} 1 \\ y_1 \end{pmatrix} + \begin{pmatrix} 1 \\ y_2 \end{pmatrix} = \begin{pmatrix} 1 \\ y_1 + y_2 \end{pmatrix}$$

$$\alpha \begin{pmatrix} 1 \\ y \end{pmatrix} = \begin{pmatrix} 1 \\ \alpha y \end{pmatrix}.$$

11. Vectors of the form $\begin{pmatrix} 3t \\ 0 \\ 5t \end{pmatrix}$ with the usual addition and scalar multiplication.

12. Vectors of the form $\begin{pmatrix} 3t \\ 1 \\ 5t \end{pmatrix}$ with the usual addition and scalar multiplication.

13. Show that the even polynomials (all exponents are even) are a vector space, but the odd polynomials (all exponents are odd) are not.
14. (a.) Show that diagonal $n \times n$ matrices are a vector space.
    (b.) The trace of a square matrix is the sum of the entries on its main diagonal. Show that $n \times n$ matrices with trace equal to zero are a vector space.
15. Show that if $\hat{u} + \hat{v} = \hat{u} + \hat{w}$, then $\hat{v} = \hat{w}$.

Section 3.3 Subspaces of a Vector Space

Definition:

Let V be a vector space and W a nonempty subset of V. W is a subspace of V if W is a vector space under the same operations as V.

The central problem of this section is to determine when a nonempty subset W of a vector space V is a subspace. We now show that some of the axioms of a vector space will always be satisfied by the elements of W by virtue of those elements being in V.

If $\hat{w}_1, \hat{w}_2 \in W$ then

$$\hat{w}_1 + \hat{w}_2 = \hat{w}_2 + \hat{w}_1$$

because $\hat{w}_1 + \hat{w}_2 = \hat{w}_2 + \hat{w}_1$ for any elements in V. However, we do not automatically know that $\hat{w}_1 + \hat{w}_2$ is an element of W, and this is required in order for W to be a vector space

Similarly, if $\hat{w}_1, \hat{w}_2, \hat{w}_3 \in W$ then

$$\left(\hat{w}_1 + \hat{w}_2\right) + \hat{w}_3 = \hat{w}_1 + \left(\hat{w}_2 + \hat{w}_3\right)$$

but again, we do not automatically know that $\left(\hat{w}_1 + \hat{w}_2\right) + \hat{w}_3$ is an element of W, and this is required in order for W to be a vector space. Thus, in order for W to be a vector space, it must be that W is closed under vector addition.

In a similar vein, if $\hat{w}_1, \hat{w}_2 \in W$, and $a, b \in \mathcal{F}$, then

$$a\left(b\hat{w}_1\right) = (ab)\hat{w}_1 \text{ and } a(\hat{w}_1 + \hat{w}_2) = a\hat{w}_1 + a\hat{w}_2$$

but there is nothing that ensures any of these are in W. Thus, in order for W to be a vector space, it must be closed under scalar multiplication.

The next theorem shows that closure under vector addition and scalar multiplication is sufficient for a nonempty set of vectors from a vector space to be a subspace.

Theorem 3.7:

A nonempty subset W of the vector space V is a subspace of V if and only if

   (i)   for every $\hat{u}, \hat{v} \in W$, we have $\hat{u} + \hat{v} \in W$ and
  (ii)   for every $\hat{u} \in W$ and $a \in \mathcal{F}$ we have $a\hat{u} \in W$.

Proof:

We must verify that the axioms of a vector space are satisfied by W.

If $\hat{0}$ is the only vector in W, then the proof is immediate.

All of the axioms except for (3.) and (4.) are immediate since the elements in the subset follow the rules of vector addition and scalar multiplication as in the vector space.

If $\hat{u} \in W$, then $-1\hat{u} = -\hat{u} \in W$ so axiom (4) is satisfied.

If $\hat{u} \in W$, then $-\hat{u} \in W$, so $\hat{u} + (-\hat{u}) = \hat{0} \in W$ and axiom (3.) is satisfied.

In showing that a set of vectors is nonempty, one can usually observe that the set contains the zero vector.

In Section 3.2, the examples of sets that were not vector spaces, failed to be closed under one of the vector operations.

    Example

    In $\mathbb{R}^3$, the only subspaces are lines through the origin, planes through the origin, {0} and $\mathbb{R}^3$.

Definition:

Let $\{\hat{v}_1, \ldots, \hat{v}_n\}$ be a collection of vectors from the vector space V. A linear combination of $\{\hat{v}_1, \ldots, \hat{v}_n\}$ is a vector of the form

$$\hat{u} = c_1\hat{v}_1 + \cdots + c_n\hat{v}_n \text{ where } c_i \in \mathcal{F}.$$

Linear combinations will play a central role in the rest of our discussions.

The next theorem highlights a very important class of vector spaces.

Theorem 3.8:

Let $\{\hat{v}_1,...,\hat{v}_n\}$ be a collection of vectors from the vector space V. The set of linear combinations of these vectors is a subspace of V.

Proof:

Note: Something that is often an impediment to beginners in proving this type of result is choosing representations. In this proof we will use the previous theorem but we need to represent $\hat{u}$ and $\hat{v}$ in that theorem as they pertain to the present case.

We first observe that

$$\hat{0} = 0\hat{v}_1 + \cdots + 0\hat{v}_n.$$

Let $\hat{u}$ and $\hat{v}$ be linear combinations of $\{\hat{v}_1,...,\hat{v}_n\}$. Then we can write

$$\hat{u} = a_1\hat{v}_1 + \cdots + a_n\hat{v}_n$$

$$\hat{v} = b_1\hat{v}_1 + \cdots + b_n\hat{v}_n$$

where $a_i, b_i \in \mathcal{F}$; $i = 1,...,n$.

So

$$\hat{u} + \hat{v} = \left(a_1\hat{v}_1 + \cdots + a_n\hat{v}_n\right) + \left(b_1\hat{v}_1 + \cdots + b_n\hat{v}_n\right) = \left(a_1 + b_1\right)\hat{v}_1 + \cdots + \left(a_n + b_n\right)\hat{v}_n$$

which is a linear combination of $\{\hat{v}_1,...,\hat{v}_n\}$.

Thus, the set is closed under vector addition.

If $c \in \mathcal{F}$, then

$$c\hat{u} = c\left(a_1\hat{v}_1 + \cdots + a_n\hat{v}_n\right) = \left(ca_1\right)\hat{v}_1 + \cdots + \left(ca_n\right)\hat{v}_n$$

which is a linear combination of $\{\hat{v}_1,...,\hat{v}_n\}$.

Thus, the set is closed under scalar multiplication.

Example

Let A be an $m \times n$ matrix and let

$$W = \left\{\hat{x} \in \mathcal{F}^n \,\middle|\, A\hat{x} = \hat{0}\right\}$$

The set W is called the null space of A. It will usually be denoted $\mathcal{N}(A)$ and will play an important role in our theory.

We show that W is a subspace of $\mathcal{F}^n$.

We have that W is nonempty since $A\hat{0} = \hat{0}$.

Suppose

$$\hat{u}, \hat{v} \in W; \text{ i.e., } A\hat{u} = \hat{0} \text{ and } A\hat{v} = \hat{0}.$$

We must show $\hat{u} + \hat{v} \in W$ and $a\hat{u} \in W$; i.e., we must show

$$A(\hat{u} + \hat{v}) = \hat{0} \text{ and } A(a\hat{u}) = \hat{0}.$$

But

$$A(\hat{u} + \hat{v}) = A(\hat{u}) + A(\hat{v}) = \hat{0} + \hat{0} = \hat{0}$$

and

$$A(a\hat{u}) = aA(\hat{u}) = a\hat{0} = \hat{0}.$$

Another note about proofs. The idea of rephrasing a hypothesis is often helpful. We rephrased the condition $\hat{v} \in W$, as $A\hat{v} = \hat{0}$. It also helps one focus on the idea that part of what needs to be shown is $A(\hat{u} + \hat{v}) = \hat{0}$.

Example

(1.) We show that vectors of the form

$$\begin{pmatrix} 2a + 5b \\ 4a - 2b \\ 3b \end{pmatrix}$$

form a subspace of $\mathbb{R}^3$ by showing such vectors are a linear combination of vectors. We have

$$\begin{pmatrix} 2a + 5b \\ 4a - 2b \\ 3b \end{pmatrix} = \begin{pmatrix} 2a \\ 4a \\ 0 \end{pmatrix} + \begin{pmatrix} 5b \\ -2b \\ 3b \end{pmatrix} = a\begin{pmatrix} 2 \\ 4 \\ 0 \end{pmatrix} + b\begin{pmatrix} 5 \\ -2 \\ 3 \end{pmatrix}.$$

Thus the set consists of the linear combinations of the vectors

$$\begin{pmatrix} 2 \\ 4 \\ 0 \end{pmatrix} \text{ and } \begin{pmatrix} 5 \\ -2 \\ 3 \end{pmatrix}$$

and the set of all linear combinations of a (nonempty) set of vectors form a subspace.

(2.) Vectors of the form

$$\begin{pmatrix} 2a \\ 4a \\ 1 \end{pmatrix}$$

do not form a subspace because the sum of two vectors of that type is not a vector of that type. In particular,

$$\begin{pmatrix} 2a \\ 4a \\ 1 \end{pmatrix} + \begin{pmatrix} 2b \\ 4b \\ 1 \end{pmatrix} = \begin{pmatrix} 2(a+b) \\ 4(a+b) \\ 2 \end{pmatrix}.$$

Also note that the zero vector is not in the set.

(3.) We show that

$$\{(x,y,z)|2x-2y-3z=0\}$$

is a subspace of $\mathbb{R}^3$.

The row reduced form of the augmented matrix associated with

$$2x-2y-3z=0$$

is

$$\begin{pmatrix} 1 & -1 & -3/2 & 0 \end{pmatrix}$$

so y and z are free variables and x is a leading variable, and we have

$$x = y + \frac{3}{2}z.$$

If $y=r$ and $z=s$, then

$$\begin{pmatrix} x \\ y \\ z \end{pmatrix} = \begin{pmatrix} r+\frac{3}{2}s \\ r \\ s \end{pmatrix} = \begin{pmatrix} r \\ r \\ 0 \end{pmatrix} + \begin{pmatrix} \frac{3}{2}s \\ 0 \\ s \end{pmatrix} = r\begin{pmatrix} 1 \\ 1 \\ 0 \end{pmatrix} + s\begin{pmatrix} \frac{3}{2} \\ 0 \\ 1 \end{pmatrix}$$

Thus the set is a linear combination of the vectors

$$\begin{pmatrix} 1 \\ 1 \\ 0 \end{pmatrix} \text{ and } \begin{pmatrix} \frac{3}{2} \\ 0 \\ 1 \end{pmatrix}$$

and so is a subspace.

Exercises

1. Tell whether the following collections of vectors form a subspace.

    (a.) Vectors of the form $\begin{pmatrix} a-b \\ 3a+2b \\ b \end{pmatrix}$.

    (b.) Vectors of the form $\begin{pmatrix} 2a-b \\ 0 \\ 3b \end{pmatrix}$.

    (c.) Vectors of the form $\begin{pmatrix} a-b \\ 3 \\ b \end{pmatrix}$.

2. Let S be the set of all 2×2 matrices for which $A = A^T$.

    Show that S is a subspace of the set of all 2×2 matrices

3. Tell whether matrices of the following form constitute a subspace of $M_{2\times2}(\mathbb{R})$

    (a.) $\begin{pmatrix} a & b \\ 0 & 0 \end{pmatrix}$

    (b.) $\begin{pmatrix} a & b \\ -b & a \end{pmatrix}$

    (c.) Invertible 2×2 matrices.

4. Let $A \in M_{2\times2}(\mathbb{R})$. Tell whether the following sets of matrices form a subspace of $M_{2\times2}(\mathbb{R})$.

    (a.) $\left\{ B \in M_{2\times2}(\mathbb{R}) | AB = \hat{0} \right\}$

    (b.) $\left\{ B \in M_{2\times2}(\mathbb{R}) | AB = BA \right\}$

    (c.) $\left\{ B \in M_{2\times2}(\mathbb{R}) | A + B = \hat{0} \right\}$

5. (a.) Show that if U and W are subspaces of the vector space V, then $U + W$ is a subspace of V where

$$U + W = \left\{ \hat{u} + \hat{v} | \hat{u} \in U, \hat{v} \in V \right\}.$$

    (b.) Give an example of nonempty subsets U and W of the vector space V for which $U + W$ is not a subspace of V.

6. Show that the set of matrices $A \in M_{2\times2}(\mathbb{R})$ for which $A^2 = A$ does not form a subspace.

7. (a.) Show that $\left\{ (x,y,z) | 2x - y - z = 0 \right\}$ is a subspace of $\mathbb{R}^3$.

    (a.) Show that $\left\{ (x,y,z) | x - 6y + 3z = 0 \right\}$ is a subspace of $\mathbb{R}^3$.

    (b.) Show that $\left\{ (x,y,z) | 3x - z = 0 \right\}$ is a subspace of $\mathbb{R}^3$.

    (c.) Show that $\left\{ (x,y,z) | 2x - y - z = 1 \right\}$ is not a subspace of $\mathbb{R}^3$.

8. Show that the solution sets for the systems of equations form subspaces of $\mathbb{R}^3$.

    (a.) $x - 3y + 4z = 0$
    $2x + y = 0$

    (b.) $7x + 5y - 3z = 0$
    $4x + y - 2z = 0$

Section 3.4 Spanning Sets, Linearly independent Sets, and Bases

We have shown that the linear combinations of a set of vectors form a subspace of a vector space. In this section we describe which sets of vectors can build (through linear combinations) the entire vector space in the most efficient way. For a given vector space, there will be many such sets of vectors. Each such set is called a basis of the vector space.

To make the central ideas more intuitive, we present the following metaphor.

Consider a toy construction set that contains parts from which you can make toys – Legos, erector sets, Lincoln logs, and Tinker Toys are some examples. Such a construction set typically comes with an instruction booklet that tells how to make particular toys from the set. The manufacturer of the construction set has two major considerations. First, the set must contain all of the parts necessary to build each of the toys in the catalog; otherwise, the customers will feel cheated. Second, there should be no surplus parts; otherwise, there is an unnecessary expense and profits will suffer.

We begin with a vector space V. We want to find a collection of vectors $\{\hat{v}_1, \ldots, \hat{v}_n\}$ in V so that

1. each vector in V can be written as a linear combination of vectors in $\{\hat{v}_1, \ldots, \hat{v}_n\}$, and
2. the set $\{\hat{v}_1, \ldots, \hat{v}_n\}$ is minimal in the sense that if any one of the vectors is removed from the set, then it will not be possible to express every vector in V as a linear combination of the reduced set.

Definition:

A vector $\hat{v} \epsilon V$ is said to be in the span of $\{\hat{v}_1, \ldots, \hat{v}_n\}$ if $\hat{v}$ can be written as a linear combination of $\{\hat{v}_1, \ldots, \hat{v}_n\}$. A set of vectors $\{\hat{v}_1, \ldots, \hat{v}_n\}$ is a spanning set for V if every vector in V can be written as a linear combination of $\{\hat{v}_1, \ldots, \hat{v}_n\}$.

Another way to say that the set of vectors $\{\hat{v}_1, \ldots, \hat{v}_n\}$ is a spanning set for V is to say that $\{\hat{v}_1, \ldots, \hat{v}_n\}$ spans V.

Definition:

Let V be vector space. A set of vectors $\{\hat{v}_1, \ldots, \hat{v}_n\}$ is linearly independent if the only way

$$a_1\hat{v}_1 + \cdots + a_n\hat{v}_n = \hat{0}$$

is for $a_i = 0$ for every $i = 1, \ldots, n$.

A set of vectors $\{\hat{v}_1, \ldots, \hat{v}_n\}$ is linearly dependent if it is not linearly independent. This means there are scalars $a_1, \ldots, a_n$ not all of which are 0, with

$$a_1\hat{v}_1 + \cdots + a_n\hat{v}_n = \hat{0}.$$

Definition:

A set of vectors $\{\hat{v}_1, \ldots, \hat{v}_n\}$ is a basis for V if it is a spanning set for V and is linearly independent.

Relating the definition of a basis to the construction toy metaphor, saying $\{\hat{v}_1,\ldots,\hat{v}_n\}$ is a spanning set is analogous to saying we have enough parts to build all the toys in the catalog and saying the set $\{\hat{v}_1,\ldots,\hat{v}_n\}$ is linearly independent is analogous to saying there are no surplus parts.

Example

The vector space $\mathcal{F}^3$ consists of the vectors

$$\left\{ \begin{pmatrix} a \\ b \\ c \end{pmatrix} \middle| a,b,c \in \mathcal{F} \right\}$$

with the operations of vector addition and scalar multiplication. The most natural basis for this vector space is the set of vectors

$$\left\{ \begin{pmatrix} 1 \\ 0 \\ 0 \end{pmatrix}, \begin{pmatrix} 0 \\ 1 \\ 0 \end{pmatrix}, \begin{pmatrix} 0 \\ 0 \\ 1 \end{pmatrix} \right\}.$$

This is called the usual (or standard) basis for $\mathcal{F}^3$ and we will use the notation

$$\hat{e}_1 = \begin{pmatrix} 1 \\ 0 \\ 0 \end{pmatrix}, \; \hat{e}_2 = \begin{pmatrix} 0 \\ 1 \\ 0 \end{pmatrix}, \; \hat{e}_3 = \begin{pmatrix} 0 \\ 0 \\ 1 \end{pmatrix}$$

for the usual basis for $\mathcal{F}^3$. The standard bases for $\mathcal{F}^n$ have similar notation.

To show that this is a basis, we must show that

1. the set spans $\mathcal{F}^3$
2. the set is linearly independent.

To show the set spans $\mathcal{F}^3$, let

$$\begin{pmatrix} a \\ b \\ c \end{pmatrix} \in \mathcal{F}^3$$

Then

$$\begin{pmatrix} a \\ b \\ c \end{pmatrix} = a \begin{pmatrix} 1 \\ 0 \\ 0 \end{pmatrix} + b \begin{pmatrix} 0 \\ 1 \\ 0 \end{pmatrix} + c \begin{pmatrix} 0 \\ 0 \\ 1 \end{pmatrix}.$$

To show the set is linearly independent, suppose that

$$a \begin{pmatrix} 1 \\ 0 \\ 0 \end{pmatrix} + b \begin{pmatrix} 0 \\ 1 \\ 0 \end{pmatrix} + c \begin{pmatrix} 0 \\ 0 \\ 1 \end{pmatrix} = \begin{pmatrix} a \\ b \\ c \end{pmatrix} = \begin{pmatrix} 0 \\ 0 \\ 0 \end{pmatrix}.$$

This is true if and only if $a = 0$, $b = 0$ and $c = 0$.
We will use analogs of this example repeatedly.

Theorem 3.9:

A set of vectors $\{\hat{v}_1,\ldots,\hat{v}_n\}$ is a basis for V if and only if every vector in V can be written as a linear combination of $\{\hat{v}_1,\ldots,\hat{v}_n\}$ in exactly one way.

Proof:

Suppose that $\{\hat{v}_1,\ldots,\hat{v}_n\}$ is a basis for V.

Since $\{\hat{v}_1,\ldots,\hat{v}_n\}$ spans V, every vector in V can be written as a linear combination of $\{\hat{v}_1,\ldots,\hat{v}_n\}$.

Next, we prove linear independence. Suppose that there is a vector $\hat{v} \in V$ that can be written as a linear combination of $\{\hat{v}_1,\ldots,\hat{v}_n\}$ in two ways, say

$$\hat{v} = a_1\hat{v}_1 + \cdots + a_n\hat{v}_n$$

and

$$\hat{v} = b_1\hat{v}_1 + \cdots + b_n\hat{v}_n.$$

Then

$$\hat{0} = \hat{v} - \hat{v} = \left(a_1\hat{v}_1 + \cdots + a_n\hat{v}_n\right) - \left(b_1\hat{v}_1 + \cdots + b_n\hat{v}_n\right)$$
$$= \left(a_1 - b_1\right)\hat{v}_1 + \cdots + \left(a_n - b_n\right)\hat{v}_n.$$

So we have written $\hat{0}$ as a linear combination of $\{\hat{v}_1,\ldots,\hat{v}_n\}$. Since $\{\hat{v}_1,\ldots,\hat{v}_n\}$ is a basis, the set is linearly independent, so $\left(a_i - b_i\right) = 0$; or $a_i = b_i$, $i = 1,\ldots,n$.

Conversely, suppose that every vector in V can be written as a linear combination of $\{\hat{v}_1,\ldots,\hat{v}_n\}$ in exactly one way.

Because every vector in V can be written as a linear combination of $\{\hat{v}_1,\ldots,\hat{v}_n\}$, the set spans V. Because

$$\hat{0} = 0\hat{v}_1 + \cdots + 0\hat{v}_n$$

and there is only one linear combination of $\{\hat{v}_1,\ldots,\hat{v}_n\}$ that gives $\hat{0}$, the set $\{\hat{v}_1,\ldots,\hat{v}_n\}$ is linearly independent.

The next theorem says that a linearly dependent set of vectors has some vectors that are, in some sense, superfluous.

Theorem 3.10:

If $\{\hat{v}_1,\hat{v}_2,\ldots,\hat{v}_n\}$ is a linearly dependent set of vectors, then one of the vectors can be written as a linear combination of the others.

The implication of this theorem is that a linearly dependent set of vectors can be reduced so that the reduced set will span the same set as the original set. This is because if

$$\hat{v} = a_1\hat{v}_1 + a_2\hat{v}_2 + \cdots + a_n\hat{v}_n$$

and

$$\hat{v}_1 = b_2\hat{v}_2 + \cdots + b_n\hat{v}_n$$

then

$$\hat{v} = a_1\hat{v}_1 + a_2\hat{v}_2 + \cdots + a_n\hat{v}_n = a_1\left(b_2\hat{v}_2 + \cdots + b_n\hat{v}_n\right) + a_2\hat{v}_2 + \cdots + a_n\hat{v}_n$$
$$= \left(a_1b_2 + a_2\right)\hat{v}_2 + \cdots + \left(a_1b_n + a_n\right)\hat{v}_n$$

and the last expression is a linear combination of $\{\hat{v}_2,\ldots,\hat{v}_n\}$.

Proof:

Since $\{\hat{v}_1,\hat{v}_2,\ldots,\hat{v}_n\}$ is a linearly dependent set of vectors, it is possible to find scalars $c_1,c_2,\ldots,c_n$, not all of which are zero so that

$$c_1\hat{v}_1 + c_2\hat{v}_2 + \cdots + c_n\hat{v}_n = \hat{0}.$$

Suppose we have arranged the vectors so that $c_1 \neq 0$. Then

$$c_1\hat{v}_1 = -c_2\hat{v}_2 - \cdots - c_n\hat{v}_n$$

and since $c_1 \neq 0$ we may divide by c_1 to get

$$\hat{v}_1 = -\left(\frac{c_2}{c_1}\right)\hat{v}_2 - \cdots - \left(\frac{c_n}{c_1}\right)\hat{v}_n.$$

Thus $\hat{v}_1$ is a linear combination of $\{\hat{v}_2,\ldots,\hat{v}_n\}$.

We will soon be precise about what is meant by the dimension of a vector space. For now, simply realize that it means something about the size of a vector space.

Every vector space has a basis. (This is a relatively deep fact.) A finite dimensional vector space is defined to be a vector space that has a finite spanning set. We will show that such a vector space has a basis. A vector space that is not finite dimensional is said to be infinite dimensional. (Showing that an infinite dimensional vector spaces has a basis is proven using Zorn's lemma or a logically equivalent axiom, which is beyond the scope of this text.)

From now on we will assume that the vector spaces we consider are finite dimensional.

For the remainder of this section and the next section, we want to prove the following facts.

1. Every basis of a given finite dimensional vector space has the same number of vectors. We call this number the dimension of the vector space.
2. Every linearly independent set of vectors that is not a basis can be expanded to be a basis. We will give an argument for why this is true and an algorithm that accomplishes the expansion.
3. Every spanning set of vectors that is not linearly independent can be reduced to a basis. We will give an argument for why this is true and an algorithm that accomplishes the reduction.

Our immediate task is to show that all bases of a given vector space have the same number of vectors. This is accomplished by Theorems 3.11 and 3.12.

Theorem 3.11:

Suppose that $\{v_1,\ldots,v_n\}$ is a basis for the vector space V. Then any set of more than n vectors will be linearly dependent.

Proof:

We demonstrate the proof in the case where a basis consists of two vectors. Suppose $\{\hat{v}_1, \hat{v}_2\}$ is a basis for the vector space V. Suppose that $\hat{w}_1, \hat{w}_2$ and $\hat{w}_3$ are vectors in V. We will show they are linearly dependent; that is, we will show that there are b_1, b_2, b_3 not all zero, with

$$b_1\hat{w}_1 + b_2\hat{w}_2 + b_3\hat{w}_3 = \hat{0}.$$

Since $\{\hat{v}_1, \hat{v}_2\}$ spans V, there are a_{11} and a_{12} with

$$\hat{w}_1 = a_{11}\hat{v}_1 + a_{12}\hat{v}_2$$

a_{21} and a_{22} with

$$\hat{w}_2 = a_{21}\hat{v}_1 + a_{22}\hat{v}_2$$

and a_{31} and a_{32} with

$$\hat{w}_3 = a_{31}\hat{v}_1 + a_{32}\hat{v}_2.$$

So

$$b_1\hat{w}_1 + b_2\hat{w}_2 + b_3\hat{w}_3 = b_1\left(a_{11}\hat{v}_1 + a_{12}\hat{v}_2\right) + b_2\left(a_{21}\hat{v}_1 + a_{22}\hat{v}_2\right) + b_3\left(a_{31}\hat{v}_1 + a_{32}\hat{v}_2\right)$$
$$= \left(b_1 a_{11} + b_2 a_{21} + b_3 a_{31}\right)\hat{v}_1 + \left(b_1 a_{12} + b_2 a_{22} + b_3 a_{32}\right)\hat{v}_2 = \hat{0}.$$

Now the a_{ij} are all fixed, and the question comes down to whether it is possible to find b_1, b_2, and b_3 not all zero, with

$$b_1 a_{11} + b_2 a_{21} + b_3 a_{31} = 0$$

and

$$b_1 a_{12} + b_2 a_{22} + b_3 a_{32} = 0.$$

Since there are two equations with three unknowns, there will be a free variable in the solution, and thus there is a solution with not all of b_1, b_2, b_3 equal to zero.

The general case uses the same ideas, and we leave the write-up as an exercise.

Theorem 3.12:

If $\{\hat{v}_1,\ldots,\hat{v}_n\}$ is a basis for the vector space V, then any set of fewer than n vectors will not span V.

Proof:

We again present a simple case that highlights the idea of the proof and leave the proof of the general case as an exercise.

Suppose that $\{\hat{v}_1,\hat{v}_2,\hat{v}_3\}$ is a basis for the vector space V and suppose that $\{\hat{w}_1,\hat{w}_2\}$ spans V. Then

$$\hat{v}_1 = a_{11}\hat{w}_1 + a_{12}\hat{w}_2$$

$$\hat{v}_2 = a_{21}\hat{w}_1 + a_{22}\hat{w}_2$$

$$\hat{v}_3 = a_{31}\hat{w}_1 + a_{32}\hat{w}_2.$$

We will show that $\{\hat{v}_1,\hat{v}_2,\hat{v}_3\}$ is not a linearly independent set.

Now,

$$b_1\hat{v}_1 + b_2\hat{v}_1 + b_3\hat{v}_3 = b_1\left(a_{11}\hat{w}_1 + a_{12}\hat{w}_2\right) + b_2\left(a_{21}\hat{w}_1 + a_{22}\hat{w}_2\right) + b_3\left(a_{31}\hat{w}_1 + a_{32}\hat{w}_2\right)$$

$$= \left(b_1 a_{11} + b_2 a_{21} + b_3 a_{31}\right)\hat{w}_1 + \left(b_1 a_{12} + b_2 a_{22} + b_3 a_{32}\right)\hat{w}_2.$$

Consider the system of equations

$$b_1 a_{11} + b_2 a_{21} + b_3 a_{31} = 0$$

$$b_1 a_{12} + b_2 a_{22} + b_3 a_{32} = 0$$

There are two equations and three unknowns (b_1, b_2, b_3) so it is possible to find a solution such that not every $b_i = 0$. This contradicts the assumption that $\{\hat{v}_1,\hat{v}_2,\hat{v}_3\}$ is a basis.

Corollary:

In a finite dimensional vector space, all bases have the same number of vectors.

Corollary:

The number of vectors in a basis for $\mathcal{F}^n$ is n, where $\mathcal{F}$ is either the real numbers or the complex numbers.

Proof:

The standard basis for $\mathcal{F}^n$ has n vectors.

Definition:

The number of vectors in a basis of a vector space is the dimension of the vector space.

Definition:

Let U and V be vector spaces over the field $\mathcal{F}$. An isomorphism from U to V is a one-to-one and onto function

$$T : U \to V$$

for which

$$T\left(\alpha_1 \hat{u}_1 + \alpha_2 \hat{u}_2\right) = \alpha_1 T\left(\hat{u}_1\right) + \alpha_2 T\left(\hat{u}_2\right)$$

for all $\hat{u}_1, \hat{u}_2 \in U$; $\alpha_1, \alpha_2 \in \mathcal{F}$.

   If such an isomorphism exists, U and V are said to be isomorphic. Isomorphic vector spaces are structurally the same; the only real difference is how the elements are named. The next theorem gives a succinct categorization of finite dimensional vector spaces.

Theorem 3.13:

Every n-dimensional vector space V over a field $\mathcal{F}$ is isomorphic to $\mathcal{F}^n$.

Proof:

Let V be a vector space with basis $\{\hat{v}_1, \ldots, \hat{v}_n\}$ and suppose $\hat{v} \in V$. There is a unique n-tuple of scalars $\left(\alpha_1, \ldots, \alpha_n\right) \in \mathcal{F}'$ for which

$$\hat{v} = \alpha_1 \hat{v}_1 + \cdots + \alpha_n \hat{v}_n.$$

Define

$$T : V \to \mathcal{F}^n$$

By

$$T\left(\hat{v}\right) = T\left(\alpha_1 \hat{v}_1 + \cdots + \alpha_n \hat{v}_n\right) = \left(\alpha_1, \ldots, \alpha_n\right).$$

The function T is one-to-one and onto.
   We show $T\left(\gamma \hat{v} + \delta \hat{w}\right) = \gamma T\left(\hat{v}\right) + \delta T\left(\hat{w}\right)$.
   Suppose $\hat{w} = \beta_1 \hat{v}_1 + \cdots + \beta_n \hat{v}_n$. Then

$$
\begin{aligned}
T\left(\gamma \hat{v} + \delta \; \hat{w}\right) &= T\left(\gamma \left(\alpha_1 \hat{v}_1 + \cdots + \alpha_n \hat{v}_n\right) + \delta \left(\beta_1 \hat{v}_1 + \cdots + \beta_n \hat{v}_n\right)\right) \\
&= T\left(\left(\gamma \alpha_1 + \delta \beta_1\right)\hat{v}_1 + \cdots + \left(\gamma \alpha_n + \delta \beta_n\right)\hat{v}_n\right) \\
&= \left(\left(\gamma \alpha_1 + \delta \beta_1\right), \ldots, \left(\gamma \alpha_n + \delta \beta_n\right)\right) \\
&= \left(\gamma \left(\alpha_1, \ldots, \alpha_n\right) + \delta \left(\beta, \ldots, \beta_n\right)\right) \\
&= \gamma T\left(\hat{v}\right) + \delta T\left(\hat{w}\right).
\end{aligned}
$$

The question that should arise is, why bother with vector spaces such as $\mathcal{P}_n$. Part of the answer is that often an intuition that is tied to a particular vector space could be lost by considering only $\mathcal{F}^n$.

We would like to be able to easily identify when a set of vectors is a basis for a subspace. The next theorems provide some ways of doing that.

Theorem 3.14:

A set of vectors that contains the zero vector is not linearly independent, and thus is not a basis.
  The proof is left as an exercise.

The most common vector spaces that we encounter are of the form $\mathcal{F}^n$. A set of vectors that does not have exactly n vectors, cannot be a basis for $\mathcal{F}^n$ because we have identified the standard basis of $\mathcal{F}^n$ as having n vectors. If the set does have n vectors, then we must do further analysis to see if it is a basis.

The next theorem gives an easy way to tell whether a set of n vectors forms a basis of $\mathcal{F}^n$.

Theorem 3.15:

A set of n vectors of $\mathcal{F}^n$ is a basis for $\mathcal{F}^n$ if and only if the determinant of the matrix whose columns (or rows) are the given vectors is not 0.
  The essential idea of the proof is given in Exercise 8.

Example

Determine whether the following sets of vectors form a basis for the appropriate vector spaces:

(a.) $\left\{ \begin{pmatrix} 1 \\ 3 \\ -2 \end{pmatrix}, \begin{pmatrix} 5 \\ -9 \\ 0 \end{pmatrix}, \begin{pmatrix} 7 \\ -3 \\ -4 \end{pmatrix} \right\}.$

There are three vectors and the vector space is $\mathcal{F}^3$ so it is possible that the set is a basis. We form the matrix

$$A = \begin{pmatrix} 1 & 5 & 7 \\ 3 & -9 & -3 \\ -2 & 0 & -4 \end{pmatrix}$$

and find $\det(A) = 0$, so the set is not a basis.

(b.) $\left\{ \begin{pmatrix} 1 \\ -1 \\ 2 \\ 0 \end{pmatrix}, \begin{pmatrix} 0 \\ 3 \\ 3 \\ -5 \end{pmatrix}, \begin{pmatrix} 2 \\ 0 \\ -1 \\ -4 \end{pmatrix}, \begin{pmatrix} 6 \\ 1 \\ 1 \\ 1 \end{pmatrix} \right\}.$

There are four vectors and the vector space is $\mathcal{F}^4$ so it is possible that the set is a basis. We form the matrix

$$B = \begin{pmatrix} 1 & 0 & 2 & 6 \\ -1 & 3 & 0 & 1 \\ 2 & 3 & -1 & 1 \\ 0 & -5 & -4 & 1 \end{pmatrix}$$

and find $\det(B) = -302$, so the set is a basis.

Exercises

In Exercises 1–6 determine whether the sets of vectors are linearly independent.

1. $\left\{ \begin{pmatrix} 1 \\ 3 \end{pmatrix}, \begin{pmatrix} 2 \\ 7 \end{pmatrix} \right\}$ in $\mathbb{R}^2$

2. $\left\{ \begin{pmatrix} 1 \\ 3 \\ 2 \end{pmatrix}, \begin{pmatrix} 4 \\ 6 \\ 9 \end{pmatrix}, \begin{pmatrix} 9 \\ 15 \\ 20 \end{pmatrix} \right\}$ in $\mathbb{R}^3$

3. $\left\{ \begin{pmatrix} 1 \\ 1 \\ 2 \end{pmatrix}, \begin{pmatrix} 2 \\ -1 \\ 0 \end{pmatrix}, \begin{pmatrix} 1 \\ -2 \\ -2 \end{pmatrix} \right\}$ in $\mathbb{R}^3$

4. $\left\{ 2, \sin^2 x, \cos^2 x \right\}$ in continuous functions.

5. $\left\{ 1 + x, 2x + x^2, x^2 \right\}$ in $P_2(x)$

6. $\left\{ \begin{pmatrix} 2 & 0 \\ 3 & 1 \end{pmatrix}, \begin{pmatrix} -1 & 4 \\ 0 & 6 \end{pmatrix}, \begin{pmatrix} 0 & 8 \\ 3 & 13 \end{pmatrix} \right\}$ in $M_{2\times 2}(\mathbb{R})$

7. Show that if $\{\hat{v}_1, \hat{v}_2, \ldots, \hat{v}_n\}$ is a linearly independent set of vectors, then any subset of $\{\hat{v}_1, \hat{v}_2, \ldots, \hat{v}_n\}$ is a linearly independent set of vectors.

8. In this exercise we justify the claim that the determinant of a matrix is zero if and only if one of the rows is a linear combination of the other rows. This will mean that the rows are linearly dependent and thus do not form a basis. We demonstrate this in the case of a 3×3 matrix.

    (a.) Show that

$$\det \begin{pmatrix} a_1 & a_2 & a_3 \\ b_1 & b_2 & b_3 \\ 0 & 0 & 0 \end{pmatrix} = 0.$$

    (b.) A linear combination of the first two rows of the matrix in part (a.) is of the form $\alpha(a_1, a_2, a_3) + \beta(b_1, b_2, b_3) = (\alpha a_1, \alpha a_2, \alpha a_3) + (\beta b_1, \beta b_2, \beta b_3)$. Show that

$$\det\begin{pmatrix} a_1 & a_2 & a_3 \\ b_1 & b_2 & b_3 \\ 0 & 0 & 0 \end{pmatrix} = \det\begin{pmatrix} a_1 & a_2 & a_3 \\ b_1 & b_2 & b_3 \\ \alpha a_1 & \alpha a_2 & \alpha a_3 \end{pmatrix}$$

$$= \det\begin{pmatrix} a_1 & a_2 & a_3 \\ b_1 & b_2 & b_3 \\ \alpha a_1 + \beta b_1 & \alpha a_2 + \beta b_2 & \alpha a_3 + \beta b_3 \end{pmatrix}.$$

Thus, if the third row is a linear combination of the first two rows, then the matrix has a determinant of zero.

(c.) Show that if the determinant of a matrix is nonzero, then the rows are independent. One way to do this is to use the fact that when a matrix is row reduced, this is done by applying the elementary row operations. These may change the value of the determinant, but will not change whether the determinant is zero.

9. Show that if $\{\hat{v}_1, \hat{v}_2, \ldots, \hat{v}_n\}$ is a linearly dependent set of vectors, then $\{\hat{v}_1, \hat{v}_2, \ldots, \hat{v}_n, \hat{v}\}$, where $\hat{v}$ is any vector, is a linearly dependent set of vectors.

10. Let $\hat{v}_1 = (7, -4, 1, 0)$, $\hat{v}_2 = (6, -5, 0, 1)$. Find whether the following vectors are in the span $\{\hat{v}_1, \hat{v}_2\}$.

(a). $(20, -13, 2, 1)$

(b). $(-1, -1, -1, 1)$

11. Let $\hat{v}_1 = (3, 1, 0)$, $\hat{v}_2 = (6, 0, -1)$. Find whether the following vectors are in the span $\{\hat{v}_1, \hat{v}_2\}$.

(a). $(9, 4, 0)$

(b). $(12, 2, -1)$

(c). $(15, 1, -2)$

12. Let $\hat{v}_1 = (1, -1, 2)$, $\hat{v}_2 = (2, 4, 5)$. Find whether the following vectors are in the span $\{\hat{v}_1, \hat{v}_2\}$.

(a). $(1, 0, 2)$

(b). $(9, -9, 21)$

(c). $(-3, -9, -8)$

13. Describe the values that a, b, and c must satisfy for the vector

$$\begin{pmatrix} a \\ b \\ c \end{pmatrix}$$

to be in the span of the following sets:

(a.) $\left\{ \begin{pmatrix} 1 \\ 2 \\ 0 \end{pmatrix}, \begin{pmatrix} 3 \\ -2 \\ 0 \end{pmatrix} \right\}$

(b.) $\left\{ \begin{pmatrix} 1 \\ -6 \\ 4 \end{pmatrix}, \begin{pmatrix} 3 \\ 2 \\ 0 \end{pmatrix} \right\}$

(c.) $\left\{ \begin{pmatrix} 1 \\ 2 \\ 3 \end{pmatrix}, \begin{pmatrix} 2 \\ 4 \\ 6 \end{pmatrix} \right\}$

14. If V is a vector space that has a basis $\{\hat{v}_1,\ldots,\hat{v}_n\}$, what can you say about a set of vectors in V that has
    (a.) more than n vectors?
    (b.) Fewer than n vectors?
    (c.) Exactly n vectors?

15. Show that a set of vectors that contains the zero vector is linearly dependent.

16. Show that if V is a vector space of dimension n and U is a subspace of V of dimension n, then $U = V$.

17. Give an example of vectors $\hat{u}, \hat{v}$, and $\hat{w}$ for which $\{\hat{u},\hat{v}\}, \{\hat{v},\hat{w}\}$ and $\{\hat{u},\hat{w}\}$ are linearly independent sets but $\{\hat{u},\hat{v},\hat{w}\}$ is not a linearly independent set.

18. Determine which of the sets is a basis for $\mathbb{R}^3$.
    (a.) $(2,0,1),(5,-1,3)$
    (b.) $(4,1,1),(3,-2,5),(1,1,1)$
    (c.) $(2,3,6),(1,3,1),(6,12,14)$
    (d.) $(2,0,1),(-6,9,-4),(0,0,0)$

19. Find a basis for the set of vectors of the form
$$\begin{bmatrix} a+2b-c \\ 3a+4b \\ 6a-2c \end{bmatrix}.$$

Section 3.5 Converting a Set of Vectors to a Basis

Our next results say that a set of linearly independent vectors that is not a basis can be expanded to be a basis and a spanning set of vectors that is not a basis can be shrunk to be a basis.

In Theorem 3.16 we will show that it is possible to add vectors to a linearly independent set until a basis is obtained and in Theorem 3.17 we will show that it is possible to delete vectors from a spanning set until a basis is obtained. However, the proofs do not give computationally simple ways to do this. After Theorem 3.16 we give an example of how to add vectors to get a basis and after Theorem 3.17 we give an example of how to delete vectors to get a basis. In the next chapter we give a justification for why the techniques are valid.

Theorem 3.16:

Suppose that V is a finite dimensional vector space and $\{\hat{v}_1,\ldots,\hat{v}_n\}$ is a linearly independent set of vectors. If $\{\hat{v}_1,\ldots,\hat{v}_n\}$ is not a basis of V, then a finite number of vectors may be added to $\{\hat{v}_1,\ldots,\hat{v}_n\}$ so that the enlarged set will be a basis.

Proof:

Since $\{\hat{v}_1,\ldots,\hat{v}_n\}$ is linearly independent but is not a basis, it does not span V. Thus, there is a $\hat{v} \in V$, $\hat{v} \neq \hat{0}$ that is not in the span of $\{\hat{v}_1,\ldots,\hat{v}_n\}$. We claim $\{\hat{v}_1,\ldots,\hat{v}_n,\hat{v}\}$ is a linearly independent set. Suppose

$$a_0\hat{v} + a_1\hat{v}_1 + \cdots + a_n\hat{v}_n = \hat{0}.$$

If $a_0 \neq 0$, then $\hat{v}$ is in the span of $\{\hat{v}_1,\ldots,\hat{v}_n\}$, which is a contradiction. Thus $a_0 = 0$, so

$$a_1\hat{v}_1 + \cdots + a_n\hat{v}_n = \hat{0}$$

and $\{\hat{v}_1,\ldots,\hat{v}_n\}$ is linearly independent, so $a_i = 0$, $i = 1, \ldots, n$. Thus, $\{\hat{v}_1,\ldots,\hat{v}_n,\hat{v}\}$ is a linearly independent set. If this set spans V, then it is a basis. Otherwise repeat the process a finite number of times with the larger sets until a basis is obtained. If the dimension of V is k, there will be $k - n$ such additions.

Theorem 3.16 shows that a linearly independent set can be expanded to a basis and gives a theoretical method of how to accomplish this, but does not give a method for finding the additional vectors. The example below gives one method.

Example

The vectors

$$\begin{pmatrix} 3 \\ -2 \\ 0 \\ 5 \end{pmatrix}, \begin{pmatrix} 1 \\ 4 \\ 7 \\ -9 \end{pmatrix}$$

are linearly independent. Complete this set of vectors to form a basis.

The technique we exploit is to choose a basis of the vector space – and it is usually simplest to use the standard basis – and attach vectors of the chosen basis to the given vectors. In this case, we would have

$$\begin{pmatrix} 3 \\ -2 \\ 0 \\ 5 \end{pmatrix}, \begin{pmatrix} 1 \\ 4 \\ 7 \\ -9 \end{pmatrix}, \begin{pmatrix} 1 \\ 0 \\ 0 \\ 0 \end{pmatrix}, \begin{pmatrix} 0 \\ 1 \\ 0 \\ 0 \end{pmatrix}, \begin{pmatrix} 0 \\ 0 \\ 1 \\ 0 \end{pmatrix}, \begin{pmatrix} 0 \\ 0 \\ 0 \\ 1 \end{pmatrix}.$$

Next, create a matrix whose columns are the vectors above. Be sure to list the vectors that are to be part of the basis in the first columns. In this case, we have

$$\begin{pmatrix} 3 & 1 & 1 & 0 & 0 & 0 \\ -2 & 4 & 0 & 1 & 0 & 0 \\ 0 & 7 & 0 & 0 & 1 & 0 \\ 5 & -9 & 0 & 0 & 0 & 1 \end{pmatrix}.$$

Row reduce this matrix. In this case, the result is

$$\begin{pmatrix} 1 & 0 & 0 & 0 & 9/35 & 1/5 \\ 0 & 1 & 0 & 0 & 1/7 & 0 \\ 0 & 0 & 1 & 0 & -32/35 & -3/5 \\ 0 & 0 & 0 & 1 & -2/35 & 2/5 \end{pmatrix}.$$

There will be (in the case where the vector space is $\mathbb{R}^4$) four leading 1s. A basis that includes the two given vectors will be the columns in the original matrix that correspond to the columns in the row reduced matrix where the leading 1s occur (we will prove this later). So in this case, a basis that contains the two given vectors is

$$\begin{pmatrix} 3 \\ -2 \\ 0 \\ 5 \end{pmatrix}, \begin{pmatrix} 1 \\ 4 \\ 7 \\ -9 \end{pmatrix}, \begin{pmatrix} 1 \\ 0 \\ 0 \\ 0 \end{pmatrix}, \begin{pmatrix} 0 \\ 1 \\ 0 \\ 0 \end{pmatrix}.$$

Theorem 3.17:

Suppose that V is a finite dimensional vector space and $\{\hat{v}_1,\ldots,\hat{v}_n\}$ is a spanning set of vectors. If $\{\hat{v}_1,\ldots,\hat{v}_n\}$ is not a basis, then a finite number of vectors may be deleted from $\{\hat{v}_1,\ldots,\hat{v}_n\}$ until the diminished set is a basis.

Proof:

Since $\{\hat{v}_1,\ldots,\hat{v}_n\}$ is a spanning set but not a basis, then it is not linearly independent, and one of the vectors can be written as a linear combination of the others. Suppose

$$\hat{v}_1 = a_2\hat{v}_2 + \ldots + a_n\hat{v}_n.$$

Then $\{\hat{v}_2,\ldots,\hat{v}_n\}$ is a spanning set of vectors. If $\{\hat{v}_2,\ldots,\hat{v}_n\}$ is a linearly independent set of vectors, then it is a basis. If not, continue the process until a basis is obtained.

Theorem 3.17 does not show explicitly how to select the vectors that should be discarded from a linearly dependent set. There may be several choices that are valid. The next example gives a computational technique for discarding superfluous vectors. The technique will be justified in the next chapter.

Example

Select from the set of vectors

$$\left\{ \begin{pmatrix} 1 \\ 7 \\ -2 \end{pmatrix} \begin{pmatrix} 4 \\ 3 \\ 2 \end{pmatrix} \begin{pmatrix} 9 \\ 13 \\ 2 \end{pmatrix} \begin{pmatrix} 10 \\ 29 \\ 0 \end{pmatrix} \begin{pmatrix} 4 \\ 0 \\ 1 \end{pmatrix} \right\}$$

a maximal linearly independent set.

We form the matrix whose columns are the vectors; that is,

$$\begin{pmatrix} 1 & 4 & 9 & 10 & 4 \\ 7 & 3 & 13 & 29 & 0 \\ -2 & 2 & 2 & 0 & 1 \end{pmatrix}.$$

When this matrix is row reduced, the result is

$$\begin{pmatrix} 1 & 0 & 1 & 0 & 73/45 \\ 0 & 1 & 2 & 0 & 191/90 \\ 0 & 0 & 0 & 1 & -11/18 \end{pmatrix}.$$

The leading 1s occur in the first, second, and fourth columns. Return to the original matrix, and those columns will be the maximal linearly independent set. This is

$$\left\{ \begin{pmatrix} 1 \\ 7 \\ -2 \end{pmatrix} \begin{pmatrix} 4 \\ 3 \\ 2 \end{pmatrix} \begin{pmatrix} 10 \\ 29 \\ 0 \end{pmatrix} \right\}.$$

The following example demonstrates that if $\hat{v}_1, \hat{v}_2, \hat{v}_3$ are vectors in $\mathbb{R}^3$ that are do not span $\mathbb{R}^3$ then $\{\hat{v}_1, \hat{v}_2, \hat{v}_3\}$ is not linearly independent.

Example

Let

$$\hat{v}_1 = \begin{pmatrix} 1 \\ 2 \\ 3 \end{pmatrix}, \hat{v}_2 = \begin{pmatrix} 1 \\ 1 \\ 1 \end{pmatrix}, \hat{v}_3 = \begin{pmatrix} 2 \\ 3 \\ 4 \end{pmatrix}, A = \begin{pmatrix} 1 & 1 & 2 \\ 2 & 1 & 3 \\ 3 & 1 & 4 \end{pmatrix}.$$

When A is row reduced the result is

$$\begin{pmatrix} 1 & 0 & 1 \\ 0 & 1 & 1 \\ 0 & 0 & 0 \end{pmatrix}$$

Thus, the solution to

$$\begin{pmatrix} 1 & 1 & 2 \\ 2 & 1 & 3 \\ 3 & 1 & 4 \end{pmatrix} \begin{pmatrix} x \\ y \\ z \end{pmatrix} = \begin{pmatrix} 0 \\ 0 \\ 0 \end{pmatrix}$$

is

$x + z = 0, y + z = 0$ so z is a free variable and $x = -z \, y = -z$

Thus any vector

$$\begin{pmatrix} -t \\ -t \\ t \end{pmatrix}$$

has

$$\begin{pmatrix} 1 & 1 & 2 \\ 2 & 1 & 3 \\ 3 & 1 & 4 \end{pmatrix}\begin{pmatrix} -t \\ -t \\ t \end{pmatrix} = \begin{pmatrix} 0 \\ 0 \\ 0 \end{pmatrix}.$$

so

$$-t\begin{pmatrix} 1 \\ 2 \\ 3 \end{pmatrix} - t\begin{pmatrix} 1 \\ 1 \\ 1 \end{pmatrix} + t\begin{pmatrix} 2 \\ 3 \\ 4 \end{pmatrix} = \begin{pmatrix} 0 \\ 0 \\ 0 \end{pmatrix}$$

and thus $\{\hat{v}_1,\hat{v}_2,\hat{v}_3\}$ is not a linearly independent set of vectors.

The last row of zeros in the row reduced form of A suggests that any vector of the form

$$\begin{pmatrix} 0 \\ 0 \\ a \end{pmatrix} a \neq 0$$

will not be in the span of $\{\hat{v}_1,\hat{v}_2,\hat{v}_3\}$.

In fact, when

$$\begin{pmatrix} 1 & 1 & 2 & 0 \\ 2 & 1 & 3 & 0 \\ 3 & 1 & 4 & a \end{pmatrix}$$

is row reduced, the result is

$$\begin{pmatrix} 1 & 0 & 1 & 0 \\ 0 & 1 & 1 & 0 \\ 0 & 0 & 0 & 1 \end{pmatrix}$$

which confirms the conjecture.

These ideas can be modified to show that any set of n vectors in $\mathbb{R}^n$ that fails to be a basis is neither linearly independent nor a spanning set. The salient point being that when an $n \times n$ matrix is row reduced, the result is either the identity matrix (in which case the columns form a basis) or has a row of zeroes.

Theorem 3.18:

Suppose that V is a vector space of dimension n.

(a.) If $\{\hat{v}_1,\ldots,\hat{v}_n\}$ is a spanning set of vectors for V then $\{\hat{v}_1,\ldots,\hat{v}_n\}$ is a basis for V.
(b.) If $\{\hat{v}_1,\ldots,\hat{v}_n\}$ is a linearly independent set of vectors for V then $\{\hat{v}_1,\ldots,\hat{v}_n\}$ is a basis for V.

Proof:

(a.) We give a proof by contradiction. Suppose that $\{\hat{v}_1,\ldots,\hat{v}_n\}$ is a spanning set, but is not a basis for V. Since $\{\hat{v}_1,\ldots,\hat{v}_n\}$ spans V but is not a basis, it must be that $\{\hat{v}_1,\ldots,\hat{v}_n\}$ is linearly dependent. Thus there is a vector in $\{\hat{v}_1,\ldots,\hat{v}_n\}$, say, $\hat{v}_1$, so that

$$\hat{v}_1 = a_2\hat{v}_2 + \cdots + a_n\hat{v}_n.$$

Thus $\{\hat{v}_2,\ldots,\hat{v}_n\}$ is a spanning set and if it is not a basis, can be shrunk to be a basis. Thus, there is a basis for V with fewer than n vectors, contradicting the assumption that V is a vector space of dimension n.

(b.) Again, we give a proof by contradiction. Suppose that $\{\hat{v}_1,\ldots,\hat{v}_n\}$ is not a basis for V. Since $\{\hat{v}_1,\ldots,\hat{v}_n\}$ is linearly independent, then it must not span V. Thus there is a vector $\hat{v} \in V$ with $\hat{v}$ not in the span of $\{\hat{v}_1,\ldots,\hat{v}_n\}$. We showed earlier that $\{\hat{v}_1,\ldots,\hat{v}_n,\hat{v}\}$ is a linearly independent set. Thus, we have a set of $n+1$ linearly independent vectors that can be expanded to a basis for V, contradicting the assumption that V is a vector space of dimension n.

A Synopsis of Sections 3.4 and 3.5

In these sections we have derived several facts about bases of vector spaces. Here is a list of these results for the case of finite dimensional vector spaces that consist of more than the zero vector.

Suppose that V is a non-trivial vector space that has a basis of k vectors. Then

1. Every basis of V has k vectors.
2. Any subset of V that contains more than k vectors cannot be linearly independent.
3. Any subset of V that contains fewer than k vectors cannot span V.
4. Any subset of V that is linearly independent but does not span V can be expanded to form a basis.
5. Any subset of V that spans V but is not linearly independent can be contracted to form a basis.
6. Any linearly independent set of k vectors is a basis for V.
7. Any spanning set of k vectors is a basis for V.
8. Any set of k vectors that is not a basis for V is neither linearly independent nor a spanning set.

We have excluded the trivial vector space (that is, the empty set) because of what might be considered a philosophical point that we prefer to avoid, but it is common to say that the empty set is a basis for the trivial vector space.

Exercises

In Exercises 1–5 (i) tell whether the given set of vectors forms a basis; (ii) if the set of vectors does not form a basis, select a maximal linearly independent set of vectors; (iii) if the maximal linearly independent set of vectors does not form a basis, extend the set to form a basis.

1. $\left\{ \begin{pmatrix} 1 \\ -2 \\ 4 \end{pmatrix}, \begin{pmatrix} 11 \\ 14 \\ 8 \end{pmatrix}, \begin{pmatrix} 2 \\ 8 \\ -4 \end{pmatrix}, \begin{pmatrix} 15 \\ 30 \\ 0 \end{pmatrix} \right\}$

2. $\left\{ \begin{pmatrix} 1 \\ -3 \\ -2 \end{pmatrix} \begin{pmatrix} -2 \\ 6 \\ 4 \end{pmatrix} \right\}$

3. $\left\{ \begin{pmatrix} 5 \\ 6 \end{pmatrix} \right\}$

4. $\left\{ \begin{pmatrix} 0 \\ 3 \\ 2 \\ 5 \end{pmatrix} \begin{pmatrix} 4 \\ -2 \\ 1 \\ 6 \end{pmatrix} \begin{pmatrix} 8 \\ 2 \\ 4 \\ 17 \end{pmatrix} \begin{pmatrix} 1 \\ 3 \\ 5 \\ 7 \end{pmatrix} \begin{pmatrix} 7 \\ -1 \\ -1 \\ 10 \end{pmatrix} \begin{pmatrix} 1 \\ 2 \\ 3 \\ 4 \end{pmatrix} \right\}$

5. $\left\{ \begin{pmatrix} 1 \\ 2 \\ 1 \end{pmatrix} \begin{pmatrix} -3 \\ 7 \\ 5 \end{pmatrix} \begin{pmatrix} 4 \\ 0 \\ 0 \end{pmatrix} \right\}$

6. Find a maximal linearly independent set in each of the following sets of vectors. If the maximal linearly independent set is not a basis, extend the maximal set to a basis.
   (a.) $(2,1,6),(3,21,9),(1,7,3),(5,22,15)$

   (b.) $(1,6,9),(4,6,2),(1,0,0),(6,2,9)$

   (c.) $(1,5),(3,15),(2,10)$

7. Construct a basis of $\mathbb{R}^4$ that includes the vector $(1,2,0,5)$.

8. (a.) Show that $\hat{v}_1$ and $\hat{v}_2$ can be written as a linear combination of $\{\hat{v}_1 + \hat{v}_2, \hat{v}_2 + \hat{v}_3, \hat{v}_3\}$.
   (b.) Show that if $\{\hat{v}_1, \hat{v}_2, \ldots, \hat{v}_n\}$ is a basis for the vector space V, then $\{\hat{v}_1 + \hat{v}_2, \hat{v}_2 + \hat{v}_3, \ldots, \hat{v}_{n-1} + \hat{v}_n, \hat{v}_n\}$ is also a basis for V.

9. Let

$$\hat{v}_1 = \begin{pmatrix} 1 \\ 0 \\ 2 \\ 1 \end{pmatrix}, \hat{v}_2 = \begin{pmatrix} 1 \\ 1 \\ 0 \\ 0 \end{pmatrix}, \hat{v}_3 = \begin{pmatrix} 3 \\ 2 \\ 2 \\ 1 \end{pmatrix}, \hat{v}_4 = \begin{pmatrix} 6 \\ -1 \\ 2 \\ 3 \end{pmatrix}.$$

   and let U be the subspace spanned by $\{\hat{v}_1, \hat{v}_2\}$ and let W be the subspace spanned by $\{\hat{v}_3, \hat{v}_4\}$.
   (a.) Find a basis for $U + W$.
   (b.) Find a basis for $U \cap W$.

10. The vectors

$$(1,6,3),(2,-3,7),(7,12,23)$$

   are not linearly independent.
   (a.) Find a vector not in the span of $\{(1,6,3),(2,-3,7),(7,12,23)\}$
   (b.) Find a non – zero vector that can be expressed as a linear c combination of

$$\{(1,6,3),(2,-3,7),(7,12,23)\}$$

in two different ways.

11. Determine h so that

$$\begin{pmatrix} 3 \\ 6 \\ 1 \end{pmatrix}, \begin{pmatrix} -2 \\ -5 \\ 0 \end{pmatrix}, \begin{pmatrix} 8 \\ 1 \\ h \end{pmatrix}$$

will not span $\mathbb{R}^3$.

12. Show that if S is a linearly independent set of vectors and S' is a proper subset of S, then span (S') is a proper subset of span (S).

13. In finding a basis for the row space of a matrix with the method described in the text, the basis vectors you obtain may not be the rows of the matrix. Describe how you could get a basis consisting of the rows of the matrix and apply your technique to the matrix

$$\begin{pmatrix} 2 & 0 & 3 & 1 & -2 & 5 \\ 4 & 1 & 6 & 3 & 3 & 9 \\ 8 & 1 & 12 & 5 & -1 & 0 \\ 0 & 7 & 2 & -4 & 4 & 0 \end{pmatrix}$$

Section 3.6 Change of Bases

A Vector Versus the Representation of a Vector.

Different bases of a vector space describe the same vector in different ways.

There are two things we want to emphasize. The first is the relationship between a vector and a basis for the vector space in which the vector lies. Borrowing from Terrence Tao, the relationship of a vector to a basis is analogous to the relationship of an idea to words. The same idea can be described in many different languages. Likewise, a vector can be described by different bases.

The second point is that we have defined a basis to be a collection of vectors that are linearly independent and span a vector space. As sets, the order of appearance of elements in a set is immaterial. All that matters is the elements; e.g., the set $\{a, b, c\}$ is the same as the set $\{c, a, b\}$. In what follows, we will describe the representation of a vector $\hat{v}$ with respect to a basis by the scalars that are the coefficients of the basis vectors in the linear combination that makes up the vector $\hat{v}$; the basis vectors themselves will not be listed. This type of representation assumes an order of the basis vectors. *The effect of all this is that when we list a set of basis vectors in this section, it will be assumed to be an ordered set.*

Thus, if it is understood that the basis of the vector space is the ordered set $\{\hat{v}_1,\ldots,\hat{v}_n\}$, we could describe

$$\hat{v} = a_1\hat{v}_1 + \cdots + a_n\hat{v}_n$$

by the ordered n-tuple of scalars $(a_1,\ldots,a_n)$.

Definition:

We call $(a_1,\ldots,a_n)$ the coordinates of $\hat{v}$ with respect to the basis $\{\hat{v}_1,\ldots,\hat{v}_n\}$.

Notation: if $\mathcal{B} = \{\hat{u}_1,\ldots,\hat{u}_n\}$ is a basis for the vector space V and $\hat{v}$ is a vector in V, then the ordered set of coordinates of $\hat{v}$ with respect to the basis $\mathcal{B}$ is denoted $[\hat{v}]_{\mathcal{B}}$.

If we are in a situation where there is no ambiguity about the basis being used, it is usually the case that the subscript $\mathcal{B}$ will be suppressed. Also, if we are using the usual (standard) basis, the subscript will be suppressed.

Changing the Representation of a Vector From One Basis to Another

We consider the following problem: given the coordinates of a vector in one basis, find the coordinates of the vector in another basis.

We first give a simple example, then give a description of the theory of the process in the more general case and then show how to solve the problem in a practical manner.

The first problem we consider is to suppose that we have a vector that is expressed as $\begin{pmatrix} a \\ b \end{pmatrix}$ in the standard basis and we want to find the coordinates in the basis $\left\{ \begin{pmatrix} 2 \\ 4 \end{pmatrix}, \begin{pmatrix} 3 \\ 7 \end{pmatrix} \right\}$. That is, we want to find c and d for which

$$\begin{pmatrix} a \\ b \end{pmatrix} = c\begin{pmatrix} 2 \\ 4 \end{pmatrix} + d\begin{pmatrix} 3 \\ 7 \end{pmatrix}.$$

To do this, we must satisfy the equations

$$a = 2c + 3d$$

$$b = 4c + 7d$$

which is equivalent to the matrix equation

$$\begin{pmatrix} a \\ b \end{pmatrix} = \begin{pmatrix} 2 & 3 \\ 4 & 7 \end{pmatrix}\begin{pmatrix} c \\ d \end{pmatrix}.$$

Since $\left\{ \begin{pmatrix} 2 \\ 4 \end{pmatrix}, \begin{pmatrix} 3 \\ 7 \end{pmatrix} \right\}$ is a basis, $\begin{pmatrix} 2 & 3 \\ 4 & 7 \end{pmatrix}$ is invertible, so

$$\begin{pmatrix} c \\ d \end{pmatrix} = \begin{pmatrix} 2 & 3 \\ 4 & 7 \end{pmatrix}^{-1}\begin{pmatrix} a \\ b \end{pmatrix}.$$

The matrix that corresponds to $\begin{pmatrix} 2 & 3 \\ 4 & 7 \end{pmatrix}$ is crucial to our computations. It is the matrix whose ith column is the ith vector of the new basis.

We convert these ideas to the case where the dimension of the vector space is n. Here the standard basis is

$$\left\{ \begin{pmatrix} 1 \\ 0 \\ \vdots \\ 0 \end{pmatrix}, \begin{pmatrix} 0 \\ 1 \\ \vdots \\ 0 \end{pmatrix}, \cdots, \begin{pmatrix} 0 \\ \vdots \\ 0 \\ 1 \end{pmatrix} \right\}$$

and we suppose the new basis is

$$\left\{ \begin{pmatrix} b_{11} \\ b_{12} \\ \vdots \\ b_{1n} \end{pmatrix}, \begin{pmatrix} b_{21} \\ b_{22} \\ \vdots \\ b_{2n} \end{pmatrix}, \cdots, \begin{pmatrix} b_{n1} \\ b_{n2} \\ \vdots \\ b_{nn} \end{pmatrix} \right\}.$$

Suppose $\begin{pmatrix} a_1 \\ a_2 \\ \vdots \\ a_n \end{pmatrix}$ is a given vector in the standard basis, and we want to find $c_1, c_2, \ldots, c_n$ for which

$$\begin{pmatrix} a_1 \\ a_2 \\ \vdots \\ a_n \end{pmatrix} = c_1 \begin{pmatrix} b_{11} \\ b_{12} \\ \vdots \\ b_{1n} \end{pmatrix} + c_2 \begin{pmatrix} b_{21} \\ b_{22} \\ \vdots \\ b_{2n} \end{pmatrix} + \cdots + c_n \begin{pmatrix} b_{n1} \\ b_{n2} \\ \vdots \\ b_{nn} \end{pmatrix}.$$

As before, this is equivalent to the system of equations

$$a_1 = b_{11}c_1 + b_{21}c_2 + \cdots + b_{n1}c_n$$

$$a_2 = b_{12}c_1 + b_{22}c_{21} + \cdots + b_{n2}c_n$$

$$\vdots$$

$$a_n = b_{1n}c_1 + b_{2n}c_2 + \cdots + b_{nn}c_n$$

which is equivalent to the matrix equation

$$\begin{pmatrix} a_1 \\ a_2 \\ \vdots \\ a_n \end{pmatrix} = \begin{pmatrix} b_{11} & b_{21} & & b_{n1} \\ b_{12} & b_{22} & \cdots & b_{n2} \\ \vdots & \vdots & & \vdots \\ b_{1n} & b_{2n} & & b_{nn} \end{pmatrix} \begin{pmatrix} c_1 \\ c_2 \\ \vdots \\ c_n \end{pmatrix}.$$

So

$$\begin{pmatrix} c_1 \\ c_2 \\ \vdots \\ c_n \end{pmatrix} = \begin{pmatrix} b_{11} & b_{21} & & b_{n1} \\ b_{12} & b_{22} & \cdots & b_{n2} \\ \vdots & \vdots & & \vdots \\ b_{1n} & b_{2n} & & b_{nn} \end{pmatrix}^{-1} \begin{pmatrix} a_1 \\ a_2 \\ \vdots \\ a_n \end{pmatrix}.$$

Thus, we have the following result.

Theorem 3.19:

If V is a finite dimensional vector space and $[\hat{x}]$ is a vector whose representation in the standard

basis is $\begin{pmatrix} a_1 \\ a_2 \\ \vdots \\ a_n \end{pmatrix}$ and

$$\mathcal{B} = \left\{ \begin{pmatrix} b_{11} \\ b_{12} \\ \vdots \\ b_{1n} \end{pmatrix}, \begin{pmatrix} b_{21} \\ b_{22} \\ \vdots \\ b_{2n} \end{pmatrix}, \cdots, \begin{pmatrix} b_{n1} \\ b_{n2} \\ \vdots \\ b_{nn} \end{pmatrix} \right\}$$

is a basis for V, then the representation for $\hat{x}$ in the $\mathcal{B}$ basis is

$$[\hat{x}]_\mathcal{B} = P_\mathcal{B}^{-1} \begin{pmatrix} a_1 \\ a_2 \\ \vdots \\ a_n \end{pmatrix} = P_\mathcal{B}^{-1} [\hat{x}] \tag{1}$$

where $P_\mathcal{B}$ is the matrix whose ith column is $\begin{pmatrix} b_{i1} \\ b_{i2} \\ \vdots \\ b_{in} \end{pmatrix}$.

$P_\mathcal{B}$ is called the transition matrix for the basis $\mathcal{B}$.

Equation (1) could be written $[\hat{x}] = P_\mathcal{B}[\hat{x}]_\mathcal{B}$.

Using this version of the formula, we get a formula for the change of basis between non-standard bases. Namely, if $\mathcal{B}$ and $\mathcal{C}$ are two bases for a vector space, then

$$P_\mathcal{B}[\hat{x}]_\mathcal{B} = [\hat{x}] = P_\mathcal{C}[\hat{x}]_\mathcal{C}.$$

Example

Let $\mathcal{C}$ be the basis

$$\mathcal{C} = \left\{ \begin{pmatrix} 1 \\ -2 \\ 2 \end{pmatrix}, \begin{pmatrix} 3 \\ 4 \\ 0 \end{pmatrix}, \begin{pmatrix} 1 \\ 1 \\ -1 \end{pmatrix} \right\}.$$

(a.) Find the representation of

$$[\hat{v}]_{\mathcal{C}} = \begin{pmatrix} 6 \\ -9 \\ 8 \end{pmatrix}$$

in the standard basis.

We have

$$P_{\mathcal{C}} = \begin{pmatrix} 1 & 3 & 1 \\ -2 & 4 & 1 \\ 2 & 0 & -1 \end{pmatrix}$$

so

$$[\hat{v}] = P_{\mathcal{C}}[\hat{v}]_{\mathcal{C}} = \begin{pmatrix} 1 & 3 & 1 \\ -2 & 4 & 1 \\ 2 & 0 & -1 \end{pmatrix}\begin{pmatrix} 6 \\ -9 \\ 8 \end{pmatrix} = \begin{pmatrix} -13 \\ -40 \\ 4 \end{pmatrix}.$$

(b.) Find the representation of

$$[\hat{v}] = \begin{pmatrix} 3 \\ 1 \\ 2 \end{pmatrix}$$

in the $\mathcal{C}$ basis.

We have

$$[\hat{v}]_{\mathcal{C}} = P_{\mathcal{C}}^{-1}[\hat{v}] = \begin{pmatrix} 1 & 3 & 1 \\ -2 & 4 & 1 \\ 2 & 0 & -1 \end{pmatrix}^{-1}\begin{pmatrix} 3 \\ 1 \\ 2 \end{pmatrix} = \begin{pmatrix} \frac{11}{12} \\ \frac{3}{4} \\ -\frac{1}{6} \end{pmatrix}.$$

Converting Between Two Non-Standard Bases

Suppose that

$$\mathcal{C} = \left\{ \begin{pmatrix} 1 \\ 3 \end{pmatrix}, \begin{pmatrix} -2 \\ 5 \end{pmatrix} \right\} \text{ and } \mathcal{D} = \left\{ \begin{pmatrix} 0 \\ 2 \end{pmatrix}, \begin{pmatrix} 1 \\ 6 \end{pmatrix} \right\}$$

and we want to convert the representation of a vector $\hat{v}$ in one basis to a representation in the other basis.

Let $P_{\mathcal{C}}$ be the transition matrix for $\mathcal{C}$ and $P_{\mathcal{D}}$ be the transition matrix for $\mathcal{D}$; that is,

$$P_{\mathcal{C}} = \begin{pmatrix} 1 & -2 \\ 3 & 5 \end{pmatrix} \text{ and } P_{\mathcal{D}} = \begin{pmatrix} 0 & 1 \\ 2 & 6 \end{pmatrix}.$$

Then

$$P_{\mathcal{C}}[\hat{v}]_{\mathcal{C}} = [\hat{v}] = P_{\mathcal{D}}[\hat{v}]_{\mathcal{D}}$$

and the equation

$$P_C\left[\hat{v}\right]_C = P_D\left[\hat{v}\right]_D$$

provides the transition mechanism between matrices. So if

$$\left[\hat{v}\right]_C = \begin{pmatrix} 9 \\ 4 \end{pmatrix}$$

then

$$\left[\hat{v}\right]_D = P_D^{-1}P_C\left[\hat{v}\right]_C = \begin{pmatrix} 0 & 1 \\ 2 & 6 \end{pmatrix}^{-1}\begin{pmatrix} 1 & -2 \\ 3 & 5 \end{pmatrix}\begin{pmatrix} 9 \\ 4 \end{pmatrix} = \begin{pmatrix} \frac{41}{2} \\ 2 \\ 1 \end{pmatrix}.$$

Example

The standard basis of $P_2(x)$ is $\left\{1,x,x^2\right\}$. Find the representation of

$$3-2x+5x^2$$

in the basis

$$B=\left\{1+x,1-x,x^2\right\}.$$

We have

$$P_B = \begin{pmatrix} 1 & 1 & 0 \\ 1 & -1 & 0 \\ 0 & 0 & 1 \end{pmatrix}, [v]=\begin{pmatrix} 3 \\ -2 \\ 5 \end{pmatrix}$$

so

$$\left[\hat{v}\right]_B = P_B^{-1}\left[\hat{v}\right]=\begin{pmatrix} 1 & 1 & 0 \\ 1 & -1 & 0 \\ 0 & 0 & 1 \end{pmatrix}^{-1}\begin{pmatrix} 3 \\ -2 \\ 5 \end{pmatrix}=\begin{pmatrix} 1/2 \\ 5/2 \\ 5 \end{pmatrix}.$$

One can check that

$$\frac{1}{2}(1+x)+\frac{5}{2}(1-x)+5x^2 =3-2x+5x^2.$$

Example

The standard basis for $M_{2\times 2}(\mathbb{R})$ is

$$\left\{\begin{pmatrix} 1 & 0 \\ 0 & 0 \end{pmatrix},\begin{pmatrix} 0 & 1 \\ 0 & 0 \end{pmatrix},\begin{pmatrix} 0 & 0 \\ 1 & 0 \end{pmatrix},\begin{pmatrix} 0 & 0 \\ 0 & 1 \end{pmatrix}\right\}.$$

Find the transition matrix for the basis

$$\mathcal{B} = \left\{ \begin{pmatrix} 1 & 0 \\ 0 & 1 \end{pmatrix}, \begin{pmatrix} 2 & 1 \\ 0 & 0 \end{pmatrix}, \begin{pmatrix} 0 & 0 \\ 1 & 1 \end{pmatrix}, \begin{pmatrix} 0 & 0 \\ 0 & 1 \end{pmatrix} \right\}.$$

We have

$$\begin{pmatrix} 1 & 0 \\ 0 & 1 \end{pmatrix} = 1\begin{pmatrix} 1 & 0 \\ 0 & 0 \end{pmatrix} + 0\begin{pmatrix} 0 & 1 \\ 0 & 0 \end{pmatrix} + 0\begin{pmatrix} 0 & 0 \\ 1 & 0 \end{pmatrix} + 1\begin{pmatrix} 0 & 0 \\ 0 & 1 \end{pmatrix}$$

$$\begin{pmatrix} 2 & 1 \\ 0 & 0 \end{pmatrix} = 2\begin{pmatrix} 1 & 0 \\ 0 & 0 \end{pmatrix} + 1\begin{pmatrix} 0 & 1 \\ 0 & 0 \end{pmatrix} + 0\begin{pmatrix} 0 & 0 \\ 1 & 0 \end{pmatrix} + 0\begin{pmatrix} 0 & 0 \\ 0 & 1 \end{pmatrix}$$

$$\begin{pmatrix} 0 & 0 \\ 1 & 1 \end{pmatrix} = 0\begin{pmatrix} 1 & 0 \\ 0 & 0 \end{pmatrix} + 0\begin{pmatrix} 0 & 1 \\ 0 & 0 \end{pmatrix} + 1\begin{pmatrix} 0 & 0 \\ 1 & 0 \end{pmatrix} + 1\begin{pmatrix} 0 & 0 \\ 0 & 1 \end{pmatrix}$$

$$\begin{pmatrix} 0 & 0 \\ 0 & 1 \end{pmatrix} = 0\begin{pmatrix} 1 & 0 \\ 0 & 0 \end{pmatrix} + 0\begin{pmatrix} 0 & 1 \\ 0 & 0 \end{pmatrix} + 0\begin{pmatrix} 0 & 0 \\ 1 & 0 \end{pmatrix} + 1\begin{pmatrix} 0 & 0 \\ 0 & 1 \end{pmatrix}$$

so

$$\mathcal{P}_\mathcal{B} = \begin{pmatrix} 1 & 2 & 0 & 0 \\ 0 & 1 & 0 & 0 \\ 0 & 0 & 1 & 0 \\ 1 & 0 & 1 & 1 \end{pmatrix}.$$

If

$$A = \begin{pmatrix} a & b \\ c & d \end{pmatrix} = a\begin{pmatrix} 1 & 0 \\ 0 & 0 \end{pmatrix} + b\begin{pmatrix} 0 & 1 \\ 0 & 0 \end{pmatrix} + c\begin{pmatrix} 0 & 0 \\ 1 & 0 \end{pmatrix} + d\begin{pmatrix} 0 & 0 \\ 0 & 1 \end{pmatrix},$$

then

$$[A]_\mathcal{B} = \mathcal{P}_\mathcal{B}^{-1}\begin{pmatrix} a \\ b \\ c \\ d \end{pmatrix} = \begin{pmatrix} 1 & 2 & 0 & 0 \\ 0 & 1 & 0 & 0 \\ 0 & 0 & 1 & 0 \\ 1 & 0 & 1 & 1 \end{pmatrix}^{-1}\begin{pmatrix} a \\ b \\ c \\ d \end{pmatrix} = \begin{pmatrix} a - 2b \\ b \\ c \\ -a + 2b - c + d \end{pmatrix}.$$

Exercises

In Exercises 1–10 find the coordinates of the vector $\hat{u}$ relative to the given basis.

1. $V = \mathbb{R}^2$, $\mathcal{B} = \left\{ \begin{pmatrix} 1 \\ 4 \end{pmatrix}, \begin{pmatrix} 0 \\ 3 \end{pmatrix} \right\}$, $u = \begin{pmatrix} 6 \\ 2 \end{pmatrix}$

2. $V = \mathbb{R}^2$, $\mathcal{B} = \left\{ \begin{pmatrix} -2 \\ 1 \end{pmatrix}, \begin{pmatrix} 1 \\ 1 \end{pmatrix} \right\}$, $u = \begin{pmatrix} 0 \\ 5 \end{pmatrix}$

3. $V = \mathbb{R}^3$, $\mathcal{B} = \left\{ \begin{pmatrix} 0 \\ 3 \\ 1 \end{pmatrix}, \begin{pmatrix} -2 \\ 1 \\ 4 \end{pmatrix}, \begin{pmatrix} 8 \\ 1 \\ 3 \end{pmatrix} \right\}$, $\hat{u} = \begin{pmatrix} 2 \\ 2 \\ 5 \end{pmatrix}$

4. $V = \mathbb{R}^3$, $\mathcal{B} = \left\{ \begin{pmatrix} 4 \\ 7 \\ -2 \end{pmatrix}, \begin{pmatrix} 0 \\ 0 \\ 1 \end{pmatrix}, \begin{pmatrix} -4 \\ 1 \\ 6 \end{pmatrix} \right\}$, $u = \begin{pmatrix} -1 \\ 0 \\ 3 \end{pmatrix}$

5. $V = \mathbb{R}^4$, $\mathcal{B} = \left\{ \begin{pmatrix} 1 \\ 0 \\ 0 \\ 2 \end{pmatrix}, \begin{pmatrix} 2 \\ 1 \\ 1 \\ 0 \end{pmatrix}, \begin{pmatrix} 3 \\ -1 \\ 4 \\ 2 \end{pmatrix}, \begin{pmatrix} 0 \\ 0 \\ 1 \\ -2 \end{pmatrix} \right\}$, $u = \begin{pmatrix} 4 \\ -2 \\ 0 \\ 1 \end{pmatrix}$

6. $V = \mathbb{R}^4$, $\mathcal{B} = \left\{ \begin{pmatrix} -1 \\ 4 \\ 1 \\ 1 \end{pmatrix}, \begin{pmatrix} 1 \\ -1 \\ 2 \\ 3 \end{pmatrix}, \begin{pmatrix} 0 \\ 1 \\ 0 \\ 1 \end{pmatrix}, \begin{pmatrix} 1 \\ 0 \\ 1 \\ 1 \end{pmatrix} \right\}$, $u = \begin{pmatrix} 5 \\ 4 \\ 1 \\ 1 \end{pmatrix}$

7. $V = P_2(x)$, $\mathcal{B} = \{1 + 2x, x - x^2, 3 + x^2\}$, $u = 3 - 4x + x^2$

8. $V = P_2(x)$, $\mathcal{B} = \{3 - x, 1 + 5x - x^2, 2x + x^2\}$, $u = 1 + 4x - 3x^2$

9. $V = \mathcal{M}_{2 \times 2}(\mathbb{R})$, $\mathcal{B} = \left\{ \begin{pmatrix} 1 & 1 \\ 1 & 1 \end{pmatrix}, \begin{pmatrix} 1 & 1 \\ 1 & 0 \end{pmatrix}, \begin{pmatrix} 1 & 0 \\ 0 & 1 \end{pmatrix}, \begin{pmatrix} 0 & 2 \\ 1 & 1 \end{pmatrix} \right\}$, $u = \begin{pmatrix} 3 & -1 \\ 0 & 6 \end{pmatrix}$

10. $V = \mathcal{M}_{2 \times 2}(\mathbb{R})$, $\mathcal{B} = \left\{ \begin{pmatrix} 2 & 0 \\ 1 & 3 \end{pmatrix}, \begin{pmatrix} 0 & 4 \\ 1 & -1 \end{pmatrix}, \begin{pmatrix} 3 & 1 \\ 0 & 1 \end{pmatrix}, \begin{pmatrix} 1 & 0 \\ -2 & 1 \end{pmatrix} \right\}$, $u = \begin{pmatrix} 2 & 3 \\ 7 & 3 \end{pmatrix}$

11. Find a basis for the symmetric 3×3 matrices.
12. (a.) Show that the set of $n \times n$ matrices whose trace is 0 is a vector space.
    (b.) Find a basis for 2×2 matrices whose trace is 0.
    (c.) Find a basis for 3×3 matrices whose trace is 0.
    In Exercises 13–16 find
    (a.) the transition matrix from $\mathcal{B}_1$ to $\mathcal{B}_2$,
    (b.) the transition matrix from $\mathcal{B}_2$ to $\mathcal{B}_1$.
    (c.) For the given $[\hat{v}]_{\mathcal{B}_1}$ find $[\hat{v}]_{\mathcal{B}_2}$,
    (d.) For the given $[\hat{v}]_{\mathcal{B}_2}$ find $[\hat{v}]_{\mathcal{B}_1}$

13. $\mathcal{B}_1 = \left\{ \begin{pmatrix} 0 \\ 3 \end{pmatrix}, \begin{pmatrix} 1 \\ 2 \end{pmatrix} \right\}$ $\mathcal{B}_2 = \left\{ \begin{pmatrix} 4 \\ -1 \end{pmatrix}, \begin{pmatrix} 1 \\ 1 \end{pmatrix} \right\}$

$$[\hat{v}]_{\mathcal{B}_1} = \begin{pmatrix} 2 \\ 5 \end{pmatrix}, \quad [\hat{v}]_{\mathcal{B}_2} = \begin{pmatrix} 6 \\ -2 \end{pmatrix}$$

14. $\mathcal{B}_1 = \left\{ \begin{pmatrix} 1 \\ -1 \end{pmatrix}, \begin{pmatrix} 2 \\ 3 \end{pmatrix} \right\} \quad \mathcal{B}_2 = \left\{ \begin{pmatrix} 1 \\ 5 \end{pmatrix}, \begin{pmatrix} 2 \\ -3 \end{pmatrix} \right\}$

$$[\hat{v}]_{\mathcal{B}_1} = \begin{pmatrix} 4 \\ 3 \end{pmatrix}, \quad [\hat{v}]_{\mathcal{B}_2} = \begin{pmatrix} 3 \\ 8 \end{pmatrix}$$

15. $\mathcal{B}_1 = \left\{ \begin{pmatrix} 1 \\ 5 \\ -2 \end{pmatrix}, \begin{pmatrix} -3 \\ 0 \\ 2 \end{pmatrix}, \begin{pmatrix} 4 \\ 1 \\ 4 \end{pmatrix} \right\} \quad \mathcal{B}_2 = \left\{ \begin{pmatrix} 2 \\ 2 \\ -2 \end{pmatrix}, \begin{pmatrix} 1 \\ 6 \\ 3 \end{pmatrix}, \begin{pmatrix} 0 \\ 1 \\ 0 \end{pmatrix} \right\}$

$$[\hat{v}]_{\mathcal{B}_1} = \begin{pmatrix} 2 \\ -3 \\ 1 \end{pmatrix}, \quad [\hat{v}]_{\mathcal{B}_2} = \begin{pmatrix} 2 \\ 0 \\ 5 \end{pmatrix},$$

16. $\mathcal{B}_1 = \left\{ \begin{pmatrix} 0 \\ 3 \\ 1 \end{pmatrix}, \begin{pmatrix} 4 \\ 5 \\ 2 \end{pmatrix}, \begin{pmatrix} 1 \\ 1 \\ 1 \end{pmatrix} \right\} \quad \mathcal{B}_2 = \left\{ \begin{pmatrix} 3 \\ 1 \\ 0 \end{pmatrix}, \begin{pmatrix} 2 \\ -1 \\ -1 \end{pmatrix}, \begin{pmatrix} 1 \\ 1 \\ 2 \end{pmatrix} \right\}$

$$[\hat{v}]_{\mathcal{B}_1} = \begin{pmatrix} 6 \\ 1 \\ 1 \end{pmatrix}, \quad [\hat{v}]_{\mathcal{B}_2} = \begin{pmatrix} 0 \\ 4 \\ -2 \end{pmatrix},$$

Section 3.7 The Null Space, Row Space, and Column Space of a Matrix

Let A be an $m \times n$ matrix. Associated with each matrix are three vector spaces.

1. The null space of A defined by

$$N(A) = \left\{ \hat{x} \in \mathcal{F}^n \,\middle|\, A\hat{x} = \hat{0} \right\}$$

2. The row space of A, which is the vector space of linear combinations of the rows of A. If A is an $m \times n$ matrix, then the row space of A is a subspace of $\mathcal{F}^n$.
3. The column space of A, which is the vector space of linear combinations of the columns of A. If A is an $m \times n$ matrix, then the column space of A is a subspace of $\mathcal{F}^m$.

In this section we will show that for a given $m \times n$ matrix A

1. The number of vectors in a basis for the column space of A is equal to the number of vectors in a basis for the row space of A. This number is called the rank of A.
2. The number of vectors in a basis for the null space of A is called the nullity of A. We will show that nullity of A + rank of A = number of columns of A.

It follows that the nullity of A is the number of free variables in the row reduced form of A and the rank of A is the number of leading variables in the row reduced form of A.

Among other things, in this section we will validate the algorithms we used in Section 3.5 to reduce/expand a given set of vectors to a basis.

Theorem 3.20:

Let A be an $m \times n$ matrix. A vector can be written as a linear combination of the columns of A if and only if that vector can be written as $A\hat{c}$ for some

$$\hat{c} \in \mathcal{F}^n$$

Proof:
Let

$$\hat{v} = A\hat{c} = \begin{pmatrix} a_{11} & a_{12} & \cdots & a_{1n} \\ a_{21} & a_{22} & \cdots & a_{2n} \\ \vdots & \vdots & & \vdots \\ a_{m1} & a_{m2} & \cdots & a_{mn} \end{pmatrix} \begin{pmatrix} c_1 \\ c_2 \\ \vdots \\ c_n \end{pmatrix}$$

$$= \begin{pmatrix} a_{11}c_1 + a_{12}c_2 + \cdots + a_{1n}c_n \\ a_{21}c_1 + a_{22}c_2 + \cdots + a_{2n}c_n \\ \vdots \\ a_{m1}c_1 + a_{m2}c_2 + \cdots + a_{mn}c_n \end{pmatrix} = \begin{pmatrix} a_{11}c_1 \\ a_{21}c_1 \\ \vdots \\ a_{m1}c_1 \end{pmatrix} + \cdots + \begin{pmatrix} a_{1n}c_n \\ a_{2n}c_n \\ \vdots \\ a_{mn}c_n \end{pmatrix}$$

$$= c_1 \begin{pmatrix} a_{11} \\ a_{21} \\ \vdots \\ a_{m1} \end{pmatrix} + \cdots + c_n \begin{pmatrix} a_{1n} \\ a_{2n} \\ \vdots \\ a_{mn} \end{pmatrix}.$$

Corollary:

The equation $A\hat{x} = \hat{b}$ has a solution if and only if $\hat{b}$ is in the column space of A.

Recall that for any function

$$f : X \to Y$$

the range of f is $\{ y \in Y \mid y = f(x) \text{ for some } x \in X \}$.

For $A : \mathbb{R}^n \to \mathbb{R}^m$, Theorem 3.20 says that the range of A is the column space of A.

Example

Let

$$A = \begin{pmatrix} 1 & -3 & 0 \\ 2 & 2 & -1 \\ 6 & 0 & 4 \\ 5 & 5 & 1 \end{pmatrix}.$$

Determine if there is a vector

$$\hat{x} = \begin{pmatrix} x_1 \\ x_2 \\ x_3 \end{pmatrix}$$

for which

$$A\hat{x} = \hat{y} = \begin{pmatrix} y_1 \\ y_2 \\ y_3 \\ y_4 \end{pmatrix}$$

if

$$(i)\ \hat{y} = \begin{pmatrix} -10 \\ 0 \\ 10 \\ 14 \end{pmatrix} \quad (ii) \begin{pmatrix} 2 \\ 1 \\ 7 \\ -6 \end{pmatrix}.$$

A useful way to think of this problem is that we are asking if $\hat{y}$ is in the column space of A. Thus, we want to know if there is a vector

$$\hat{x} = \begin{pmatrix} x_1 \\ x_2 \\ x_3 \end{pmatrix} \text{ for which}$$

$$A\hat{x} = \begin{pmatrix} 1 & -3 & 0 \\ 2 & 2 & -1 \\ 6 & 0 & 4 \\ 5 & 5 & 1 \end{pmatrix} \begin{pmatrix} x_1 \\ x_2 \\ x_3 \end{pmatrix} = x_1 \begin{pmatrix} 1 \\ 2 \\ 6 \\ 5 \end{pmatrix} + x_2 \begin{pmatrix} -3 \\ 2 \\ 0 \\ 5 \end{pmatrix} + x_3 \begin{pmatrix} 0 \\ -1 \\ 4 \\ 1 \end{pmatrix} = \hat{y} = \begin{pmatrix} y_1 \\ y_2 \\ y_3 \\ y_4 \end{pmatrix}.$$

This is most conveniently solved by converting the vector equation to the system of linear equations

$$x_1 - 3x_2 = y_1$$

$$2x_1 + 2x_2 - x_3 = y_2$$

$$6x_1 + 4x_3 = y_3$$

$$5x_1 + 5x_2 + x_3 = y_4$$

(i) For

$$\hat{y} = \begin{pmatrix} -10 \\ 0 \\ 10 \\ 14 \end{pmatrix}$$

the system has the augmented matrix

$$\begin{pmatrix} 1 & -3 & 0 & -10 \\ 2 & 2 & -1 & 0 \\ 6 & 0 & 4 & 10 \\ 5 & 5 & 1 & 14 \end{pmatrix}.$$

When row reduced the result is

$$\begin{pmatrix} 1 & 0 & 0 & -1 \\ 0 & 1 & 0 & 3 \\ 0 & 0 & 1 & 4 \\ 0 & 0 & 0 & 0 \end{pmatrix}$$

and we have $x_1 = -1$, $x_2 = 3$, $x_3 = 4$ so if

$$\hat{x} = \begin{pmatrix} -1 \\ 3 \\ 4 \end{pmatrix}$$

we have

$$A\hat{x} = \begin{pmatrix} 1 & -3 & 0 \\ 2 & 2 & -1 \\ 6 & 0 & 4 \\ 5 & 5 & 1 \end{pmatrix} \begin{pmatrix} -1 \\ 3 \\ 4 \end{pmatrix} = \begin{pmatrix} -10 \\ 0 \\ 10 \\ 14 \end{pmatrix}.$$

(ii) For

$$\hat{y} = \begin{pmatrix} 2 \\ 1 \\ 7 \\ -6 \end{pmatrix}$$

the system has the augmented matrix

$$\begin{pmatrix} 1 & -3 & 0 & 2 \\ 2 & 2 & -1 & 1 \\ 6 & 0 & 4 & 7 \\ 5 & 5 & 1 & -6 \end{pmatrix}.$$

When row reduced the result is

$$\begin{pmatrix} 1 & 0 & 0 & 0 \\ 0 & 1 & 0 & 0 \\ 0 & 0 & 1 & 0 \\ 0 & 0 & 0 & 1 \end{pmatrix}$$

and there is no solution.

The major goal of this section is, given a matrix A, find bases for the null space of A, the row space of A and the column space of A. We will first demonstrate how this is done when the matrix is in row reduced form and then see what changes occur in going from a matrix that is not in row reduced form to a matrix that is in row reduced form. We will see that the row space and null space are unchanged but there is a change in the column space.

Example

Consider the matrix A that is in row reduced form:

$$A = \begin{pmatrix} 1 & 4 & 0 & 0 & 0 \\ 0 & 0 & 1 & 0 & 5 \\ 0 & 0 & 0 & 1 & 0 \\ 0 & 0 & 0 & 0 & 0 \end{pmatrix}.$$

The non-zero rows all have leading 1s and these rows are linearly independent. This is because the leading 1s occur in columns where there is only one non-zero entry. In this case, the leading 1s occur at the first, third, and fourth columns.

To expand what we have said, in this example, the first three rows have non-zero elements. Let

$$\hat{v}_1 = (1,4,0,0,0), \hat{v}_2 = (0,0,1,0,5), \hat{v}_3 = (0,0,0,1,0).$$

So if

$$a\hat{v}_1 + b\hat{v}_2 + c\hat{v}_3 = (a,*,b,c,*) = (0,0,0,0,0)$$

then $a = b = c = 0$, so $\{\hat{v}_1, \hat{v}_2, \hat{v}_3\}$ is a linearly independent set.

Likewise, the columns where leading 1s occur form a basis for the column space. In the example, these columns are

$$\begin{pmatrix} 1 \\ 0 \\ 0 \\ 0 \end{pmatrix}, \begin{pmatrix} 0 \\ 1 \\ 0 \\ 0 \end{pmatrix}, \begin{pmatrix} 0 \\ 0 \\ 1 \\ 0 \end{pmatrix}.$$

In the matrix A the second column is a multiple of the first column and the fifth column is a multiple of the third column. For all matrices that are in row reduced form, a column that does not have a leading 1 is a multiple of a column that has a leading 1.

The three vectors above form a basis for the subspace

$$\left\{ \begin{pmatrix} x \\ y \\ z \\ 0 \end{pmatrix} \middle| x, y, z \in \mathbb{R} \right\}.$$

Thus, in the case of a row reduced matrix, the number of leading 1s is equal to the dimension of the row space and is also equal to the dimension of the column space.

For the null space of this example, the free variables are x_2 and x_5, and we have

$$x_1 + 4x_2 = 0 \text{ so } x_1 = -4x_2$$

$$x_3 + 5x_5 = 0, \text{ so } x_3 = -4x_5$$

$$x_4 = 0.$$

Thus, a vector in the null space of A is of the form

$$\hat{x} = \begin{pmatrix} x_1 \\ x_2 \\ x_3 \\ x_4 \\ x_5 \end{pmatrix} = \begin{pmatrix} -4x_2 \\ x_2 \\ -4x_5 \\ 0 \\ x_5 \end{pmatrix} = \begin{pmatrix} -4x_2 \\ x_2 \\ 0 \\ 0 \\ 0 \end{pmatrix} + \begin{pmatrix} 0 \\ 0 \\ -4x_5 \\ 0 \\ x_5 \end{pmatrix} = x_2 \begin{pmatrix} -4 \\ 1 \\ 0 \\ 0 \\ 0 \end{pmatrix} + x_5 \begin{pmatrix} 0 \\ 0 \\ -4 \\ 0 \\ 1 \end{pmatrix}$$

and a basis for the null space is

$$\left\{ \begin{pmatrix} -4 \\ 1 \\ 0 \\ 0 \\ 0 \end{pmatrix}, \begin{pmatrix} 0 \\ 0 \\ -4 \\ 0 \\ 1 \end{pmatrix} \right\}.$$

Note that each free variable gives rise to a basis vector of the null space.

Thus, if the matrix is row reduced, we have

number of free variables = dimension of the null space.

We will later show this is true even if the matrix is not in row reduced form.

Since in row reduced form, each variable is a free variable or a leading variable, it follows that the number of leading variables plus the number of free variables is equal to the number of columns of the matrix.

We now explore the changes that occur in the row, column, and null spaces when elementary row operations are applied to a matrix. We found earlier that elementary row operations do not change the null space of a matrix.

Theorem 3.21:

Applying an elementary row operation to a matrix does not change the row space of the matrix. That is, if E is a matrix for which multiplying the matrix A on the left by E results in having the matrix A changed by an elementary row operation, then the matrix EA has the same row space as A.

Note that if E is an elementary matrix, then it satisfies the hypothesis of the theorem.

Proof:

We consider each of the elementary row operations.

1. Interchanging two rows of a matrix only changes the order in the vectors whose linear combinations make up the row space.

2. Multiplying a row of a matrix by a non-zero constant does not change the possible linear combinations of the rows.
3. For the effect of replacing a row by itself plus a non-zero multiple of another row, suppose two rows are $\hat{r}_1$ and $\hat{r}_2$ and $\hat{r}_1$ is replaced by $\hat{r}_1 + a\hat{r}_2$. We show that any linear combination of $\hat{r}_1$ and $\hat{r}_2$ can be written as a linear combination of $\hat{r}_1 + a\hat{r}_2$ and $\hat{r}_2$.

We suppose that

$$c\hat{r}_1 + d\hat{r}_2 = e\left(\hat{r}_1 + a\hat{r}_2\right) + f\hat{r}_2.$$

We know a, c and d, and want to find e and f.

Observe that if

$$e = c \text{ and } f = d - ac$$

then we have

$$e\left(\hat{r}_1 + a\hat{r}_2\right) + f\hat{r}_2 = c\left(\hat{r}_1 + a\hat{r}_2\right) + \left(d - ac\right)\hat{r}_2 = c\hat{r}_1 + \left(ac + d - ac\right)\hat{r}_2 = c\hat{r}_1 + d\hat{r}_2$$

Also, any vector in the span of $\left\{\left(\hat{r}_1 + a\hat{r}_2\right), \hat{r}_2\right\}$ is in the span of $\left\{\hat{r}_1, \hat{r}_2\right\}$.

Corollary:

If A is a matrix with row reduced form B, then A and B have the same row space.

This result is useful because a basis of the row space of a row reduced matrix consists of the non-zero rows.

Example

Let

$$A = \begin{pmatrix} 5 & 10 & 2 & 8 & 0 & 12 \\ 1 & 2 & 0 & 0 & 0 & 2 \\ 2 & 4 & 1 & 4 & 0 & 5 \\ 1 & 2 & 0 & 0 & 1 & 5 \\ 2 & 4 & 1 & 4 & 2 & 2 \\ 3 & 6 & 1 & 4 & 0 & 7 \end{pmatrix}$$

When A is row reduced, the result is

$$B = \begin{pmatrix} 1 & 2 & 0 & 0 & 0 & 0 \\ 0 & 0 & 1 & 4 & 0 & 0 \\ 0 & 0 & 0 & 0 & 1 & 0 \\ 0 & 0 & 0 & 0 & 0 & 1 \\ 0 & 0 & 0 & 0 & 0 & 0 \\ 0 & 0 & 0 & 0 & 0 & 0 \end{pmatrix}$$

so a basis for the row space of A is

$$\left\{(1\ 2\ 0\ 0\ 0\ 0), (0\ 0\ 1\ 4\ 0\ 0), (0\ 0\ 0\ 0\ 1\ 0), (0\ 0\ 0\ 0\ 0\ 1)\right\}.$$

While elementary row operations do not change the row space or null space of a matrix, they can change the column space of a matrix as the next example shows.

Example

Consider a 2×2 matrix where the second row is a non-zero multiple of the first, say

$$A=\begin{pmatrix}1 & 2\\ 4 & 8\end{pmatrix}.$$

The rows and columns are each multiples of the other. The column space is spanned by the vector

$$\begin{pmatrix}1\\ 4\end{pmatrix}$$

and the row space is spanned by the vector $(1,2)$.

When the matrix A is row reduced, the result is

$$\begin{pmatrix}1 & 2\\ 0 & 0\end{pmatrix}.$$

The column space of the row reduced matrix is spanned by the vector

$$\begin{pmatrix}1\\ 0\end{pmatrix}$$

which is different from the space spanned by

$$\begin{pmatrix}1\\ 4\end{pmatrix}$$

So while the elementary row operations do not change the space spanned by the rows of a matrix, the span of the column space can change. In this example, while the column space changed with the elementary row operations, the dimension of the column space did not. The next theorem shows that this is true in general.

Theorem 3.22:

If A and B are row equivalent matrices, then the columns of A are linearly independent if and only if the corresponding columns of B are linearly independent. In particular, a set of columns of A forms a basis for the column space of A if and only if the corresponding columns of B form a basis for the column space of B.

Proof:

We use the fact that if A and B are row equivalent matrices, then the null space of A is equal to the null space of B. We also use the fact that the columns of A are linearly dependent if and only if there is a non-zero vector $\hat{x}$ for which $A\hat{x}=\hat{0}$.

Thus, we have

$$A\hat{x}=\hat{0}\text{ if and only if }B\hat{x}=\hat{0}\text{ for some }\hat{x}\neq\hat{0}$$

if and only if the columns of A are dependent if and only if the columns of B are dependent. This gives a way to find a basis for the column space of a matrix. Namely,

(1.) put the matrix in row reduced form.
(2.) Identify the columns of the row reduced form where the leading 1s occur.
(3.) The corresponding columns in the original matrix will be a basis for the column space of the original matrix.

Example

We find a basis for column space of the matrix

$$A = \begin{pmatrix} 5 & 10 & 2 & 8 & 0 & 12 \\ 1 & 2 & 0 & 0 & 0 & 2 \\ 2 & 4 & 1 & 4 & 0 & 5 \\ 1 & 2 & 0 & 0 & 1 & 5 \\ 2 & 4 & 1 & 4 & 2 & 2 \\ 3 & 6 & 1 & 4 & 0 & 7 \end{pmatrix}.$$

When A is row reduced, the result is

$$B = \begin{pmatrix} 1 & 2 & 0 & 0 & 0 & 0 \\ 0 & 0 & 1 & 4 & 0 & 0 \\ 0 & 0 & 0 & 0 & 1 & 0 \\ 0 & 0 & 0 & 0 & 0 & 1 \\ 0 & 0 & 0 & 0 & 0 & 0 \\ 0 & 0 & 0 & 0 & 0 & 0 \end{pmatrix}.$$

In the row reduced form, the leading 1s occurred in the first, third, fifth, and sixth columns. These do not constitute a basis for the column space of A, but if we take the corresponding columns from the original matrix, these will be a basis for the column space. Thus, a basis for the column space of the matrix A is

$$\left\{ \begin{pmatrix} 5 \\ 1 \\ 2 \\ 1 \\ 2 \\ 3 \end{pmatrix}, \begin{pmatrix} 2 \\ 0 \\ 1 \\ 0 \\ 1 \\ 1 \end{pmatrix}, \begin{pmatrix} 0 \\ 0 \\ 0 \\ 1 \\ 2 \\ 0 \end{pmatrix}, \begin{pmatrix} 12 \\ 2 \\ 5 \\ 5 \\ 2 \\ 7 \end{pmatrix} \right\}.$$

Summarizing our results, we have that for any matrix $A : \mathbb{R}^n \to \mathbb{R}^m$ each non-zero row in row reduced form gives rise to a distinct vector in a basis for the row space. Each non-zero row in row reduced form is headed by a leading 1. Thus,

the number of leading variables in row reduced form
= dimension of the row space = dimension of the column space

so

number of free variables + number of leading variables
= number of variables in the domain = number of columns of the matrix

Another method to determine a basis for the column space of a matrix is to take the transpose of the matrix, and a basis for the row space of the transposed matrix will be a basis for the column space of the original matrix (when the rows are converted to columns).

Definition:

The rank of a matrix is the dimension of the row space (which is also the dimension of the column space).

Definition:

The dimension of the null space of a matrix is called the nullity of the matrix.
    Theorem 3.23 gives the following result which is the restatement of an earlier result in a slightly different context.

Theorem 3.23:

For a matrix A

$$\text{rank of } A + \text{nullity of } A = \text{number of columns of } A.$$

Example

Suppose that A is a 3×8 matrix. Find the minimum possible value of the nullity of A.
    Since A is a 3×8 matrix, the maximum possible value of the rank of A is of 3. Since

$$\text{rank of } A + \text{nullity of } A = \text{number of columns of } A = 8$$

and the minimum value of the nullity of A occurs when the rank of A is the largest, the minimum value of the nullity of A is

$$8 - 3 = 5.$$

Exercises

1. Express the product $A\hat{u}$ as a linear combination of the columns of A.

(a.) $\begin{pmatrix} 3 & -1 \\ 4 & 6 \end{pmatrix}\begin{pmatrix} -2 \\ 5 \end{pmatrix}$ (b.) $\begin{pmatrix} 1 & 0 \\ -3 & 2 \end{pmatrix}\begin{pmatrix} 3 \\ 6 \end{pmatrix}$

(c.) $\begin{pmatrix} 2 & 2 & 0 \\ -1 & -3 & -6 \\ 1 & -2 & 4 \end{pmatrix}\begin{pmatrix} -2 \\ 0 \\ 3 \end{pmatrix}$ (d.) $\begin{pmatrix} 8 & -2 & 2 \\ 0 & 3 & 1 \\ 5 & 4 & -4 \end{pmatrix}\begin{pmatrix} 4 \\ -1 \\ 1 \end{pmatrix}$

(e.) $\begin{pmatrix} 2 & 1 & 1 \\ 0 & -2 & 3 \end{pmatrix}\begin{pmatrix} 2 \\ 3 \\ -5 \end{pmatrix}$ (f.) $\begin{pmatrix} 1 & 6 \\ -2 & 2 \\ 3 & 4 \end{pmatrix}\begin{pmatrix} 2 \\ -5 \end{pmatrix}$

2. Determine whether $\hat{b}$ is in the column space of A. If that is the case, then express $\hat{b}$ as a linear combination of the columns of A.

(a.) $A = \begin{pmatrix} 2 & 3 \\ -1 & 0 \end{pmatrix}$, $b = \begin{pmatrix} 0 \\ 5 \end{pmatrix}$ (b.) $A = \begin{pmatrix} 1 & 2 \\ 3 & 6 \end{pmatrix}$, $b = \begin{pmatrix} 1 \\ -2 \end{pmatrix}$

(c.) $A = \begin{pmatrix} 1 & 0 & 2 \\ 3 & 3 & 1 \\ 0 & -2 & 4 \end{pmatrix}$, $b = \begin{pmatrix} 1 \\ 1 \\ 0 \end{pmatrix}$ (d.) $A = \begin{pmatrix} 1 & 3 & 5 \\ -1 & 4 & 2 \\ 2 & 3 & 7 \end{pmatrix}$, $b = \begin{pmatrix} 4 \\ 3 \\ -9 \end{pmatrix}$

3. Determine bases for the row space, column space, and null space for the matrices below.

(a.) $\begin{pmatrix} 1 & 3 \\ 2 & 6 \end{pmatrix}$ (b.) $\begin{pmatrix} 2 & 0 & -1 \\ 3 & 1 & 1 \end{pmatrix}$

(c.) $\begin{pmatrix} 0 & 2 & 1 \\ 3 & -4 & 3 \\ 6 & -6 & 7 \end{pmatrix}$ (d.) $\begin{pmatrix} 3 & 2 & 0 & 2 & -3 & 1 \\ -1 & 2 & 0 & 1 & -1 & 4 \\ 0 & 1 & 3 & 1 & -2 & 0 \\ 1 & 4 & 1 & 5 & 0 & 3 \end{pmatrix}$

4. Let A and B be square matrices so that AB is defined.
   (a.) Show that the null space of AB contains the null space of B. Is it necessarily true that the null space of AB contains the null space of A? Can you add a hypothesis that will make that true?
   (b.) Show that the row space of B contains the row space of AB.

5. Show that if A and B are matrices for which $AB = 0$, then the column space of B is contained in the null space of A.

Section 3.8 Sums and Direct Sums of Vector Spaces (Optional)

The topics covered in this section deal with ways to decompose a vector space and ways to put vector spaces together to form a new vector space.

Suppose that V is a vector space and A and B are non – empty subsets of V. We define

$$A + B = \left\{ \hat{v} \in V | v = \hat{a} + \hat{b}, \hat{a} \in A, \hat{b} \in B \right\}.$$

It is not necessarily the case that $A + B$ is a subspace of V, but if A and B are subspaces of V, then $A + B$ is a subspace of V as we show in Theorem 3.24.

Example

Let $V = \mathbb{R}^3$ and $U_1 = \left\{ (x,0,0) | x \in \mathbb{R} \right\}, U_2 = \left\{ (y,y,0) | y \in \mathbb{R} \right\}$. Then

$$U_1 + U_2 = \left\{ (w,z,0) | w, z \in \mathbb{R} \right\}.$$

This is because we can take $y = z$ and $x = w - z$, so that

$$(w-z,0,0) + (z,z,0) = (w,z,0).$$

Theorem 3.24:

Suppose that V is a vector space and $U_1,\ldots,U_n$ are subspaces of V. Then $U_1+\cdots+U_n$ is the smallest subspace of V containing $U_1,\ldots,U_n$.

Proof:

We first show $U_1+\cdots+U_n$ is a subspace of V. Let $\hat{v},\hat{w} \in U_1+\cdots+U_n$. So

$$\hat{v}=\hat{v}_1+\cdots\hat{v}_n,\ \hat{w}=\hat{w}_1+\cdots\hat{w}_n;\ \hat{v}_i,\hat{w}_i \in U_i, i=1,\ldots,n.$$

Then

$$\hat{v}+\hat{w}=\left(\hat{v}_1+\cdots\hat{v}_n\right)+\left(\hat{w}_1+\cdots\hat{w}_n\right)$$
$$=\left(\hat{v}_1+\hat{w}_1\right)+\cdots+\left(\hat{v}_n+\hat{w}_n\right)\varepsilon\, U_1+\cdots+U_n$$

and

$$\alpha\hat{v}=\alpha\left(\hat{v}_1+\cdots\hat{v}_n\right)=\alpha\hat{v}_1+\cdots+\alpha\hat{v}_n\,\varepsilon\, U_1+\cdots+U_n$$

so $U_1+\cdots+U_n$ is a subspace of V.

Note that $U_1+\cdots+U_n$ contains each U_i, because if $\hat{u}_i \in U_i$, then

$$\hat{u}_i=\hat{0}+\cdots+\hat{0}+\hat{u}_i+\hat{0}+\cdots+\hat{0}\,\varepsilon\, U_1+\cdots+U_n\ .$$

Also, any subspace containing $U_1,\cdots,U_n$ contains all finite sums of elements from those sets and thus contains $U_1+\cdots+U_n$.

More important than sums of subspaces to applications in linear algebra is the idea of the direct sum of subspaces, because it provides a way of decomposing a vector space.

Definition:

Suppose that V is a vector space and $U_1,\ldots,U_n$ are subspaces of V. We say that V is the direct sum of $U_1,\ldots,U_n$ if each vector in V can be written as the sum of elements from $U_1,\ldots,U_n$ in exactly one way. In this case, we write

$$V=U_1\oplus\cdots\oplus U_n.$$

Example

Suppose that V is a vector space with basis $\{\hat{v}_1,\ldots,\hat{v}_k\}$. Let

$$U_i=\{\alpha\hat{v}_i\,|\,\alpha\in\mathcal{F}\}.$$

Then U_i is a subspace of V and

$$V=U_1\oplus\cdots\oplus U_n.$$

This is because if $\hat{v} \in V$ and if $\{\hat{v}_1, \ldots, \hat{v}_k\}$ is a basis for $\{\hat{v}_1, \ldots, \hat{v}_k\}$, then there is exactly one way to write $\hat{v}$ as a linear combination of $\{\hat{v}_1, \ldots, \hat{v}_k\}$.

Theorem 3.25:

Suppose that V is a vector space and $U_1, \ldots, U_n$ are subspaces of V. We have

$$V = U_1 \oplus \cdots \oplus U_n.$$

if and only if both of the following conditions hold:
   (a.) $V = U_1 + \cdots + U_n$

   (b.) If $\hat{u}_i \in U_i$, $i = 1, \ldots, n$ and $\hat{u}_1 + \cdots + \hat{u}_n = \hat{0}$, then $\hat{u}_i = 0, i = 1, \ldots, n$.

Proof:
Suppose that

$$V = U_1 \oplus \cdots \oplus U_n.$$

Then $V = U_1 + \cdots + U_n$. Suppose that $\hat{u}_1 + \cdots + \hat{u}_n = \hat{0}$.

One way this can happen is $\hat{u}_i = \hat{0}$ for $i = 1, \ldots, n$. Since

$$V = U_1 \oplus \cdots \oplus U_n,$$

this is the only way that it can happen.

Suppose that conditions (a.) and (b.) hold. By condition (a.) every vector in V can be expressed as a sum of vectors $\hat{u}_1 + \cdots + \hat{u}_n$ with $\hat{u}_i \in U_i$. We must show the expression is unique. Suppose there is a vector $\hat{v} \in V$ with

$$\hat{v} = \hat{u}_1 + \cdots + \hat{u}_n, \ \hat{u}_i \in U_i$$

and

$$\hat{v} = \hat{w}_1 + \cdots + \hat{w}_n, \ \hat{w}_i \in U_i.$$

Then

$$\hat{0} = \hat{v} - \hat{v} = \left(\hat{u}_1 + \cdots + \hat{u}_n \right) - \left(\hat{w}_1 + \cdots + \hat{w}_n \right) = \left(\hat{u}_1 - \hat{w}_1 \right) + \cdots + \left(\hat{u}_n - \hat{w}_n \right)$$

and $\left(\hat{u}_i - \hat{w}_i \right) \in U_i$ so by condition (b.) $\hat{u}_i - \hat{w}_i = \hat{0}$ or $\hat{u}_i = \hat{w}_i$.

Corollary:
Suppose that V is a vector space and U and W are subspaces of V. Then

$$V = U \oplus W$$

if and only if

$$V = U + W \text{ and } U \cap W = \{\hat{0}\}$$

Proof:

We need to show that $U \cap W = \{\hat{0}\}$ is equivalent to the statement if $\hat{u} \epsilon U, \hat{w} \epsilon W$ and $\hat{u} + \hat{w} = \hat{0}$ then $\hat{u} = \hat{w} = \hat{0}$.

Now since $\hat{u} + \hat{w} = \hat{0}$, we have $\hat{u} = -\hat{w}$, so $-\hat{w} \epsilon U$ and thus $\hat{w} \epsilon U$. Since $U \cap W = \{\hat{0}\}$ and $\hat{w} \epsilon U \cap W$ we must have $\hat{w} = \hat{0}$. Thus

$$\hat{u} = -\hat{w} = \hat{0}.$$

Theorem 3.26:

Suppose that V is a vector space of dimension n and U is a subspace of V of dimension k where $k < n$. Then there is a subspace W of V of dimension $n - k$ with $V = U \oplus W$.

Proof:

Let $\{\hat{v}_1, , ., \hat{v}_k\}$ be a basis for U. Extend $\{\hat{v}_1, , ., \hat{v}_k\}$ to be a basis for V, say $\{\hat{v}_1, \ldots, \hat{v}_k, \hat{v}_{k+1}, \ldots, \hat{v}_n\}$. Let W be the subspace of V formed by linear combinations of $\{\hat{v}_{k+1}, \ldots, \hat{v}_n\}$. If $\hat{v} \epsilon U \cap W$, then

$$\hat{v} = c_1 \hat{v}_1 + \cdots c_k \hat{v}_k \epsilon U \text{ and } \hat{v} = c_{k+1} \hat{v}_{k+1} + \cdots c_n \hat{v}_n \epsilon W.$$

Thus

$$\hat{0} = \hat{v} - \hat{v} = \left(c_1 \hat{v}_1 + \cdots c_k \hat{v}_k \right) - \left(c_{k+1} \hat{v}_{k+1} + \cdots c_n \hat{v}_n \right)$$

and since $\{\hat{v}_1, , ., \hat{v}_k, \hat{v}_{k+1}, \ldots, \hat{v}_n\}$ is a basis, $c_i = 0$, $i = 1, \ldots, n$ so $v = 0$.

Also $U + W$ contains all linear combinations of $\{\hat{v}_1, \ldots, \hat{v}_k, \hat{v}_{k+1}, \ldots, \hat{v}_n\}$ so $U \oplus W = V$.

Corollary:
If $V = U \oplus W$, then $\dim V = \dim U + \dim W$.

Proof:

Suppose that $\{\hat{u}_1, , ., \hat{u}_k\}$ is a basis for U and $\{\hat{w}_1, , ., \hat{w}_j\}$ is a basis for W.
Since $V = U + W$, then $\{\hat{u}_1, , ., \hat{u}_k, \hat{w}_1, , ., \hat{w}_j\}$ spans V. Also

$$\{\hat{u}_1, , ., \hat{u}_k\} \cap \{\hat{w}_1, , ., \hat{w}_j\} = \varnothing.$$

We show $\left\{ \hat{u}_1, , ., \hat{u}_k, \hat{w}_1, , ., \hat{w}_j \right\}$ is a linearly independent set. Suppose

$$c_1 \hat{u}_1 + \cdots + c_k \hat{u}_k + d_1 \hat{w}_1 + \cdots + d_j \hat{w}_j = \hat{0}.$$

Now

$$c_1 \hat{u}_1 + \cdots + c_k \hat{u}_k = \hat{u} \in U \quad \text{and} \quad d_1 \hat{w}_1 + \cdots + d_j \hat{w}_j = \hat{w} \in W$$

and

$$\hat{u} + \hat{w} = \hat{0} \quad \text{so} \quad \hat{u} = -\hat{w}.$$

But U and W are subspaces of V with $U \cap W = \left\{ \hat{0} \right\}$, so $\hat{u} = \hat{w} = \hat{0}$.
Thus

$$c_1 \hat{u}_1 + \cdots + c_k \hat{u}_k = \hat{0} \quad \text{and} \quad d_1 \hat{w}_1 + \cdots + d_j \hat{w}_j = \hat{0}$$

and since $\left\{ \hat{u}_1, , ., \hat{u}_k \right\}$ and $\left\{ \hat{w}_1, , ., \hat{w}_j \right\}$ are linearly independent sets,

$$c_1 = \cdots = c_k = 0, \quad \text{and} \quad d_1 = \cdots = d_j = 0.$$

Exercises

1. Let U be the subspace of $\mathbb{R}^4$ spanned by $\hat{v}_1 = (3, 0, 1, 1)$ and $\hat{v}_2 = (2, 2, 2, 1)$. Find a basis for a subspace W of $\mathbb{R}^4$ for which

   $$\mathbb{R}^4 = U \oplus W.$$

2. Let $\mathcal{P}_n(\mathbb{R})$ be the polynomials of degree n or less with coefficients in $\mathbb{R}$. Let $\mathcal{E}(\mathbb{R})$ be the polynomials in $\mathcal{P}_n(\mathbb{R})$ for which the odd powers of x have coefficients equal to 0 and $\mathcal{O}(\mathbb{R})$ be the polynomials in $\mathcal{P}_n(\mathbb{R})$ for which the even powers of x have coefficients equal to 0. Show that

   $$\mathcal{P}_n(\mathbb{R}) = \mathcal{E}(\mathbb{R}) \oplus \mathcal{O}(\mathbb{R}).$$

3. Let S be the set of symmetric 2×2 matrices; that is, matrices for which $A = A^T$. These are matrices of the form

   $$\begin{pmatrix} a & c \\ c & b \end{pmatrix}.$$

   Let T be the set of antisymmetric 2×2 matrices; that is, matrices for which $A = -A^T$. These are matrices of the form

$$\begin{pmatrix} 0 & c \\ -c & 0 \end{pmatrix}.$$

(a). Show that T and S are subspaces of $M_{2\times2}(\mathbb{R})$.

(b). Show that $M_{2\times2}(\mathbb{R}) = S \oplus T$.

4. (a.) Show that if U and W are subspaces of the vector space V, then $U + W$ is a subspace of V.

(b.) Give an example of nonempty subsets U and W of the vector space V for which $U + W$ is not a subspace of V.

5. Suppose that V is a vector space and A and B are subspaces of V with bases $\{\hat{a}_1,\ldots,\hat{a}_j\}$ and $\{\hat{b}_1,\ldots,\hat{b}_k\}$. Show that

$$V = A \oplus B$$

If and only if $\{\hat{a}_1,\ldots,\hat{a}_j,\hat{b}_1,\ldots,\hat{b}_k\}$ is a basis for V.

6. Use the result of Exercise 5 to determine whether $V = A \oplus B$.

If $V \neq A \oplus B$ find a non-zero vector in $A \cap B$ or a vector in V that is not in $A + B$

(a.) $V = \mathbb{R}^4$, A is the subspace spanned by

$$\{(1,-2,0,6),(3,7,1,-1)\}$$

and B is the subspace spanned by

$$\{(1,0,0,0),(5,3,1,11)\}.$$

(b). $V = \mathbb{R}^5$, A is the subspace spanned by

$$\{(3,-2,6,6,4),(2,5,1,-1,0)\}$$

and B is the subspace spanned by

$$\{(1,2,3,0,0),(6,5,-3,1,1)\}.$$

(c.) $V = \mathbb{R}^4$, A is the subspace spanned by

$$\{(1,-2,0,6),(3,7,1,-1)\}$$

and B is the subspace spanned by

$$\{(1,0,0,0),(3,2,1,5)\}.$$

7. Show that if

V is a vector space with subspaces W_1 and W_2, then $W_1 \cup W_2$ is a subspace if and only if $W_1 \subset W_2$ or $W_2 \subset W_1$.

Chapter 4

Linear Transformations

Historical Note

In 1848 in England, J.J. Sylvester first introduced the term "matrix", as a name for an array of numbers. Matrix algebra was furthered by the work of Arthur Cayley circa 1855. Cayley studied compositions of linear transformations and was led to define matrix multiplication so that the matrix of coefficients for the composite transformation ST is the product of the matrix for S times the matrix for T. He went on to study the algebra of these compositions including matrix inverses. The famous Cayley–Hamilton theorem which asserts that a square matrix is a root of its characteristic polynomial was given by Cayley in his 1858 *Memoir on the Theory of Matrices*.

The modern and more precise definition of a vector space was introduced by Giuseppe Peano in 1888. By 1900, a theory of linear transformations of finite-dimensional vector spaces had emerged. Linear algebra took its modern form in the first half of the twentieth century, when many ideas and methods of previous centuries were generalized as abstract algebra.

In this chapter we study linear transformations, which are the most important functions between vector spaces. We will see that for U and V vector spaces with bases $\mathcal{B}_1$ and $\mathcal{B}_2$ respectively and $T : U \to V$ a linear transformation, there is a matrix $A(\mathcal{B}_1, \mathcal{B}_2)$ such that T is expressed by multiplying vectors in U by the matrix $A(\mathcal{B}_1, \mathcal{B}_2)$. The matrix $A(\mathcal{B}_1, \mathcal{B}_2)$ will depend on the choice of bases. This means we can apply the theory we have developed for matrices to the study of linear transformations.

Section 4.1 Properties of a Linear Transformation

Definition:

Suppose that U and V are vector spaces over the same scalar field $\mathcal{F}$. A linear transformation (linear function) T from U to V is a function

$$T : U \to V$$

that satisfies

$$T\left(\hat{u}_1 + \hat{u}_2\right) = T\left(\hat{u}_1\right) + T\left(\hat{u}_2\right) \text{ for all } \hat{u}_1, \hat{u}_2 \in U$$

and

$$T\left(a\hat{u}\right) = aT\left(\hat{u}\right) \text{ for all } \hat{u} \in U, a \in \mathcal{F}.$$

These two conditions are often combined into the single condition

$$T\left(a_1\hat{u}_1 + a_2\hat{u}_2\right) = a_1T\left(\hat{u}_1\right) + a_2T\left(\hat{u}_2\right) \text{ for all } \hat{u}_1, \hat{u}_2 \in U, a_1, a_2 \in \mathcal{F}.$$

If

$$T:V \rightarrow V$$

is a linear transformation, then T is said to be a linear operator on V.

Notation:

The set of linear transformations from U to V is denoted $L(U,V)$ and the set of linear operators from V to V is denoted $L(V)$.

The properties described in the definition are saying that linear transformations preserve vector addition and scalar multiplication.

The diagram in Figure 4.1(a.) illustrates what the definition is saying for vector addition.

One could begin with $\hat{u}_1, \hat{u}_2 \in U$ and first apply T, to each vectors obtaining the vectors $T\left(\hat{u}_1\right), T\left(\hat{u}_2\right) \in V$.

One could then add $T\left(\hat{u}_1\right), T\left(\hat{u}_2\right)$ to get $T\left(\hat{u}_1\right) + T\left(\hat{u}_2\right) \in V$. This is the path α in Figure 4.1(a.).

A second choice would be to first add $\hat{u}_1, \hat{u}_2 \in U$, to obtain $\hat{u}_1 + \hat{u}_2 \in U$, then apply T to $\hat{u}_1 + \hat{u}_2$ obtaining $T(\hat{u}_1 + \hat{u}_2) \in V$. This is the path β in Figure 4.1(a.).

Depending on the function, one could get different answers following the different paths. For example if

$$f\left(x\right) = x^2$$

then

$$f\left(x_1\right) + f\left(x_2\right) = x_1^2 + x_2^2$$

and

$$f\left(x_1 + x_2\right) = \left(x_1 + x_2\right)^2 = x_1^2 + 2x_1x_2 + x_2^2$$

We could follow the same ideas with scalar multiplication. If a is a scalar and $\hat{u} \in U$, one could form the vector $a\hat{u} \in U$, then apply T to $a\hat{u}$ to obtain $T(a\hat{u}) \in V$. This is the path α in Figure 4.1(b.).

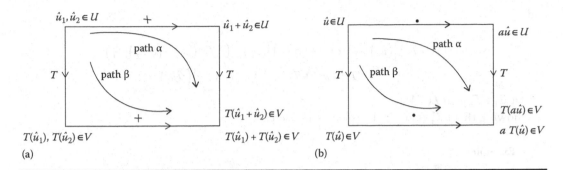

Figure 4.1   A linear transformation preserves vector addition and scalar multiplication.

One could have first applied T to $\hat{u} \in U$ to obtain $T(\hat{u}) \in V$, then multiplied by a to obtain $aT(u) \in V$. This is the path β in Figure 4.1(b.).

The function

$$f(x) = x^2$$

again provides an example where the different paths give different answers. We have

$$f(2x) = (2x)^2 = 4x^2$$

and

$$2f(x) = 2(x^2) = 2x^2.$$

The definition says that if T is a linear transformation, then we end up in the same place with either path.

Definition:

If $T_1, T_2 \in L(U,V)$ and $\alpha \in \mathcal{F}$, we define $T_1 + T_2$ and αT_1 by

$$(T_1 + T_2)(\hat{u}) = T_1(\hat{u}) + T_2(\hat{u}),$$

$$(\alpha T_1)(\hat{u}) = \alpha\left(T_1(\hat{u})\right) \text{ for all } \hat{u} \in U, \alpha \in \mathcal{F}.$$

We note that $T_1 + T_2$ and αT_1 are in $L(U,V)$. This is because if $\hat{u}_1, \hat{u}_2 \in U$ and $\alpha \in \mathcal{F}$ then

$$
\begin{aligned}
(T_1 + T_2)(\hat{u}_1 + \hat{u}_2) &= T_1(\hat{u}_1 + \hat{u}_2) + T_2(\hat{u}_1 + \hat{u}_2) \\
&= T_1(\hat{u}_1) + T_1(\hat{u}_2) + T_2(\hat{u}_1) + T_2(\hat{u}_2) \\
&= T_1(\hat{u}_1) + T_2(\hat{u}_1) + T_1(\hat{u}_2) + T_2(\hat{u}_2) \\
&= (T_1 + T_2)(\hat{u}_1) + (T_1 + T_2)(\hat{u}_2)
\end{aligned}
$$

and

$$\left[\alpha\left(T_1+T_2\right)\right](\hat{u}_1)=\left[\left(\alpha T_1\right)+\left(\alpha T_2\right)\right](\hat{u}_1)=\left(\alpha T_1\right)(\hat{u}_1)+\left(\alpha T_2\right)(\hat{u}_1)$$
$$=T_1\left(\alpha\hat{u}_1\right)+T_2\left(\alpha\hat{u}_1\right)=\left(T_1+T_2\right)\left(\alpha\hat{u}_1\right).$$

so that $T_1 + T_2$ is in $L(U,V)$.

Showing that αT_1 is in $L(U,V)$ is left as an exercise.

Example

The linear transformation $T:U\rightarrow V$ defined by $T\left(\hat{u}\right)=\hat{0}_v$ for every $\hat{u}\in U$ is the zero transformation, which we will denote by **0**.

The linear operator $T:U\rightarrow V$ defined by $T\left(\hat{u}\right)=\hat{u}$ for every $\hat{u}\in U$ is the identity operator on U, which we will denote by I_U.

Theorem 4.1:

With the notation above, $L(U,V)$ and $L(V)$ are vector spaces over $\mathcal{F}$.

The proof is left as an exercise.

Theorem 4.2:

Let $T\in L\left(U,V\right)$ and $\hat{0}_U$ and $\hat{0}_V$ be the zero vectors for U and V, respectively. Then

1. $T\left(\hat{0}_U\right)=\hat{0}_V$.

2. For all $\hat{u}\in U, T\left(-\hat{u}\right)=-T\left(\hat{u}\right)$.

Proof:

    1.   We have

$$T\left(\hat{0}_U\right)=T\left(0\hat{0}_U\right)=0T\left(\hat{0}_U\right)=\hat{0}_V$$

since $T\left(\hat{0}_U\right)$ is a vector in V and $0v=\hat{0}_V$ for all $v\in V$.

2. We have

$$\hat{0}_V =T\left(\hat{0}_U\right)=T\left[\hat{u}+\left(-\hat{u}\right)\right]=T\left(\hat{u}\right)+T\left(-\hat{u}\right),$$

so $T\left(-\hat{u}\right)$ is the additive inverse of $T\left(\hat{u}\right)$, and since additive inverses are unique, $T\left(-\hat{u}\right)=-T\left(\hat{u}\right)$.

Example

We show that the function

$$T:\mathbb{R}^2\rightarrow\mathbb{R}^3$$

defined by

$$T\begin{pmatrix} x \\ y \end{pmatrix} = \begin{pmatrix} 2x+y \\ x-y \\ 4x \end{pmatrix}$$

is a linear transformation.

Let

$$\hat{u}_1 = \begin{pmatrix} x_1 \\ y_1 \end{pmatrix}, \quad \hat{u}_2 = \begin{pmatrix} x_2 \\ y_2 \end{pmatrix}.$$

Then

$$T\left(\hat{u}_1 + \hat{u}_2\right) = T\left[\begin{pmatrix} x_1 \\ y_1 \end{pmatrix} + \begin{pmatrix} x_2 \\ y_2 \end{pmatrix}\right] = T\begin{pmatrix} x_1+x_2 \\ y_1+y_2 \end{pmatrix} = \begin{pmatrix} 2(x_1+x_2)+(y_1+y_2) \\ (x_1+x_2)-(y_1+y_2) \\ 4(x_1+x_2) \end{pmatrix}$$

$$= \begin{pmatrix} (2x_1+y_1)+(2x_2+y_2) \\ (x_1-y_1)+(x_2-y_2) \\ 4x_1+4x_2 \end{pmatrix} = \begin{pmatrix} 2x_1+y_1 \\ x_1-y_1 \\ 4x_1 \end{pmatrix} + \begin{pmatrix} 2x_2+y_2 \\ x_2-y_2 \\ 4x_2 \end{pmatrix}$$

$$= T\begin{pmatrix} x_1 \\ y_1 \end{pmatrix} + T\begin{pmatrix} x_2 \\ y_2 \end{pmatrix} = T\left(\hat{u}_1\right) + T\left(\hat{u}_2\right).$$

Also,

$$T\left(a\hat{u}_1\right) = T\left[a\begin{pmatrix} x_1 \\ y_1 \end{pmatrix}\right] = T\begin{pmatrix} ax_1 \\ ay_1 \end{pmatrix} = \begin{pmatrix} 2ax_1+ay_1 \\ ax_1-ay_1 \\ 4ax_1 \end{pmatrix} = a\begin{pmatrix} 2x_1+y_1 \\ x_1-y_1 \\ 4x_1 \end{pmatrix} = aT\left(\hat{u}_1\right)$$

Example

We show the function

$$T : \mathbb{R}^2 \to \mathbb{R}^3$$

defined by

$$T\begin{pmatrix} x \\ y \end{pmatrix} = \begin{pmatrix} x^2 \\ x-y \\ 4x \end{pmatrix}$$

is not a linear transformation.

Using the notation of the previous example,

$$T\left(\hat{u}_1 + \hat{u}_2\right) = T\left[\begin{pmatrix} x_1 \\ y_1 \end{pmatrix} + \begin{pmatrix} x_2 \\ y_2 \end{pmatrix}\right] = T\begin{pmatrix} x_1+x_2 \\ y_1+y_2 \end{pmatrix}$$

$$= \begin{pmatrix} (x_1+x_2)^2 \\ (x_1+x_2)-(y_1+y_2) \\ 4(x_1+x_2) \end{pmatrix} = \begin{pmatrix} x_1^2+2x_1x_2+x_1^2 \\ (x_1+x_2)-(y_1+y_2) \\ 4(x_1+x_2) \end{pmatrix}$$

but

$$T\left(\hat{u}_1\right)+T\left(\hat{u}_2\right)=T\begin{pmatrix}x_1\\y_1\end{pmatrix}+T\begin{pmatrix}x_2\\y_2\end{pmatrix}=\begin{pmatrix}x_1^2\\x_1-y_1\\4x_1\end{pmatrix}+\begin{pmatrix}x_2^2\\x_2-y_2\\4x_2\end{pmatrix}$$

$$=\begin{pmatrix}x_1^2+x_1^2\\(x_1+x_2)-(y_1+y_2)\\4(x_1+x_2)\end{pmatrix}$$

so

$$T\left(\hat{u}_1+\hat{u}_2\right)\ne T\left(\hat{u}_1\right)+T\left(\hat{u}_2\right).$$

The two examples above illustrate the fact that any linear transformation

$$T:\mathbb{R}^m\to\mathbb{R}^n$$

can be represented as

$$T\begin{pmatrix}x_1\\\vdots\\x_m\end{pmatrix}=\begin{pmatrix}T_1(x_1,..,x_m)\\\vdots\\T_n(x_1,..,x_m)\end{pmatrix}$$

where

$$T_i:\mathbb{R}^m\to\mathbb{R}.$$

The function T will be linear if and only if every T_i, $i=1,\ldots,n$, is linear.

Example
Consider

$$\frac{d}{dx}:\mathcal{P}_n(\mathbb{R})\to\mathcal{P}_{n-1}(\mathbb{R}).$$

We have

$$\frac{d}{dx}\left(af(x)+bg(x)\right)=\frac{d}{dx}\left(af(x)\right)+\frac{d}{dx}\left(bg(x)\right)=a\frac{d}{dx}f(x)+b\frac{d}{dx}g(x)$$

so $\frac{d}{dx}$ is a linear transformation.

Example
Consider

$$\int_\alpha^\beta:\mathcal{P}_n(\mathbb{R})\to\mathbb{R}.$$

We have

$$\int_\alpha^\beta \big(af(x)+bg(x)\big)dx = \int_\alpha^\beta \big(af(x)\big)dx + \int_\alpha^\beta \big(bg(x)\big)dx$$

$$= a\int_\alpha^\beta f(x)dx + b\int_\alpha^\beta g(x)dx$$

so $\int_\alpha^\beta$ is a linear transformation.

Theorem 4.3:

If $T \in L(U,V)$ is a one-to-one and onto linear transformation, then $T^{-1} \in L(V,U)$.

Proof:

For

$$T^{-1}:V \to U$$

to exist, it is necessary and sufficient for T to be a one-to-one and onto function.

It is necessary to show T^{-1} is a linear transformation.

Suppose $\hat{v}_1,\hat{v}_2 \in V$. Then there are unique $\hat{u}_1,\hat{u}_2 \in U$ with

$$T(\hat{u}_1)=\hat{v}_1 \text{ and } T(\hat{u}_2)=\hat{v}_2.$$

Thus

$$\hat{u}_1 = T^{-1}(\hat{v}_1) \text{ and } \hat{u}_2 = T^{-1}(\hat{v}_2)$$

So

$$T^{-1}(\hat{v}_1+\hat{v}_2)=T^{-1}\big(T(\hat{u}_1)+T(\hat{u}_2)\big)=T^{-1}\big(T(\hat{u}_1+\hat{u}_2)\big)$$

$$= \hat{u}_1+\hat{u}_2 = T^{-1}\hat{v}_1 + T^{-1}\hat{v}_2.$$

For α a scalar, we have

$$T^{-1}(\alpha\hat{v}_1)=T^{-1}\big(\alpha T(\hat{u}_1)\big)=T^{-1}\big(T(\alpha\hat{u}_1)\big)=\alpha\hat{u}_1=\alpha T^{-1}(\hat{v}_1).$$

Null Space and Range (Image) of a Linear Transformation

Definition:

If

$$T:U \to V$$

is a linear transformation then the null space of T (also called the kernel of T), denoted $\mathcal{N}(T)$, is defined by

$$\mathcal{N}(T) = \left\{ \hat{u} \in U \mid T(\hat{u}) = \hat{0}_V \right\}.$$

Thus $\mathcal{N}(T) \subset U$.

Each linear transformation T also has another subspace associated with it, called the range of T.

Definition:

Let

$$T : U \to V$$

be a linear transformation. The range of T, (also called the image of T) denoted $\mathcal{R}(T)$, is defined by

$$\mathcal{R}(T) = \left\{ \hat{w} \in W \mid \hat{w} = T(\hat{v}) \text{ for some } \hat{v} \in V \right\} = \left\{ T(\hat{v}) \mid \hat{v} \in V \right\}.$$

Theorem 4.4:

Let

$$T : U \to V$$

be a linear transformation. Then the range of T is a subspace of V.

Proof:

Since $T(\hat{0}_U) = \hat{0}_V$, we have $\hat{0}_V \in \mathcal{R}(T)$.

Suppose that $\hat{v}_1, \hat{v}_2 \in \mathcal{R}(T)$. Then there are $\hat{u}_1, \hat{u}_2 \in U$ with

$$T(\hat{u}_1) = \hat{v}_1, \quad T(\hat{u}_2) = \hat{v}_2.$$

Thus,

$T(\hat{u}_1 + \hat{u}_2) = T(\hat{u}_1) + T(\hat{u}_2) = \hat{v}_1 + \hat{v}_2$ and so $\hat{v}_1 + \hat{v}_2 \in \mathcal{R}(T)$.

If $\alpha \in \mathcal{F}$ and $\hat{v} \in \mathcal{R}(T)$, then there is a $\hat{u} \in U$ with $\hat{v} = T(\hat{u})$.
Then

$$T(\alpha\hat{u}) = \alpha T(\hat{u}) = \alpha\hat{v} \in \mathcal{R}(T).$$

Later in this chapter we will demonstrate how to easily find a basis for the null space and the basis of the range space of a linear transformation.

Exercises

1. For $T : \mathbb{R}^3 \to \mathbb{R}^3$ defined below, tell whether T is a linear transformation.

(a.) $T\begin{pmatrix} x \\ y \\ z \end{pmatrix} = \begin{pmatrix} 3y - z \\ 2x + y \\ 0 \end{pmatrix}$

(b.) $T\begin{pmatrix} x \\ y \\ z \end{pmatrix} = \begin{pmatrix} 3y - z \\ 2x + y \\ 1 \end{pmatrix}$

(c.) $T\begin{pmatrix} x \\ y \\ z \end{pmatrix} = \begin{pmatrix} 0 \\ 2x + y^2 \\ z \end{pmatrix}$

2. Is $T : P_n(\mathbb{R}) \to P_{n+2}(\mathbb{R})$ defined by

$$T(f(x)) = x^2(f(x))$$

a linear transformation?

3. Tell whether the following transformations

$$T : M_{2 \times 2}(\mathbb{R}) \to \mathbb{R}$$

are linear.

(a). $T\left(\begin{pmatrix} a & b \\ c & d \end{pmatrix}\right) = ad - bc$

(b). $T\left(\begin{pmatrix} a & b \\ c & d \end{pmatrix}\right) = 3a + 2b - c$

4. Show that $T : M_{n \times m}(\mathbb{R}) \to M_{m \times n}(\mathbb{R})$ defined by

$$T(A) = A^T$$

is a linear transformation.

5. Show that $T : \mathbb{R}^2 \to \mathbb{R}^2$ defined by

$$T\begin{pmatrix} x \\ y \end{pmatrix} = \begin{pmatrix} \cos\theta & -\sin\theta \\ \sin\theta & \cos\theta \end{pmatrix}\begin{pmatrix} x \\ y \end{pmatrix} = \begin{pmatrix} x\cos\theta - y\sin\theta \\ x\sin\theta + y\cos\theta \end{pmatrix}$$

is a linear transformation. This is the function that rotates a vector counterclockwise through the angle θ.

6. (a.) Show that the function $T : C[x] \to C[x]$ defined by

$$T\big(f(x)\big) = f\big(x^2 + 1\big)$$

is a linear transformation.

    (b.) Show that the function $T : C[x] \to C[x]$ defined by

$$T\big(f(x)\big) = \big(f(x)\big)^2 + 1$$

        is not a linear transformation

7. Let $T : U \to V$ be a linear transformation. Show that
    (a.) T is one-to-one if and only if the image of a linearly independent set is a linearly independent set.
    (b.) T is a 1–1 function if and only if $\ker(T) = \{\hat{0}\}$.

8. If T is a linear transformation with

$$T(\hat{v}_1) = (4,7,-1), \ T(\hat{v}_2) = (3,0,5), \ T(\hat{v}_3) = (1,1,1),$$

    find

$$T\big(2\hat{v}_1 - 7\hat{v}_2 + \hat{v}_3\big)$$

9. Suppose $T : V \to V$ is a linear transformation.
    (a.) Show that $T^2 = 0$ if and only if $\mathcal{R}(T) \subset \mathcal{N}(T)$.
    (b.) Give an example of such a $T \neq 0$.

10. Show that the zero transformation and identity transformation are linear transformations.

11. Show that a reflection $(x, y) \to (x, -y)$ in $\mathbb{R}^2$ is a linear transformation.

12. Show that permutation of components in $\mathbb{R}^2$ is a linear transformation.

13. Show that a projection $(x, y) \to (x, 0)$ in $\mathbb{R}^2$ is a linear transformation.

14. Show that $L(U,V)$ and $L(V)$ are vector spaces over F.

Section 4.2 Representing a Linear Transformation

We now prove the fundamental fact that a linear transformation can be expressed as multiplication of a vector by a matrix. However, for any particular linear transformation, the matrix will be determined by both the linear transformation and the choice of bases for the domain vector space and the range vector space.

    We return to the idea of a basis for a vector space. A vector space has many different bases. A fixed vector has a unique expression for a given basis, but the same vector will usually have different representations in different bases.

Theorem 4.5:

A linear transformation is determined by its effect on the basis elements.
    What this theorem is saying is if

$$T : U \to V$$

is a linear transformation and $\{\hat{u}_1,\ldots,\hat{u}_n\}$ is a basis of U and we know $T(\hat{u}_1),\ldots,T(\hat{u}_n)$ then we know $T(\hat{u})$ for any vector $\hat{u} \in U$. This is not true for an arbitrary function. For example if

$$f : \mathbb{R} \to \mathbb{R}$$

and we know the value of $f(x)$ for any finite set $\{x_1, x_2, \ldots, x_n\}$ of x's that is not enough to determine the value of $f(x)$ for a value of x that is not in $\{x_1, x_2, \ldots, x_n\}$.

Proof:

Let

$$T \in L(U,V),$$

and let $\{\hat{e}_1,\ldots,\hat{e}_n\}$ be a basis for U. If $\hat{u} \in U$, then there are unique scalars $a_1,\ldots,a_n$ for which

$$\hat{u} = a_1\hat{e}_1 + \cdots + a_n\hat{e}_n.$$

Thus,

$$T(\hat{u}) = T(a_1\hat{e}_1 + \cdots + a_n\hat{e}_n) = a_1 T(\hat{e}_1) + \cdots + a_n T(\hat{e}_n).$$

So if $\{\hat{e}_1,\ldots,\hat{e}_n\}$ is a basis for U, then any vector $T(\hat{u})$ in the range of T is a linear combination of $\{T(\hat{e}_1),\ldots,T(\hat{e}_n)\}$.

A slightly different way to express the previous sentence is $\{T(\hat{e}_1),\ldots,T(\hat{e}_n)\}$ spans the range of T.

This does not say that $\{T(\hat{e}_1),\ldots,T(\hat{e}_n)\}$ is a basis for the range of T because $\{T(\hat{e}_1),\ldots,T(\hat{e}_n)\}$ is not necessarily linearly independent, as we show in the exercises.

The Representation of a Linear Transformation in the Usual Basis

As we stated, the major result of this section is that a linear transformation can be expressed as multiplication by a matrix. The example below will show how to do this. *However, the matrix will depend on the basis, and this important fact may not be clear until later.*

Example

Let $T : \mathbb{R}^2 \to \mathbb{R}^3$ be defined by

$$T\begin{pmatrix} x \\ y \end{pmatrix} = \begin{pmatrix} 3x - 4y \\ x + y \\ 2y \end{pmatrix}. \tag{1}$$

We will use the standard bases for $\mathbb{R}^2$ and $\mathbb{R}^3$. According to Theorem 4.5, if we know the effect of T on $\hat{e}_1$ and $\hat{e}_2$ we know what T does to every vector.

Equation (1) can be written as

$$T\begin{pmatrix} x \\ y \end{pmatrix} = \begin{pmatrix} 3x-4y \\ x+y \\ 2y \end{pmatrix} = \begin{pmatrix} 3x \\ x \\ 0 \end{pmatrix} + \begin{pmatrix} -4y \\ y \\ 2y \end{pmatrix} = x\begin{pmatrix} 3 \\ 1 \\ 0 \end{pmatrix} + y\begin{pmatrix} -4 \\ 1 \\ 2 \end{pmatrix}$$

and in Chapter 1, we learned the right-hand side can be expressed as the matrix equation

$$x\begin{pmatrix} 3 \\ 1 \\ 0 \end{pmatrix} + y\begin{pmatrix} -4 \\ 1 \\ 2 \end{pmatrix} = \begin{pmatrix} 3 & -4 \\ 1 & 1 \\ 0 & 2 \end{pmatrix}\begin{pmatrix} x \\ y \end{pmatrix}.$$

So we have

$$T\begin{pmatrix} x \\ y \end{pmatrix} = \begin{pmatrix} 3 & -4 \\ 1 & 1 \\ 0 & 2 \end{pmatrix}\begin{pmatrix} x \\ y \end{pmatrix}$$

and we have shown that evaluating T at the vector $\begin{pmatrix} x \\ y \end{pmatrix}$ is the same as multiplying the vector $\begin{pmatrix} x \\ y \end{pmatrix}$ by the matrix

$$\begin{pmatrix} 3 & -4 \\ 1 & 1 \\ 0 & 2 \end{pmatrix}.$$

We next want to determine how to easily construct the matrix associated with a linear transformation with respect to the standard bases. Consider

$$T\begin{pmatrix} x \\ y \end{pmatrix} = \begin{pmatrix} 3 & -4 \\ 1 & 1 \\ 0 & 2 \end{pmatrix}\begin{pmatrix} x \\ y \end{pmatrix}$$

If we let $\begin{pmatrix} x \\ y \end{pmatrix}$ be the first basis vector; that is, $\begin{pmatrix} 1 \\ 0 \end{pmatrix}$, then we get

$$T\begin{pmatrix} 1 \\ 0 \end{pmatrix} = \begin{pmatrix} 3 & -4 \\ 1 & 1 \\ 0 & 2 \end{pmatrix}\begin{pmatrix} 1 \\ 0 \end{pmatrix} = \begin{pmatrix} 3 \\ 1 \\ 0 \end{pmatrix}$$

which is the first column of the matrix. Likewise, $T\begin{pmatrix} 0 \\ 1 \end{pmatrix}$ gives the second column of the matrix. So, at least when we use the standard basis vectors, we have found the following useful principle.

Theorem 4.6:

Given a linear transformation $T : \mathbb{R}^n \to \mathbb{R}^m$, the matrix associated with T with the standard bases will have $T(\hat{e}_k)$ as its kth column where $\hat{e}_k$ is the kth standard basis vector in $\mathbb{R}^n$.

This is saying that the matrix of T with respect to the usual bases is

$$\left[T\left(\hat{e}_1\right),T\left(\hat{e}_2\right)...,T\left(\hat{e}_n\right)\right].$$

Example

Find the matrix A for the following linear transformations with respect to the standard bases.

(a.) $\quad T\begin{pmatrix} x \\ y \\ z \end{pmatrix} = \begin{pmatrix} 4y-6z \\ 2x+y \\ 5z \end{pmatrix}$

We have

$$T\begin{pmatrix} 1 \\ 0 \\ 0 \end{pmatrix} = \begin{pmatrix} 0 \\ 2 \\ 0 \end{pmatrix};\quad T\begin{pmatrix} 0 \\ 1 \\ 0 \end{pmatrix} = \begin{pmatrix} 4 \\ 1 \\ 0 \end{pmatrix};\quad T\begin{pmatrix} 0 \\ 0 \\ 1 \end{pmatrix} = \begin{pmatrix} -6 \\ 0 \\ 5 \end{pmatrix}$$

so

$$A = \begin{pmatrix} 0 & 4 & -6 \\ 2 & 1 & 0 \\ 0 & 0 & 5 \end{pmatrix}.$$

(b.) T rotates each vector in $\mathbb{R}^2$ through the angle θ.

See Figure 4.2. The vector $(1,0)$ is moved to $(\cos\theta, \sin\theta)$, and $(0,1)$ is moved to $(-\sin\theta, \cos\theta)$, so

$$T\begin{pmatrix} 1 \\ 0 \end{pmatrix} = \begin{pmatrix} \cos\theta \\ \sin\theta \end{pmatrix};\; T\begin{pmatrix} 0 \\ 1 \end{pmatrix} = \begin{pmatrix} -\sin\theta \\ \cos\theta \end{pmatrix}$$

so

$$A = \begin{pmatrix} \cos\theta & -\sin\theta \\ \sin\theta & \cos\theta \end{pmatrix}.$$

(c.) For the linear transformation

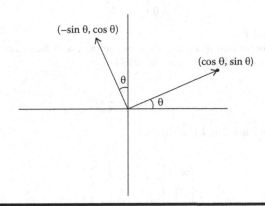

Figure 4.2   Rotation about the origin through angle θ.

$$T\begin{pmatrix} x \\ y \end{pmatrix} = \begin{pmatrix} x-y \\ 4x+3y \end{pmatrix} = x\begin{pmatrix} 1 \\ 4 \end{pmatrix} + y\begin{pmatrix} -1 \\ 3 \end{pmatrix}$$

We might (correctly) infer from part (a.) that

$$A = \begin{pmatrix} 1 & -1 \\ 4 & 3 \end{pmatrix}$$

in the standard basis.

Example

Let $T : \mathbb{R}^3 \to \mathbb{R}^2$ be defined by

$$T\begin{pmatrix} x \\ y \\ z \end{pmatrix} = \begin{pmatrix} 2x-6y+2z \\ 3y+z \end{pmatrix}.$$

We find a basis for the range of T and the null space of T. We have

$$T\begin{pmatrix} 1 \\ 0 \\ 0 \end{pmatrix} = \begin{pmatrix} 2 \\ 0 \end{pmatrix}, \quad T\begin{pmatrix} 0 \\ 1 \\ 0 \end{pmatrix} = \begin{pmatrix} -6 \\ 3 \end{pmatrix}, \quad T\begin{pmatrix} 0 \\ 0 \\ 1 \end{pmatrix} = \begin{pmatrix} 2 \\ 1 \end{pmatrix}$$

so a spanning set for the range of T is

$$\left\{ \begin{pmatrix} 2 \\ 0 \end{pmatrix}, \begin{pmatrix} -6 \\ 3 \end{pmatrix}, \begin{pmatrix} 2 \\ 1 \end{pmatrix} \right\}.$$

This cannot be a basis for the range of T (there are three vectors that have two components) so we use the methods in Chapter 3 to find a maximal linearly independent subset. With row reduction, we obtain

$$\begin{pmatrix} 2 & -6 & 2 \\ 0 & 3 & 1 \end{pmatrix} \to \begin{pmatrix} 1 & 0 & 2 \\ 0 & 1 & 1/3 \end{pmatrix}$$

and so

$$\left\{ \begin{pmatrix} 2 \\ 0 \end{pmatrix}, \begin{pmatrix} -6 \\ 3 \end{pmatrix} \right\}$$

is a basis for the range of T.

To obtain a basis for the null space of T, we note that the row reduced form of the matrix gives that x and y are leading variables and z is a free variable and

$$x = -2z, y = -\frac{1}{3}z.$$

Thus, a vector in the null space of T is of the form

$$\begin{pmatrix} x \\ y \\ z \end{pmatrix} = \begin{pmatrix} -2z \\ -\frac{1}{3}z \\ z \end{pmatrix} = z\begin{pmatrix} -2 \\ -\frac{1}{3} \\ 1 \end{pmatrix}$$

so

$$\left\{ \begin{pmatrix} -2 \\ -\dfrac{1}{3} \\ 1 \end{pmatrix} \right\}$$

is a basis for the null space of T.

Theorem 4.7:

Let

$$T : V \to V$$

be a linear operator. Then

(a.) T is invertible if and only if the matrix of T with respect to any basis is invertible.
(b.) If T is invertible, and A is the matrix representation of T with respect to a particular basis, then the matrix representation of T^{-1} with respect to the same basis is A^{-1}.

The proof is left to the exercises.
    The next theorem was given in a different context in Chapter 3.

Theorem 4.8:

If V is a finite dimensional vector space and $T : V \to W$ is a linear transformation, then

$$\dim \mathcal{R}(T) + \dim \mathcal{N}(T) = \dim(V).$$

Exercises

In Exercises 1–4 give the matrix associated with the linear transformation.

1. $T \begin{pmatrix} x \\ y \end{pmatrix} = \begin{pmatrix} 4x - 3y \\ x \\ 5y \end{pmatrix}$

2. $T \begin{pmatrix} x \\ y \\ z \end{pmatrix} = \begin{pmatrix} x - y + 2z \\ 7x + 5z \end{pmatrix}$

3. $T(x) = (3x)$

4. $T\begin{pmatrix} z \\ y \\ z \end{pmatrix} = \begin{pmatrix} x - 2y + 9z \\ y + 4z \\ x - 5y \\ 0 \end{pmatrix}$

In Exercises 5–8 give the linear transformation associated with the matrix.

5. $\begin{pmatrix} 1 & -3 \\ 7 & 0 \\ 5 & 9 \end{pmatrix}$

6. $\begin{pmatrix} 2 & 0 & -3 & 1 \\ 5 & -2 & 4 & 0 \end{pmatrix}$

7. (2)

8. $\begin{pmatrix} 5 \\ 0 \\ 0 \\ 6 \\ 2 \end{pmatrix}$

9. Suppose $T : \mathbb{R}^2 \to \mathbb{R}^3$ is a linear transformation with

$$T\begin{pmatrix} 1 \\ -3 \end{pmatrix} = \begin{pmatrix} 3 \\ 7 \\ 2 \end{pmatrix} \text{ and } T\begin{pmatrix} -2 \\ 2 \end{pmatrix} = \begin{pmatrix} -1 \\ -4 \\ 0 \end{pmatrix}.$$

(a.) Find $T\begin{pmatrix} 1 \\ 0 \end{pmatrix}$ and $T\begin{pmatrix} 0 \\ 1 \end{pmatrix}$

(b.) Find $T\begin{pmatrix} 5 \\ -6 \end{pmatrix}$

(c.) Find $T\begin{pmatrix} x \\ y \end{pmatrix}$

10. Suppose that $T_1 : U \to V$ and $T_2 : U \to V$ are linear transformations. Show $T_1 + T_2$ is a linear transformation.

11. The linear transformation T that reflects a vector about the line $y = mx$ is represented in the standard basis by the matrix

$$\begin{pmatrix} \dfrac{5}{13} & \dfrac{12}{13} \\ \dfrac{12}{13} & -\dfrac{5}{13} \end{pmatrix}.$$

In this problem, we want to find m.

(a.) What are

$$T\begin{pmatrix}1\\0\end{pmatrix} \text{ and } T\begin{pmatrix}0\\1\end{pmatrix}?$$

(b.) What is the slope of the line between $\begin{pmatrix}1\\0\end{pmatrix}$ and $T\begin{pmatrix}1\\0\end{pmatrix}?$

(c.) Use the answer to part (b.) to find m.

12. Find the value of m if the linear transformation T that reflects a vector about the line $y = mx$ is represented in the standard basis by the matrix

$$\begin{pmatrix} \dfrac{3}{5} & \dfrac{2}{5} \\[2mm] \dfrac{2}{5} & -\dfrac{3}{5} \end{pmatrix}.$$

13. Let $T : \mathbb{R}^3 \to \mathbb{R}^2$ be the linear transformation that reflects a vector about the line $y = x$.
    (a). Find the matrix of T with respect to the standard bases.
    (b). Find a basis for the null space (kernel) of T.
    (c). Find a basis for the range of T.

14. Let $T : \mathbb{R}^3 \to \mathbb{R}^2$ be defined by

$$T\begin{pmatrix}x\\y\\z\end{pmatrix} = \begin{pmatrix} 3x + 2y - 4z \\ x - 5z \end{pmatrix}.$$

(a.) Find the matrix of T with respect to the standard basis.
(b.) Find a basis for the null space (kernel) of T.
(c.) Find a basis for the range of T.

15. Let $T : P_n(\mathbb{R}) \to P_{n-1}(\mathbb{R})$ be defined by

$$T(f(x)) = \frac{d}{dx}(f(x)).$$

The standard basis for $P_k(\mathbb{R})$ is $\{1, x, x^2, \ldots, x^k\}$.

(a.) Find the matrix of T with respect to the standard bases.
(b.) Find a basis for the null space (kernel) of T.
(c.) Find a basis for the range of T.

16. Give an example to demonstrate the following statement is false.

   If $T : \mathbb{R}^n \to \mathbb{R}^m$ is a linear transformation, and $\{\hat{v}_1, \ldots, \hat{v}_n\}$ is a basis for $\mathbb{R}^n$, then $\{T(\hat{v}_1), \ldots, T(\hat{v}_n)\}$ is a basis for the image of T.

   (d.) Add a hypothesis that will make the above statement true.

17. Let $T : P_2(\mathbb{R}) \to P_2(\mathbb{R})$ be defined by

$$T\left(a + bx + cx^2\right) = b + \left(a + 2c\right)x + ax^2.$$

(a.) Find the matrix representation of T with respect to the basis {1, x, x^2}.
(b.) Find a basis for the null space of T.
(c.) Find a basis for the image of T.

18. Let $T : P_2\left(\mathbb{R}\right) \to P_2\left(\mathbb{R}\right)$ be defined by

$$T\left(a + bx + cx^2\right) = \left(a - 2b\right) + \left(3a + c\right)x + \left(a + b\right)x^2.$$

(a.) Find the matrix representation of T with respect to the basis {1, x, x^2}.
(b.) Find a basis for the null space of T.
(c.) Find a basis for the image of T.

19. Let V be the vector space spanned by $\left\{e^{3x}\sin x, e^{3x}\cos x\right\}$
    Let $T : V \to V$ be defined by

$$T\left(f\left(x\right)\right) = 2f'\left(x\right) + f\left(x\right).$$

(a.) Show that T is a linear transformation.
(b.) Give the matrix representation of V with respect to the given basis.
(c.) Solve the equation

$$2y' + y = e^{3x}\sin x.$$

20. Let T:VV be a linear operator. Prove
   (a.) T is invertible if and only if the matrix of T with respect to any basis is invertible.
   (b.) If T is invertible and A is the matrix representation of T with respect to a particular basis, then the matrix representation of T^{-1} with respect to the same basis is A^{-1}.

Historical Note

Hermann Grassmann (born 15 April 1809, died 26 September 1877)

Beginning in 1827, Grassmann studied theology at the University of Berlin, also taking classes in classical languages, philosophy, and literature. He does not appear to have taken courses in mathematics or physics.

Although lacking university training in mathematics, it was the field that most interested him after completing his studies in Berlin. Around this time (1834), he made his first significant mathematical discoveries, which led him to the important ideas he set out in his 1844 paper *The Ausdehnungslehre*.

In 1844, Grassmann published his masterpiece, *Lineale Ausdehnungslehre*, commonly referred to as *The Ausdehnungslehre*, which translates as "theory of extension" or "theory of extensive magnitudes". In this work, which is considered a masterpiece of originality, he developed the idea of an algebra in which the symbols representing geometric entities such as points, lines, and planes, are manipulated using certain rules.

Since *The Ausdehnungslehre* proposed a new foundation for all of mathematics, the work began with general definitions of a philosophical nature. Grassmann then showed that once geometry is put into the algebraic form he advocated, the number three has no privileged role as the number of spatial dimensions; the number of possible dimensions is in fact unbounded.

The Ausdehnungslehre was a revolutionary text, too far ahead of its time to be appreciated. When Grassmann submitted it to apply for a professorship in 1847, the ministry asked Ernst Kummer for a report. Kummer noted that there were good ideas in it, but found the exposition deficient and advised against giving Grassmann a university position. Over the next ten or so years, Grassmann wrote a variety of work applying his theory of extension and several papers on algebraic curves and surfaces, in the hope that these applications would lead others to take his theory seriously.

Fearnley-Sander writes: "

All mathematicians stand, as Newton said he did, on the shoulders of giants, but few have come closer than Hermann Grassmann to creating, single-handedly, a new subject".

The *Ausdehnungslehre* introduces two revolutionary concepts. First is the notion of a general n-dimensional vector space. Grassmann introduces the notion of linearly dependent vectors and develops the "elementary" theory of finite-dimensional vector spaces, as can be found in all of today's books on linear algebra. The second idea is the introduction of multilinear algebra, which involves the product of vectors or "multivectors". This leads to a complete theory of the "exterior algebra", an area of mathematics whose importance could not have been anticipated at that time.

Section 4.3 Finding the Representation of a Linear Operator with Respect to Different Bases

Suppose we know the representation of a linear transformation in the usual basis, and are given a second basis for the input vector space, and a third basis as a basis for the output vector space. We want to find the output vector of the linear transformation in the third basis if the input vector is given in the second basis.

Suppose

$$T : U \to V$$

is a linear transformation and we have a basis $\mathcal{B}_1 = \left\{ \hat{u}_1, \ldots, \hat{u}_m \right\}$ for U and a basis $\mathcal{B}_2 = \left\{ \hat{v}_1, \ldots, \hat{v}_n \right\}$ for V. Let T denote the matrix representation of T with respect to the standard basis. To find the matrix representation of T with respect to the bases $\mathcal{B}_1$ and $\mathcal{B}_2$, that is, where the input vector is expressed in the basis $\mathcal{B}_1$ and the output vector is expressed in the basis $\mathcal{B}_2$, we execute the following steps:

1. Map $\left[\hat{u} \right]_{\mathcal{B}_1} \to \hat{u}$ using $P_{\mathcal{B}_1} \left[\hat{u} \right]_{\mathcal{B}_1} = \hat{u}$ where $P_{\mathcal{B}_1}$ is the transition matrix defined in Chapter 3.

2. $\hat{u} \to T\hat{u}$ Map

3. Map $T\hat{u} \to \left[T\hat{u} \right]_{\mathcal{B}_2}$ using $\left[T\hat{u} \right]_{\mathcal{B}_2} = P_{\mathcal{B}_2}^{-1} \left(T\hat{u} \right)$

4. If we let $\left[T \right]_{\mathcal{B}_1}^{\mathcal{B}_2}$ denote the matrix representation of the linear transformation T where the input vector is represented in the basis $\mathcal{B}_1$ and the output vector is represented in the basis $\mathcal{B}_2$, we have

$$\left[T \right]_{\mathcal{B}_1}^{\mathcal{B}_2} = P_{\mathcal{B}_2}^{-1} T P_{\mathcal{B}_1}$$

Example

Let $T : \mathbb{R}^2 \to \mathbb{R}^3$ be defined by

$$T\begin{pmatrix} x \\ y \end{pmatrix} = \begin{pmatrix} x + 2y \\ 3x - 4y \\ y \end{pmatrix}$$

and suppose the bases for $\mathbb{R}^2$ and $\mathbb{R}^3$ are

$$\mathcal{B}_1 = \left\{ \begin{pmatrix} 1 \\ 2 \end{pmatrix}, \begin{pmatrix} 3 \\ 0 \end{pmatrix} \right\} \text{ and } \mathcal{B}_2 = \left\{ \begin{pmatrix} 1 \\ 1 \\ 0 \end{pmatrix}, \begin{pmatrix} 0 \\ 1 \\ 1 \end{pmatrix}, \begin{pmatrix} 1 \\ 0 \\ 1 \end{pmatrix} \right\}$$

respectively.

Here

$$P_{\mathcal{B}_1} = \begin{pmatrix} 1 & 3 \\ 2 & 0 \end{pmatrix} \quad P_{\mathcal{B}_2} = \begin{pmatrix} 1 & 0 & 1 \\ 1 & 1 & 0 \\ 0 & 1 & 1 \end{pmatrix}$$

and the matrix for T in the standard basis is

$$\begin{pmatrix} 1 & 2 \\ 3 & -4 \\ 0 & 1 \end{pmatrix}.$$

The matrix representation for the given bases is

$$\left(P_{\mathcal{B}_2} \right)^{-1} TP_{\mathcal{B}_1} = \begin{pmatrix} 1 & 0 & 1 \\ 1 & 1 & 0 \\ 0 & 1 & 1 \end{pmatrix}^{-1} \begin{pmatrix} 1 & 2 \\ 3 & -4 \\ 0 & 1 \end{pmatrix} \begin{pmatrix} 1 & 3 \\ 2 & 0 \end{pmatrix} = \begin{pmatrix} -1 & 6 \\ -4 & 3 \\ 6 & -3 \end{pmatrix}$$

Suppose the coordinates of $\hat{u}$ in the $\mathcal{B}_1$ basis are

$$[\hat{u}]_{\mathcal{B}_1} = \begin{pmatrix} 5 \\ 7 \end{pmatrix}$$

and we want to find the output in the $\mathcal{B}_2$ basis. We have

$$[T\hat{u}]_{\mathcal{B}_2} = \left(\left(P_{\mathcal{B}_2} \right)^{-1} TP_{\mathcal{B}_1} \right)[\hat{u}]_{\mathcal{B}_1} = [T]_{\mathcal{B}_1}^{\mathcal{B}_2}[\hat{u}]_{\mathcal{B}_1} = \begin{pmatrix} -1 & 6 \\ -4 & 3 \\ 6 & -3 \end{pmatrix}\begin{pmatrix} 5 \\ 7 \end{pmatrix} = \begin{pmatrix} 37 \\ 1 \\ 9 \end{pmatrix}.$$

Exercises

1. The set

$$\mathcal{B} = \left\{ \begin{pmatrix} 2 \\ 1 \end{pmatrix}, \begin{pmatrix} 3 \\ 5 \end{pmatrix} \right\}$$

is a basis for $\mathbb{R}^2$. Suppose T is a linear transformation with

$$T\begin{pmatrix}1\\0\end{pmatrix}=\begin{pmatrix}4\\0\end{pmatrix}, \ T\begin{pmatrix}0\\1\end{pmatrix}=\begin{pmatrix}6\\-2\end{pmatrix}.$$

(a.) Find the matrix of T with respect to the basis $\mathcal{B}$.

(b.) Find a basis for the null space (kernel) of T with respect to the basis $\mathcal{B}$.

(c.) Find a basis for the range of T with respect to the basis $\mathcal{B}$.

2. For $T : \mathbb{R}^3 \rightarrow \mathbb{R}^3$ given by

$$T\begin{pmatrix}x\\y\\z\end{pmatrix}=\begin{pmatrix}3x-4z\\4x\\x+y+z\end{pmatrix}$$

find the matrix representation of T with respect to the basis for the domain

$$\mathcal{B}_1 = \left\{ \begin{pmatrix}1\\0\\0\end{pmatrix}, \begin{pmatrix}-1\\1\\2\end{pmatrix}, \begin{pmatrix}3\\2\\0\end{pmatrix} \right\}$$

and the basis for the range space

$$\mathcal{B}_2 = \left\{ \begin{pmatrix}1\\1\\0\end{pmatrix}, \begin{pmatrix}0\\1\\0\end{pmatrix}, \begin{pmatrix}0\\1\\1\end{pmatrix} \right\}$$

3. For $T : \mathbb{R}^3 \rightarrow \mathbb{R}^3$ given by

$$T\begin{pmatrix}x\\y\\z\end{pmatrix}=\begin{pmatrix}y-3z\\x-4y\\x+2y-z\end{pmatrix}$$

find the matrix representation of T with respect to

$$\mathcal{B}_1 = \left\{ \begin{pmatrix}1\\0\\1\end{pmatrix}, \begin{pmatrix}1\\1\\2\end{pmatrix}, \begin{pmatrix}0\\2\\1\end{pmatrix} \right\}$$

the basis for the domain and

$$\mathcal{B}_2 = \left\{ \begin{pmatrix}1\\1\\0\end{pmatrix}, \begin{pmatrix}0\\1\\0\end{pmatrix}, \begin{pmatrix}0\\1\\1\end{pmatrix} \right\}$$

the basis for the range space.

4. For $T : \mathbb{R}^3 \to \mathbb{R}^3$ given by

$$T\begin{pmatrix} x \\ y \\ z \end{pmatrix} = \begin{pmatrix} 3x - 4z \\ 4x \\ x + y + z \end{pmatrix}$$

find the matrix representation of T with respect to

$$\mathcal{B}_1 = \left\{ \begin{pmatrix} 1 \\ 0 \\ 1 \end{pmatrix}, \begin{pmatrix} -1 \\ 1 \\ 0 \end{pmatrix}, \begin{pmatrix} 1 \\ 2 \\ 0 \end{pmatrix} \right\}$$

the basis for the domain and

$$\mathcal{B}_2 = \left\{ \begin{pmatrix} 1 \\ 1 \\ 0 \end{pmatrix}, \begin{pmatrix} 0 \\ 1 \\ 0 \end{pmatrix}, \begin{pmatrix} 0 \\ 1 \\ 1 \end{pmatrix} \right\}$$

the basis for the range space.

Section 4.4 Composition (Multiplication) of Linear Transformations

The definition of matrix multiplication we gave earlier probably made little sense intuitively. Why was it so constructed? We will give a partial answer now.

Earlier, we showed there is an algebra connected with linear transformations. Two elements of $L(U,V)$ can be added to give an element of $L(U,V)$. That is, $L(U,V)$ is closed under addition. Also if a is a scalar and $T \in L(U,V)$ then $aT \in L(U,V)$.

If we take the matrix representation of elements of $L(U,V)$, then the method of matrix addition gives the matrix of the sum of the corresponding linear transformations. Similar ideas hold for scalar multiples of linear transformations.

There is not a "multiplication" of linear transformations. Instead, linear transformations can be combined by function composition provided the domains and ranges are suitable. If $T \in L(V)$ then $T \circ T \in L(V)$ and we will use the notation

$$T \circ T = T^2.$$

Similarly, for any positive integer n,

$$T^n = T \circ \cdots \circ T \quad (n \text{ factors}).$$

In the next few paragraphs we will represent a linear transformation T that has the matrix A as its representation by T_A.

If T_A and $T_B \in L(V)$, then $T_B \circ T_A \in L(V)$ and if $T_A \in L(U,V)$ and $T_B \in L(V,W)$ then $T_B \circ T_A \in L(U,W)$.

We would like to define matrix multiplication so that if A is the matrix associated with T_A and B is the matrix associated with T_B, then BA is the matrix associated with $T_B \circ T_A$.

The following example gives some insight into why the definition we gave for matrix multiplication is right for this purpose.

Example

Suppose that U, V and W are vector spaces with bases $\{\hat{u}_1, \hat{u}_2, \hat{u}_3\}$, $\{\hat{v}_1, \hat{v}_2, \hat{v}_3\}$ and $\{\hat{w}_1, \hat{w}_2, \hat{w}_3\}$, respectively. Further suppose that

$$T_A : U \to V$$

with

$$T_A(\hat{u}_1) = a_{11}\hat{v}_1 + a_{21}\hat{v}_2 + a_{31}\hat{v}_3$$

$$T_A(\hat{u}_2) = a_{12}\hat{v}_1 + a_{22}\hat{v}_2 + a_{32}\hat{v}_3$$

$$T_A(\hat{u}_3) = a_{13}\hat{v}_1 + a_{23}\hat{v}_2 + a_{33}\hat{v}_3$$

so that the matrix of T_A with respect to the given bases is

$$A = \begin{pmatrix} a_{11} & a_{12} & a_{13} \\ a_{21} & a_{22} & a_{23} \\ a_{31} & a_{32} & a_{33} \end{pmatrix}.$$

Also, suppose that

$$T_B : V \to W$$

with

$$T_B(\hat{v}_1) = b_{11}\hat{w}_1 + b_{21}\hat{w}_2 + b_{31}\hat{w}_3$$

$$T_B(\hat{v}_2) = b_{12}\hat{w}_1 + b_{22}\hat{w}_2 + b_{32}\hat{w}_3$$

$$T_B(\hat{v}_3) = b_{13}\hat{w}_1 + b_{23}\hat{w}_2 + b_{33}\hat{w}_3$$

so that the matrix of T_B with respect to the given bases is

$$B = \begin{pmatrix} b_{11} & b_{12} & b_{13} \\ b_{21} & b_{22} & b_{23} \\ b_{31} & b_{32} & b_{33} \end{pmatrix}.$$

We want to find the matrix associated with $T_B \circ T_A$.

Now

$$
\begin{aligned}
u_1 &\to a_{11}\hat{v}_1 + a_{21}\hat{v}_2 + a_{31}\hat{v}_3 \\
&\to a_{11}\left(b_{11}\hat{w}_1 + b_{21}\hat{w}_2 + b_{31}\hat{w}_3\right) + a_{21}\left(b_{12}\hat{w}_1 + b_{22}\hat{w}_2 + b_{32}\hat{w}_3\right) \\
&\quad + a_{31}\left(b_{13}\hat{w}_1 + b_{23}\hat{w}_2 + b_{33}\hat{w}_3\right) \\
&= \hat{w}_1\left(a_{11}b_{11} + a_{21}b_{12} + a_{31}b_{13}\right) + \hat{w}_2\left(a_{11}b_{21} + a_{21}b_{22} + a_{31}b_{23}\right) \\
&\quad + \hat{w}_3\left(a_{11}b_{31} + a_{21}b_{32} + a_{31}b_{33}\right).
\end{aligned}
\tag{2}
$$

We rewrite the last terms in Equation (2) listing the b_{ij} factors first. This gives

$$\hat{w}_1\left(b_{11}a_{11}+b_{12}a_{21}+b_{13}a_{31}\right)+\hat{w}_2\left(b_{21}a_{11}+b_{22}a_{21}+b_{23}a_{31}\right)$$
$$+\hat{w}_3\left(b_{31}a_{11}+b_{32}a_{21}+b_{33}a_{31}\right).$$

Thus, the first column of the matrix associated with $T_B \circ T_A$ is

$$\begin{pmatrix} b_{11}a_{11}+b_{12}a_{21}+b_{13}a_{31} \\ b_{21}a_{11}+b_{22}a_{21}+b_{23}a_{31} \\ b_{31}a_{11}+b_{32}a_{21}+b_{33}a_{31} \end{pmatrix} = \begin{pmatrix} \displaystyle\sum_{k=1}^{3} b_{1k}a_{k1} \\ \displaystyle\sum_{k=1}^{3} b_{2k}a_{k1} \\ \displaystyle\sum_{k=1}^{3} b_{3k}a_{k1} \end{pmatrix},$$

and the other columns are expressed in the analogous manner.

If we want to define the matrix product BA so that BA will be the matrix associated with $T_B \circ T_A$, the computations above suggest that

$$(BA)_{ij} = \sum_{k=1}^{3} b_{ik}a_{kj}$$

which is the formula we gave for matrix multiplication.

Exercises

1. Suppose

$$A = \begin{pmatrix} 1 & -1 \\ 2 & 3 \end{pmatrix} \text{ and } B = \begin{pmatrix} 0 & 2 \\ -4 & 5 \end{pmatrix}$$

Let

$$T : \mathbb{R}^2 \to \mathbb{R}^2 \text{ and } S : \mathbb{R}^2 \to \mathbb{R}^2$$

be defined by

$$T\begin{pmatrix} x \\ y \end{pmatrix} = A\begin{pmatrix} x \\ y \end{pmatrix} \text{ and } S\begin{pmatrix} x \\ y \end{pmatrix} = B\begin{pmatrix} x \\ y \end{pmatrix}.$$

Compute

$$T\begin{pmatrix} 1 \\ 0 \end{pmatrix}, \ T\begin{pmatrix} 0 \\ 1 \end{pmatrix}, \ S\left(T\begin{pmatrix} 1 \\ 0 \end{pmatrix}\right), \text{ and } S\left(T\begin{pmatrix} 0 \\ 1 \end{pmatrix}\right)$$

(a.) Compute BA

(b.) Compute $BA\begin{pmatrix} 1 \\ 0 \end{pmatrix}$ and $BA\begin{pmatrix} 0 \\ 1 \end{pmatrix}$.

2. Let $T: P_2(\mathbb{R}) \to P_2(\mathbb{R})$ be defined by $T(p(x)) = p(x-1)$ and $S: P_2(\mathbb{R}) \to P_2(\mathbb{R})$ be defined by $S(p(x)) = p(x+1)$.

Find $(S \circ T)$ and $(T \circ S)$.

(a.) The standard basis for $P_2(\mathbb{R})$ is $\{1, x, x^2\}$. Find the matrices of S and T with respect to the standard bases.

(b.) Let A and B be the matrices for S and T, respectively.

Find AB and BA.

3. Let $S: \mathbb{R}^2 \to \mathbb{R}^2$ and $T: \mathbb{R}^2 \to \mathbb{R}^3$ be defined by

$$S\begin{pmatrix} a \\ b \end{pmatrix} = \begin{pmatrix} a+b \\ a-b \end{pmatrix}; \; T\begin{pmatrix} a \\ b \end{pmatrix} = \begin{pmatrix} 2b \\ a+4b \\ -3a \end{pmatrix}.$$

(a.) Find $(T \circ S)\begin{pmatrix} a \\ b \end{pmatrix}$

(b.) Find the matrix representation of S, T, and $T \circ S$ with respect to the standard bases.

(c.) Compare the product of the matrix representation of T and the matrix representation of S with the matrix representation $T \circ S$.

4. Show that if $T_1: U \to V$ and $T_2: V \to W$ are linear transformations then $(T_2 \circ T_1): U \to W$ is a linear transformation.

5. Recall that the matrix representation of the linear transformation that rotates a vector counterclockwise through an angle θ in $\mathbb{R}^2$ with respect to the standard basis is

$$\begin{pmatrix} \cos\theta & -\sin\theta \\ \sin\theta & \cos\theta \end{pmatrix}.$$

(a.) Find the matrices of the linear transformations so that the first transformation rotates a vector counterclockwise through an angle θ and the second transformation rotates a vector counterclockwise through an angle α with respect to the standard basis.

(b.) Should the product of these matrices commute?

(c.) Use the product of these matrices to find a formula for $\sin(\theta + \alpha)$ and $\cos(\theta + \alpha)$.

(d.) Show that $T(-\theta) = [T(\theta)]^{-1}$

(e.) Let

$$A(\theta) = \begin{pmatrix} \cos\theta & -\sin\theta \\ \sin\theta & \cos\theta \end{pmatrix}.$$

Give a geometric argument for why

$$A(2\theta) = \left[A(\theta)\right]^2$$

and use this to derive formulas for sin (2θ) and cos (2θ).

6. Let V be the vector space of functions that have continuous derivatives of all orders, and let $In : V \to V$ and $D : V \to V$ be given by

$$In(f) = \int_0^x f(t)\,dt; \quad D(f) = f'(x).$$

Find $(D \circ In)(f)$ and $(In \circ D)(f)$ for

(a.) $f(x) = 5x^3 + 3x^2$

(b.) $f(x) = 3x + 4$

(c.) $f(x) = e^x$

Chapter 5

Eigenvalues and Eigenvectors

Section 5.1 Determining Eigenvalues and Eigenvectors

Suppose we have a finite dimensional vector space V and a linear operator

$$T : V \rightarrow V.$$

If V has dimension n, then T can be expressed as multiplication by an $n \times n$ matrix, but the matrix will depend on the basis for V. If the scalar field is $\mathbb{C}$, there will be certain "distinguished" vectors, which we call eigenvectors. If the scalar field is $\mathbb{R}$, this is a possibility, but not a certainty.

We begin our investigation where the scalar field is $\mathbb{R}$ and the vector space is $\mathbb{R}^n$.

Definition:

Let V be the vector space $\mathbb{R}^n$ and let A be an $n \times n$ matrix. A non-zero vector $\hat{v} \in V$ is an eigenvector of A with eigenvalue λ if

$$A\hat{v} = \lambda\hat{v}.$$

Note that $\hat{0}$ cannot be an eigenvector, but 0 can be an eigenvalue. Also, if zero is an eigenvalue for A then A is not a one to one function. (Why?)

The effect of A on an eigenvector depends on the value of λ.

Value of λ	Effect of A on the eigenvector $\hat{v}$
$0 < \lambda < 1$	Direction of $\hat{v}$ is unchanged. Length of $\hat{v}$ is decreased
$\lambda = 1$	$\hat{v}$ is unchanged
$1 < \lambda$	Direction of $\hat{v}$ is unchanged. Length of $\hat{v}$ is increased

Value of λ	Effect of A on the eigenvector $\hat{v}$
$\lambda = 0$	$\hat{v}$ becomes the zero vector
$-1 < \lambda < 0$	Direction of $\hat{v}$ is reversed. Length of $\hat{v}$ is decreased
$\lambda < -1$	Direction of $\hat{v}$ is reversed. Length of $\hat{v}$ is increased

Our first task is, given the matrix A, find the eigenvalues and eigenvectors of A.
Consider

$$A\hat{v} = \lambda\hat{v}, \quad \hat{v} \neq \hat{0}.$$

Then

$$A\hat{v} - \lambda\hat{v} = \hat{0}. \tag{1}$$

We replace $\lambda\hat{v}$ by $\lambda I \hat{v}$ in Equation (1) to get

$$A\hat{v} - \lambda I \hat{v} = \hat{0}.$$

This is done so that factoring out the $\hat{v}$ will be legitimate. The expression $(A - \lambda I)$ makes sense, but $(A - \lambda)$ does not.

Now

$$A\hat{v} - \lambda I \hat{v} = (A - \lambda I)\hat{v} = \hat{0}$$

so we have the non-zero vector $\hat{v}$ in the null space of $(A - \lambda I)$. This occurs if and only if $(A - \lambda I)$ is not invertible, which occurs if and only if the determinant of $(A - \lambda I)$ is 0. Thus we have the following result.

Theorem 5.1:

The eigenvalues of the matrix A are the values of λ for which

$$\det(A - \lambda I) = 0.$$

Example

Find the eigenvalues for

$$A = \begin{pmatrix} 2 & 1 \\ 0 & -1 \end{pmatrix}.$$

We have

$$A - \lambda I = \begin{pmatrix} 2-\lambda & 1 \\ 0 & -1-\lambda \end{pmatrix}$$

$$\det(A - \lambda I) = (2-\lambda)(-1-\lambda) - 0 = \lambda^2 - \lambda - 2.$$

The eigenvalues of this particular matrix are the values of λ for which

$$\lambda^2 - \lambda - 2 = (\lambda + 1)(\lambda - 2) = 0$$

which are $\lambda = 2$ and $\lambda = -1$.

The expression $\det(A - \lambda I)$, when expanded, is called the characteristic polynomial of A, which is denoted $P_A(\lambda)$. If A is an $n \times n$ matrix, then the characteristic polynomial of A will have degree n. Expanding $\det(A - \lambda I)$ and finding the values for which $\det(A - \lambda I) = 0$ (that is, finding the eigenvalues) can be computationally challenging and should be done using a computer in most cases. In the real numbers, the characteristic polynomial may not factor completely, but in the complex numbers, we have the following result.

Theorem 5.2:

In the complex numbers, the characteristic polynomial factors into linear factors and

$$P_A(\lambda) = (\lambda - \lambda_1)(\lambda - \lambda_2)\cdots(\lambda - \lambda_n)$$

where λ_i are eigenvalues repeated according to their multiplicity.

Example

The matrix that rotates a vector through an angle θ in two dimensions is

$$A(\theta) = \begin{pmatrix} \cos\theta & -\sin\theta \\ \sin\theta & \cos\theta \end{pmatrix}.$$

If $\theta = 90°$, then there will not be an eigenvector. (What is the geometric reason there is not?)
    To see what happens in this case note that

$$A(90°) = \begin{pmatrix} 0 & -1 \\ 1 & 0 \end{pmatrix}$$

so

$$A(90°) - \lambda I = \begin{pmatrix} -\lambda & -1 \\ 1 & -\lambda \end{pmatrix}$$

and

$$\det\left[A(90°) - \lambda I\right] = \lambda^2 + 1.$$

Thus, no eigenvalues of $A(90°)$ exist in the real numbers.

This example shows an important difference between the cases when the scalar field is $\mathbb{R}$ and when the scalar field is $\mathbb{C}$. If the scalar field is $\mathbb{C}$, the characteristic polynomial will always factor into linear factors.

Finding the Eigenvectors after the Eigenvalues Have Been Found

Suppose that we have found the eigenvalues for a matrix A. The next task is to find the eigenvectors associated with each eigenvalue. Suppose that λ_1 is an eigenvalue for A and we want to find the associated eigenvectors. That is, we want to find the non-zero vectors $\hat{v}$ for which

$$A\hat{v} = \lambda_1\hat{v} \text{ or } \left(A - \lambda_1 I\right)\hat{v} = \hat{0}.$$

This is exactly the null space of $\left(A - \lambda_1 I\right)$ *except for the zero vector.*

Definition:

If A is an $n \times n$ matrix with eigenvalue λ, the eigenspace of λ is the null space of $A - \lambda I$.

Example

We recall how we found the null space of a matrix B.
Suppose

$$B = \begin{pmatrix} 1 & -1 & 3 \\ -10 & 8 & 8 \\ 14 & -12 & 4 \end{pmatrix}$$

and we want to solve $B\hat{x} = \hat{0}$.

When B is row reduced, the result is

$$\begin{pmatrix} 1 & 0 & -16 \\ 0 & 1 & -19 \\ 0 & 0 & 0 \end{pmatrix}.$$

If we denote the variables as x_1, x_2, x_3, then we have the equations

$$x_1 - 16x_3 = 0, \text{ or } x_1 = 16x_3$$

$$x_2 - 19x_3 = 0, \text{ or } x_2 = 19x_3.$$

So x_3 is the free variable and if $x_3 = t$, then $x_1 = 16t$ and $x_2 = 19t$.
Thus a vector in the null space of B is

$$\begin{pmatrix} x_1 \\ x_2 \\ x_3 \end{pmatrix} = \begin{pmatrix} 16t \\ 19t \\ t \end{pmatrix} = t \begin{pmatrix} 16 \\ 19 \\ 1 \end{pmatrix}$$

and

$$\left\{ \begin{pmatrix} 16 \\ 19 \\ 1 \end{pmatrix} \right\}.$$

is a basis for the null space of B.

We apply this technique to find a basis for each eigenspace.

Example

$$A = \begin{pmatrix} 2 & 1 \\ 0 & -1 \end{pmatrix}.$$

To find a basis for the eigenspace for the eigenvalue $\lambda = 2$, we find

$$A - 2I = \begin{pmatrix} 2 & 1 \\ 0 & -1 \end{pmatrix} - 2 \begin{pmatrix} 1 & 0 \\ 0 & 1 \end{pmatrix} = \begin{pmatrix} 0 & 1 \\ 0 & -3 \end{pmatrix}.$$

When $A - 2I$ is row reduced, the result is

$$\begin{pmatrix} 0 & 1 \\ 0 & 0 \end{pmatrix}.$$

So $x_2 = 0$, and x_1 is the free variable. This gives the eigenvector

$$\begin{pmatrix} x_1 \\ 0 \end{pmatrix} = x_1 \begin{pmatrix} 1 \\ 0 \end{pmatrix}.$$

Putting this in standard form, we have the eigenvectors for $\lambda = 2$ are

$$t \begin{pmatrix} 1 \\ 0 \end{pmatrix}$$

and a basis for the eigenspace for $\lambda = 2$ is

$$\left\{ \begin{pmatrix} 1 \\ 0 \end{pmatrix} \right\}.$$

If we repeat this procedure for $\lambda = -1$, we get that a basis for the eigenspace for $\lambda = -1$ is

$$\left\{ \begin{pmatrix} 1 \\ -3 \end{pmatrix} \right\}.$$

Theorem 5.3:

A set of eigenvectors, each of which has a different eigenvalue from the others, is a linearly independent set.

Proof:

We give the proof in the case of three vectors, and leave the general case as an exercise.

Suppose that $\hat{v}_1, \hat{v}_2$, and $\hat{v}_3$ are eigenvectors for A with distinct eigenvalues $\lambda_1, \lambda_2, \lambda_3$ and suppose that

$$c_1\hat{v}_1 + c_2\hat{v}_2 + c_3\hat{v}_3 = \hat{0}. \tag{2}$$

We will show that each $c_i = 0$.

Multiply Equation (2) by $(A - \lambda_3 I)(A - \lambda_2 I)$. Note that

$$(A - \lambda_2 I)\hat{v}_1 = A\hat{v}_1 - \lambda_2\hat{v}_1 = \lambda_1\hat{v}_1 - \lambda_2\hat{v}_1 = (\lambda_1 - \lambda_2)\hat{v}_1$$

so

$$(A - \lambda_3 I)(A - \lambda_2 I)\hat{v}_1 = (A - \lambda_3 I)\left[(A - \lambda_2 I)\hat{v}_1\right] = (A - \lambda_3 I)(\lambda_1 - \lambda_2)\hat{v}_1$$
$$= (\lambda_1 - \lambda_2)(A - \lambda_3 I)\hat{v}_1 = (\lambda_1 - \lambda_2)(\lambda_1 - \lambda_3)\hat{v}_1.$$

Similarly,

$$(A - \lambda_2 I)\hat{v}_2 = A\hat{v}_2 - \lambda_2\hat{v}_2 = \lambda_2\hat{v}_2 - \lambda_2\hat{v}_2 = \hat{0}$$

so

$$(A - \lambda_3 I)(A - \lambda_2 I)\hat{v}_2 = (A - \lambda_3 I)\hat{0} = \hat{0}.$$

Finally,

$$(A - \lambda_2 I)\hat{v}_3 = A\hat{v}_3 - \lambda_2\hat{v}_3 = \lambda_3\hat{v}_3 - \lambda_2\hat{v}_3 = (\lambda_3 - \lambda_2)\hat{v}_3$$

so

$$(A - \lambda_3 I)(A - \lambda_2 I)\hat{v}_3 = (A - \lambda_3 I)(\lambda_3 - \lambda_2)\hat{v}_3 = (\lambda_3 - \lambda_2)\left[(A - \lambda_3 I)\hat{v}_3\right]$$
$$= (\lambda_3 - \lambda_2)(\lambda_3 - \lambda_3)\hat{v}_3 = \hat{0}.$$

Thus,

$$(A - \lambda_3 I)(A - \lambda_2 I)\left[c_1\hat{v}_1 + c_2\hat{v}_2 + c_3\hat{v}_3\right]$$
$$= c_1(\lambda_1 - \lambda_2)(\lambda_1 - \lambda_3)\hat{v}_1 + c_2\hat{0} + c_3\hat{0} = c_1(\lambda_1 - \lambda_2)(\lambda_1 - \lambda_3)\hat{v}_1 = \hat{0}.$$

So,

$$c_1(\lambda_1 - \lambda_2)(\lambda_1 - \lambda_3)\hat{v}_1 = \hat{0}$$

and since $(\lambda_1 - \lambda_2)(\lambda_1 - \lambda_3) \neq 0$ and $\hat{v}_1 \neq \hat{0}$, it must be that $c_1 = 0$.

Similarly, multiplying Equation (2) by $(A - \lambda_3 I)(A - \lambda_1 I)$ yields $c_2 = 0$ and multiplying Equation (2) by $(A - \lambda_2 I)(A - \lambda_1 I)$ yields $c_3 = 0$.

Observations:

1. If A is a matrix with eigenvector $\hat{v}$ whose eigenvalue is λ, then $A^n \hat{v} = \lambda^n \hat{v}$.
2. If A is a matrix with eigenvector $\hat{v}$ whose eigenvalue is λ, then

for any scalar α, $\alpha\hat{v}$ is also an eigenvector of A with eigenvalue λ.

This is because

$$A(\alpha\hat{v}) = \alpha(A\hat{v}) = \alpha(\lambda\hat{v}) = \lambda(\alpha\hat{v}),$$

It is important to realize that a single eigenvalue may have more than one linearly independent eigenvector since the null space of $(A - \lambda I)$ may have dimensions greater than one.

Example

Consider the two matrices

$$A = \begin{pmatrix} 2 & 0 \\ 0 & 2 \end{pmatrix} \text{ and } B = \begin{pmatrix} 2 & 1 \\ 0 & 2 \end{pmatrix}.$$

We first find the eigenvalues and eigenvectors for A.

The characteristic polynomial for A is $(2-\lambda)^2$ so $\lambda = 2$ is the only eigenvalue. Now

$$A - 2I = \begin{pmatrix} 0 & 0 \\ 0 & 0 \end{pmatrix}$$

So there are two free variables, x_1 and x_2 and there are two linearly independent eigenvectors

$$\begin{pmatrix} 1 \\ 0 \end{pmatrix} \text{ and } \begin{pmatrix} 0 \\ 1 \end{pmatrix}.$$

The characteristic polynomial for B is also $(2-\lambda)^2$ so $\lambda = 2$ is again the only eigenvalue. Now

$$B - 2I = \begin{pmatrix} 0 & 1 \\ 0 & 0 \end{pmatrix}$$

so $x_1 = 0$ and x_2 is free. Thus, there is only one linearly independent eigenvector

$$\begin{pmatrix} 0 \\ 1 \end{pmatrix}.$$

It is important to realize that even if the characteristic polynomials of different matrices are the same, their eigenspaces may be different.

Exercises

In Exercises 1–10 (i) find the characteristic polynomial; (ii) find the eigenvalues; (iii) find a basis for each eigenspace.

1. $\begin{pmatrix} 3 & 0 & 0 \\ 0 & 2 & 1 \\ -1 & 0 & 2 \end{pmatrix}$

2. $\begin{pmatrix} 1 & 0 & 0 \\ -2 & -1 & -2 \\ 3 & 6 & 6 \end{pmatrix}$

3. $\begin{pmatrix} 2 & 1 & 0 \\ 0 & 2 & 1 \\ 0 & 0 & 2 \end{pmatrix}$

4. $\begin{pmatrix} 3 & 1 & 0 \\ 0 & 3 & 0 \\ 0 & 0 & 3 \end{pmatrix}$

5. $\begin{pmatrix} 3 & 3 & -2 \\ 2 & -3 & 2 \\ -4 & 1 & 1 \end{pmatrix}$

6. $\begin{pmatrix} 1 & -3 & 3 \\ 3 & -5 & 3 \\ 6 & -6 & 6 \end{pmatrix}$

7. $\begin{pmatrix} 0 & 0 & -2 \\ 1 & 2 & 1 \\ 0 & 0 & 3 \end{pmatrix}$

8. $\begin{pmatrix} 12 & 0 & 0 \\ -16 & -4 & 0 \\ -16 & 0 & -4 \end{pmatrix}$

9. $\begin{pmatrix} 3 & 6 & 0 & 0 \\ 1 & -2 & 0 & 0 \\ 0 & 0 & 3 & -1 \\ 0 & 0 & 0 & 3 \end{pmatrix}$

10. $\begin{pmatrix} 1 & 1 & 0 & -1 \\ 0 & -3 & 2 & 6 \\ 0 & 0 & 0 & 5 \\ 0 & 0 & 0 & 4 \end{pmatrix}$

11. Show that if A is an invertible matrix and λ is an eigenvalue of A with eigenvector $\hat{v}$, then $\dfrac{1}{\lambda}$ is an eigenvalue of A^{-1} with eigenvector $\hat{v}$.

12. (a.) Show that if A is a square matrix, then A and A^T, have the same eigenvalues.
    (b.) Give an example of a 2×2 matrix A for which A and A^T have different eigenspaces.

13. (a.) Suppose that A and B are $n \times n$ matrices with α an eigenvalue of A and β an eigenvalue of B. Must $\alpha + \beta$ be an eigenvalue of $A + B$?
    (b.) Suppose that A and B are $n \times n$ matrices with $\hat{v}$ an eigenvalue of A and B. Must $\hat{v}$ be an eigenvalue of $A + B$?

14. Show that if α is an eigenvalue of A, then α^2 is an eigenvalue of A^2.

15. If A is a matrix for which A^k is the zero matrix for some positive integer k show that every eigenvalue of A must be 0.

16. Show that if A is a square matrix then A is invertible if and only if 0 is not an eigenvalue of A.

17. Find the characteristic polynomial of

$$\begin{pmatrix} 3 & 9 & 5 & 1 \\ 0 & -4 & 0 & 0 \\ 0 & 0 & 2 & 4 \\ 0 & 0 & 0 & 6 \end{pmatrix}$$

18. Give a geometric argument for finding two independent eigenvectors for the matrix that reflects a point about the line $y = mx$ in $\mathbb{R}^2$.

19. Without doing major calculations, find an eigenvector for the matrices below. Give the eigenvalue of the vector you selected.

(a.) $$\begin{pmatrix} 2 & 2 & 2 & 2 \\ 2 & 2 & 2 & 2 \\ 2 & 2 & 2 & 2 \\ 2 & 2 & 2 & 2 \end{pmatrix}$$

(b.) $$\begin{pmatrix} 1 & 2 & 3 & 4 \\ 4 & 2 & 3 & 1 \\ 2 & 1 & 4 & 3 \\ 3 & 1 & 2 & 4 \end{pmatrix}$$

(c.) A^6 where $A = \begin{pmatrix} 2 & 2 \\ 2 & 2 \end{pmatrix}$.

20. Show that for a 2×2 matrix

$$A = \begin{pmatrix} a & b \\ c & d \end{pmatrix}$$

the characteristic polynomial is

$$\lambda^2 - \lambda Tr(A) + \det(A)$$

where $Tr(A)$ is the trace of A.

This is a special case of the formula for the characteristic polynomial of an $n \times n$ matrix which is

$$\lambda^n - a_1 \lambda^{n-1} + a_2 \lambda^{n-2} - \cdots + (-1)^n a_n$$

where $a_1 = Tr(A)$ and $a_n = \det(A)$.

21. If the characteristic polynomial of a matrix A is $(\lambda - 1)^2 (\lambda - 3)$
    - (a.) What is the size of the matrix A?
    - (b.) What is det (A)?
    - (c.) What is $Tr(A)$?
    - (d.) Is A invertible?
22. Show that if λ is an eigenvalue of A with eigenvector $\hat{v}$, then for any number c, $(\lambda + c)$ is an eigenvalue of $A + cI$ with eigenvector $\hat{v}$.

Section 5.2 Diagonalizing a Matrix

If we have a basis for $\mathcal{F}^n$ that consists of eigenvectors of an $n \times n$ matrix A, then the representation of A with respect to that basis is diagonal. The advantages of this from a computational viewpoint are huge.

For example, if

$$A = \begin{pmatrix} 0 & 0 & 0 \\ 0 & -1 & 0 \\ 0 & 0 & -.5 \end{pmatrix}$$

then

$$A^n = \begin{pmatrix} 0 & 0 & 0 \\ 0 & (-1)^n & 0 \\ 0 & 0 & (-.5)^n \end{pmatrix}.$$

It is sometimes possible to make sense of a transcendental function of a matrix using Taylor series expansions. We show in the exercises that for certain matrices, expressions such as e^A and sin A make sense.

The next result is immediate from Theorem 5.3.

Theorem 5.4:

If the characteristic polynomial of A factors into distinct linear factors, then there is a basis of V that consists of eigenvectors of A.

The Algebraic and Geometric Multiplicities of an Eigenvalue

Definition:

The geometric multiplicity of an eigenvalue is the dimension of the eigenspace of the eigenvalue.

The algebraic multiplicity of an eigenvalue λ is the exponent of the factor $(x-\lambda)$ in the characteristic polynomial.

In Chapter 9 we show that the geometric multiplicity of an eigenvalue is less than or equal to the algebraic multiplicity. As an earlier example shows, one can determine the algebraic multiplicity of an eigenvalue from the characteristic polynomial, but not the geometric multiplicity unless the algebraic multiplicity is one.

Example

Suppose the matrix A has characteristic polynomial

$$P_A(x) = (x)^3 (x-3)^4 (x+1)^2.$$

What conclusions can we draw?

1. A is a 9×9 matrix, because the sum of the exponents is 9.
2. The eigen values of A are $\lambda = 0, 3$ and -1.
3. The algebraic multiplicity of $\lambda = 0$ is 3. All we can say about the geometric multiplicity of $\lambda = 0$ is that it is 1, 2, or 3.
4. The algebraic multiplicity of $\lambda = 3$ is 4. All we can say about the geometric multiplicity of $\lambda = 3$ is that it is 1, 2, 3, or 4.
5. The algebraic multiplicity of $\lambda = -1$ is 2. All we can say about the geometric multiplicity of $\lambda = -1$ is that it is 1 or 2.
6. There will be a basis consisting of eigenvectors of A if and only if the geometric multiplicity of each eigenvalue is equal to the algebraic multiplicity of that eigenvalue.
7. The matrix is not invertible because 0 is an eigenvalue.

Diagonalizing a Matrix

A crucial thing to remember in the next discussion is that if A and B are matrices so that AB is defined, then

$$AB = \left[A\hat{b}_1, \ldots, A\hat{b}_k \right]$$

where $A\hat{b}_i$ is the column vector obtained by multiplying the ith column of B on the left by A. We review this in the 2×2 case. If

$$A = \begin{pmatrix} a_{11} & a_{12} \\ a_{21} & a_{22} \end{pmatrix} \quad B = \begin{pmatrix} b_{11} & b_{12} \\ b_{21} & b_{22} \end{pmatrix}$$

then

$$A\hat{b}_1 = \begin{pmatrix} a_{11} & a_{12} \\ a_{21} & a_{22} \end{pmatrix} \begin{pmatrix} b_{11} \\ b_{21} \end{pmatrix} = \begin{pmatrix} a_{11}b_{11} + a_{12}b_{21} \\ a_{21}b_{11} + a_{22}b_{21} \end{pmatrix}$$

$$A\hat{b}_2 = \begin{pmatrix} a_{11} & a_{12} \\ a_{21} & a_{22} \end{pmatrix} \begin{pmatrix} b_{12} \\ b_{22} \end{pmatrix} = \begin{pmatrix} a_{11}b_{12} + a_{12}b_{22} \\ a_{21}b_{12} + a_{22}b_{22} \end{pmatrix}$$

$$AB = \begin{pmatrix} a_{11} & a_{12} \\ a_{21} & a_{22} \end{pmatrix} \begin{pmatrix} b_{11} & b_{12} \\ b_{21} & b_{22} \end{pmatrix} = \begin{pmatrix} a_{11}b_{11} + a_{12}b_{21} & a_{11}b_{12} + a_{12}b_{22} \\ a_{21}b_{11} + a_{22}b_{21} & a_{21}b_{12} + a_{22}b_{22} \end{pmatrix}$$

Example

Suppose that A is an $n \times n$ matrix that has n linearly independent eigenvectors $\{\hat{v}_1, \ldots, \hat{v}_n\}$ and λ_i is the eigenvalue of $\hat{v}_i$. We are not saying that all the λ_i's are distinct.

We have

$$A[\hat{v}_1, \hat{v}_2, \ldots, \hat{v}_n] = [A\hat{v}_1, A\hat{v}_2, \ldots, A\hat{v}_n] = [\lambda_1 \hat{v}_1, \lambda_2 \hat{v}_2, \ldots, \lambda_n \hat{v}_n]$$

$$= [\hat{v}_1, \hat{v}_2, \ldots, \hat{v}_n] \begin{pmatrix} \lambda_1 & 0 & \cdots & 0 \\ 0 & \lambda_2 & \cdots & 0 \\ \vdots & \vdots & \ddots & \vdots \\ 0 & 0 & \cdots & \lambda_n \end{pmatrix}.$$

Let P be the matrix whose columns are the eigenvectors $\hat{v}_1, \ldots, \hat{v}_n$ and D the diagonal matrix

$$D = \begin{pmatrix} \lambda_1 & 0 & \cdots & 0 \\ 0 & \lambda_2 & \cdots & 0 \\ \vdots & \vdots & \cdots & \vdots \\ 0 & 0 & \cdots & \lambda_n \end{pmatrix}.$$

Then

$$AP = PD. \tag{3}$$

Note: If Equation (3) is not clear, consider the 2×2 case where A has eigenvectors

$$\hat{v}_1 = \begin{pmatrix} v_{11} \\ v_{21} \end{pmatrix} \quad \hat{v}_2 = \begin{pmatrix} v_{12} \\ v_{22} \end{pmatrix}$$

with eigenvalues λ_1 and λ_2, respectively. Then

$$AP = A[\hat{v}_1 \ \hat{v}_2] = [A\hat{v}_1 \ A\hat{v}_2] = [\lambda_1 \hat{v}_1 \ \lambda_2 \hat{v}_2] = \begin{pmatrix} \lambda_1 v_{11} & \lambda_2 v_{12} \\ \lambda_1 v_{21} & \lambda_2 v_{22} \end{pmatrix}$$

and

$$PD = \begin{pmatrix} v_{11} & v_{12} \\ v_{21} & v_{22} \end{pmatrix} \begin{pmatrix} \lambda_1 & 0 \\ 0 & \lambda_2 \end{pmatrix} = \begin{pmatrix} v_{11}\lambda_1 + 0 & 0 + v_{12}\lambda_2 \\ v_{21}\lambda_1 + 0 & 0 + v_{22}\lambda_2 \end{pmatrix} = \begin{pmatrix} \lambda_1 v_{11} & \lambda_2 v_{12} \\ \lambda_1 v_{21} & \lambda_2 v_{22} \end{pmatrix}.$$

Typically, we will be given the matrix A and want to find the matrix D. Rearranging Equation (3) gives

$$D = P^{-1}AP.$$

We know that P^{-1} exists because $\{\hat{v}_1, \ldots, \hat{v}_n\}$ is a basis.

Recapping, we have the following result.

Theorem 5.5:

Let A be a matrix for which there is a basis of eigenvectors $\{\hat{v}_1,\ldots,\hat{v}_n\}$. Then

$$P^{-1}AP = D$$

where P is the matrix whose columns are the eigenvectors of A and D is the diagonal matrix whose diagonal entries are the eigenvalues listed in the same order as the corresponding eigenvectors in P.

Not every square matrix can be diagonalized. Theorem 5.5 gives the method of accomplishing a diagonalization when it can be done. In most cases, it is not immediately obvious when a matrix can be diagonalized. We give several examples and exercises showing matrices that can and matrices that cannot be diagonalized.

There is a special category of matrices that can always be diagonalized and are easily recognized. In $\mathbb{R}^n$ these are the symmetric matrices – matrices for which $A^T = A$. We will see in Chapter 7 that a matrix is symmetric if and only if there is a matrix P for which

$$P = P^T = P^{-1} \quad \text{and} \quad P^{-1}AP = D.$$

Such a matrix P is called an orthogonal matrix.

Example

Diagonalize, if possible, the matrix

$$A = \begin{pmatrix} 2 & 0 & 0 \\ -3 & 0 & 1 \\ 0 & 1 & 0 \end{pmatrix},$$

The characteristic polynomial is

$$(2-x)(x-1)(x+1)$$

so there are three eigenvalues; $\lambda = 2$, $\lambda = 1$ and $\lambda = -1$.

Since the eigenvalues have algebraic dimension one, the matrix can be diagonalized.

For

$$\lambda = 2, \text{eigenvector} = \begin{pmatrix} -1 \\ 2 \\ 1 \end{pmatrix}; \ \lambda = 1, \text{eigenvector} = \begin{pmatrix} 0 \\ 1 \\ 1 \end{pmatrix};$$

$$\lambda = -1, \text{eigenvector} = \begin{pmatrix} 0 \\ 1 \\ -1 \end{pmatrix}.$$

Then, we can take

$$P = \begin{pmatrix} -1 & 0 & 0 \\ 2 & 1 & 1 \\ 1 & 1 & -1 \end{pmatrix}$$

and

$$P^{-1}AP = \begin{pmatrix} -1 & 0 & 0 \\ 2 & 1 & 1 \\ 1 & 1 & -1 \end{pmatrix}^{-1} \begin{pmatrix} 2 & 0 & 0 \\ -3 & 0 & 1 \\ 0 & 1 & 0 \end{pmatrix} \begin{pmatrix} -1 & 0 & 0 \\ 2 & 1 & 1 \\ 1 & 1 & -1 \end{pmatrix} = \begin{pmatrix} 2 & 0 & 0 \\ 0 & 1 & 0 \\ 0 & 0 & -1 \end{pmatrix}.$$

Exercises

In Exercises 1–10, the matrices are the same as those from Section 5.1. If possible, diagonalize the matrix.

1. $\begin{pmatrix} 3 & 0 & 0 \\ 0 & 2 & 1 \\ -1 & 0 & 2 \end{pmatrix}$

2. $\begin{pmatrix} 1 & 0 & 0 \\ -2 & -1 & -2 \\ 3 & 6 & 6 \end{pmatrix}$

3. $\begin{pmatrix} 2 & 1 & 0 \\ 0 & 2 & 1 \\ 0 & 0 & 2 \end{pmatrix}$

4. $\begin{pmatrix} 3 & 1 & 0 \\ 0 & 3 & 0 \\ 0 & 0 & 3 \end{pmatrix}$

5. $\begin{pmatrix} 3 & 3 & -2 \\ 2 & -3 & 2 \\ -4 & 1 & 1 \end{pmatrix}$

6. $\begin{pmatrix} 1 & -3 & 3 \\ 3 & -5 & 3 \\ 6 & -6 & 6 \end{pmatrix}$

7. $\begin{pmatrix} 0 & 0 & -2 \\ 1 & 2 & 1 \\ 0 & 0 & 3 \end{pmatrix}$

8. $\begin{pmatrix} 12 & 0 & 0 \\ -16 & -4 & 0 \\ -16 & 0 & -4 \end{pmatrix}$

9. $\begin{pmatrix} 3 & 6 & 0 & 0 \\ 1 & -2 & 0 & 0 \\ 0 & 0 & 3 & -1 \\ 0 & 0 & 0 & 3 \end{pmatrix}$

10. $\begin{pmatrix} 1 & 1 & 0 & -1 \\ 0 & -3 & 2 & 6 \\ 0 & 0 & 0 & 5 \\ 0 & 0 & 0 & 4 \end{pmatrix}$

11. If the characteristic polynomial of a matrix A is $(\lambda + 2)^3 (\lambda - 1)^4 (\lambda)$
    (a.) What is the size of the matrix A?
    (b.) What are the algebraic and geometric dimensions of each eigenspace?
    (c.) Is A invertible?

Section 5.3 Similar Matrices

If $T : V \to V$ is a linear operator, and $\mathcal{B}$ is a basis for V then there is a matrix A for which

$$T(\hat{x}) = A\hat{x}$$

where the matrix A depends on both T and $\mathcal{B}$. An important distinction between T and A is that T describes how a vector $\hat{x}$ is changed and is independent of the basis of V, whereas A gives the representation of $T(\hat{x})$ in a given coordinate system, and is basis and T dependent.

In this section, we characterize the class of matrices that represent a particular linear operator.

Definition:

If A and B are square matrices, then A is similar to B if there is an invertible matrix P with $B = P^{-1}AP$.

The main result of this section is that two matrices represent the same linear operator with respect to different bases if and only if the matrices are similar.

For the purposes of this section, it will be convenient to describe what is meant by an "equivalence relation", but that idea will not be used in the sequel. In the present setting, we consider a relation between $n \times n$ matrices. We will say that two such matrices are related if they are similar.

A relation is an equivalence relation if three conditions hold. These are:

1. A matrix is related to itself. (This is called the reflexive property.)
2. If A is related to B, then B is related to A. (Symmetric property.)
3. If A is related to B and B is related to C, then A is related to C. (Transitive property.)

Theorem 5.6:

Being similar is an equivalence relation on the set of $n \times n$ matrices.

Proof:

A is similar to A since $A = I^{-1}AI$.

If A is similar to B then B is similar to A since if $B = P^{-1}AP$, then

$$PBP^{-1} = P\left(P^{-1}AP\right)P^{-1} = \left(PP^{-1}\right)A\left(PP^{-1}\right) = A.$$

Said another way,

$$\left(P^{-1}\right)^{-1} BP^{-1} = A.$$

Also, if A is similar to B then B is similar to C, then A is similar to C because if

$$B = P^{-1}AP \quad \text{and} \quad C = Q^{-1}BQ$$

then

$$C = Q^{-1}BQ = Q^{-1}\left(P^{-1}AP\right)Q = \left(Q^{-1}P^{-1}\right)A(PQ) = (PQ)^{-1}A(PQ).$$

An equivalence relation has the effect of partitioning a set. This means it divides a set into pieces, called equivalence classes, and each element in the set is in exactly one equivalence class. The equivalence classes are determined by the condition that A and B are in the same equivalence class if and only if they are related to each other.

Similar matrices share several characteristics. The next theorems highlight some of them.

Theorem 5.7:

Similar matrices have the same characteristic polynomial.

Proof:

Suppose A is similar to B. Then there is an invertible matrix P with

$$B = P^{-1}AP$$

so

$$B - \lambda I = P^{-1}AP - \lambda I = P^{-1}AP - P^{-1}\lambda IP = P^{-1}\left(A - \lambda I\right)P$$

and thus

$$\det\left(B - \lambda I\right) = \det\left[P^{-1}\left(A - \lambda I\right)P\right] = \det\left(P^{-1}\right)\det\left(A - \lambda I\right)\det\left(P\right)$$
$$= \left[\det\left(P\right)\right]^{-1}\det\left(A - \lambda I\right)\det\left(P\right) = \det\left(A - \lambda I\right).$$

Corollary:

For similar matrices, the eigenvalues and the algebraic dimensions of their eigenspaces are the same.

While similar matrices have the same characteristic polynomial, two matrices that have the same characteristic polynomial are not necessarily similar.

Example

The matrices A and B where

$$A = \begin{pmatrix} 2 & 1 \\ 0 & 2 \end{pmatrix} \text{ and } B = \begin{pmatrix} 2 & 0 \\ 0 & 2 \end{pmatrix}$$

have the same characteristic polynomial, but are not similar.

We showed in a previous example that A has a basis of eigenvectors but B does not.

The next theorem states that two $n \times n$ matrices are similar if and only if they represent the same linear transformation with respect to different bases.

Theorem 5.8:

Let

$$T : \mathcal{F}^n \to \mathcal{F}^n$$

be a linear transformation. Let A be the matrix representation of T with respect to the standard basis. The matrix A is similar to the matrix B if and only if there is a basis of $\mathcal{F}^n$ for which B is the representation of T with respect to that basis.

Proof:

We have seen most of the ideas in this proof at the end of Section 4.2. The explanation and next example show why the result is true.

There is a one-to-one and onto correspondence between *ordered bases* of $\mathcal{F}^n$ and $n \times n$ invertible matrices with entries in $\mathcal{F}$. Let

$$T : V \to V$$

be a linear operator and let A be the matrix representation of T with respect to the standard basis. To be more concrete, we give the idea of the proof with an example. Let

$$T : \mathbb{R}^3 \to \mathbb{R}^3$$

and let A be the matrix representation of T with respect to the ordered usual basis, with

$$A = \begin{pmatrix} 1 & 2 & 5 \\ 3 & 0 & 3 \\ -2 & 4 & 6 \end{pmatrix}.$$

Choose the ordered basis

$$\mathcal{B} = \left\{ \begin{pmatrix} 1 \\ 1 \\ 1 \end{pmatrix}, \begin{pmatrix} 2 \\ 2 \\ 0 \end{pmatrix}, \begin{pmatrix} 3 \\ 0 \\ 0 \end{pmatrix} \right\}.$$

Now $\mathcal{B}$ is uniquely associated with the invertible matrix

$$S = \begin{pmatrix} 1 & 2 & 3 \\ 1 & 2 & 0 \\ 1 & 0 & 0 \end{pmatrix}.$$

We have previously shown that

$$S^{-1}AS = \left[T \right]_{\mathcal{B}}^{\mathcal{B}}$$

so that $S^{-1}AS$ is the representation of T with respect to the basis $\mathcal{B}$.

The theorems above say that each linear transformation on a vector space gives rise to an equivalence class of matrices, and that different linear transformations give rise to different equivalence classes of matrices.

We have shown that similar matrices have the same characteristic polynomial, and hence the same eigenvalues and the eigenvalues have the same algebraic multiplicity. We have now shown that similar matrices have the same geometric multiplicity.

The next example demonstrates that while similar matrices have the same correspondence of eigenvectors and eigenvalues, the representation of the eigenvectors will be different in different bases.

Example

Let $T : \mathbb{R}^3 \to \mathbb{R}^3$. Suppose the representation of T in the standard basis is

$$A = \begin{pmatrix} -2 & 3 & 1 \\ 0 & 1 & 1 \\ -3 & 4 & 1 \end{pmatrix}.$$

The eigenvalues of A are $\lambda = -2$, 0 and 2. Suppose that $\mathcal{B}$ is the basis

$$\mathcal{B} = \left\{ \begin{pmatrix} 1 \\ 1 \\ 1 \end{pmatrix}, \begin{pmatrix} 1 \\ 1 \\ 0 \end{pmatrix}, \begin{pmatrix} 1 \\ 0 \\ 0 \end{pmatrix} \right\}$$

so that the change of basis matrix is

$$P_{\mathcal{B}} = \begin{pmatrix} 1 & 1 & 1 \\ 1 & 1 & 0 \\ 1 & 0 & 0 \end{pmatrix}.$$

Now

$$B = P_{\mathcal{B}}^{-1} A P_{\mathcal{B}} = \begin{pmatrix} 2 & 1 & 3 \\ 0 & 0 & -3 \\ 0 & 0 & -2 \end{pmatrix}.$$

The eigenvalues of B are also $\lambda = -2$, 0 and 2.

The eigenvectors of A are

$$\lambda = -2, \ \hat{v}_{-2} = \begin{pmatrix} 5 \\ -3 \\ 9 \end{pmatrix}; \quad \lambda = 0, \ \hat{v}_0 = \begin{pmatrix} 1 \\ 1 \\ -1 \end{pmatrix}; \quad \lambda = 2, \ \hat{v}_2 = \begin{pmatrix} 1 \\ 1 \\ 1 \end{pmatrix}.$$

The eigenvectors of B are

$$\lambda = -2, w_{-2} = \begin{pmatrix} 9 \\ -12 \\ 8 \end{pmatrix}_{\mathcal{B}}; \lambda = 0, w_0 = \begin{pmatrix} -1 \\ 2 \\ 0 \end{pmatrix}_{\mathcal{B}}; \lambda = 2, w_2 = \begin{pmatrix} 1 \\ 0 \\ 0 \end{pmatrix}_{\mathcal{B}}.$$

Note that the representations of the eigenvectors of B are in the $\mathcal{B}$ basis. We show that the eigenvectors are the same; it is just the representations are different. We check this:

$$\hat{w}_{-2} = \begin{pmatrix} 9 \\ -12 \\ 8 \end{pmatrix}_{\mathcal{B}} = 9\begin{pmatrix} 1 \\ 1 \\ 1 \end{pmatrix} - 12\begin{pmatrix} 1 \\ 1 \\ 0 \end{pmatrix} + 8\begin{pmatrix} 1 \\ 0 \\ 0 \end{pmatrix} = \begin{pmatrix} 5 \\ -3 \\ 9 \end{pmatrix} = \hat{v}_{-2}$$

$$\hat{w}_0 = \begin{pmatrix} -1 \\ 2 \\ 0 \end{pmatrix}_{\mathcal{B}} = -1\begin{pmatrix} 1 \\ 1 \\ 1 \end{pmatrix} + 2\begin{pmatrix} 1 \\ 1 \\ 0 \end{pmatrix} + 0\begin{pmatrix} 1 \\ 0 \\ 0 \end{pmatrix} = \begin{pmatrix} 1 \\ 1 \\ -1 \end{pmatrix} = \hat{v}_0$$

$$\hat{w}_2 = \begin{pmatrix} 1 \\ 0 \\ 0 \end{pmatrix}_{\mathcal{B}} = 1\begin{pmatrix} 1 \\ 1 \\ 1 \end{pmatrix} + 0\begin{pmatrix} 1 \\ 1 \\ 0 \end{pmatrix} + 0\begin{pmatrix} 1 \\ 0 \\ 0 \end{pmatrix} = \begin{pmatrix} 1 \\ 1 \\ 1 \end{pmatrix} = \hat{v}_2.$$

If we visualize a vector as an arrow, the eigenvector of a linear operator is a vector that changes only its length (if the eigenvalue is negative, the direction is reversed) when the linear transformation acts on it. However, the basis in which the eigenvector is represented is immaterial.

Exercises

1. Suppose $T : \mathbb{R}^2 \to \mathbb{R}^2$ with

$$T\begin{pmatrix} 1 \\ 0 \end{pmatrix} = \begin{pmatrix} 2 \\ 5 \end{pmatrix}, \ T\begin{pmatrix} 0 \\ 1 \end{pmatrix} = \begin{pmatrix} -1 \\ 3 \end{pmatrix}.$$

Consider the basis $\mathcal{B} = \{\hat{b}_1, \hat{b}_2\}$ with

$$\hat{b}_1 = \begin{pmatrix} 2 \\ 1 \end{pmatrix}, \ \hat{b}_2 = \begin{pmatrix} 1 \\ 1 \end{pmatrix}.$$

Find $\left[T\right]_{\mathcal{B}}^{\mathcal{B}}$.

2. Suppose $T : \mathbb{R}^2 \to \mathbb{R}^2$ with

$$T\begin{pmatrix} 1 \\ 0 \end{pmatrix} = \begin{pmatrix} 3 \\ -2 \end{pmatrix}, \; T\begin{pmatrix} 0 \\ 1 \end{pmatrix} = \begin{pmatrix} 6 \\ 4 \end{pmatrix}.$$

Consider the basis $\mathcal{B} = \left\{\hat{b}_1, \hat{b}_2\right\}$, with

$$\hat{b}_1 = \begin{pmatrix} 1 \\ -1 \end{pmatrix}, \; \hat{b}_2 = \begin{pmatrix} -4 \\ 3 \end{pmatrix}.$$

Find $\left[T\right]_{\mathcal{B}}^{\mathcal{B}}$.

3. Suppose $T : \mathbb{R}^3 \to \mathbb{R}^3$ with

$$T\begin{pmatrix} 1 \\ 0 \\ 0 \end{pmatrix} = \begin{pmatrix} 1 \\ 1 \\ 2 \end{pmatrix}, \; T\begin{pmatrix} 1 \\ 0 \\ 0 \end{pmatrix} = \begin{pmatrix} 0 \\ 2 \\ 3 \end{pmatrix}, \; T\begin{pmatrix} 0 \\ 0 \\ 1 \end{pmatrix} = \begin{pmatrix} 1 \\ 1 \\ 0 \end{pmatrix}.$$

Consider the basis $\mathcal{B} = \left\{\hat{b}_1, \hat{b}_2, \hat{b}_3\right\}$ with

$$\hat{b}_1 = \begin{pmatrix} 2 \\ 0 \\ 1 \end{pmatrix}, \; \hat{b}_2 = \begin{pmatrix} 0 \\ 4 \\ -1 \end{pmatrix}, \; \hat{b}_3 = \begin{pmatrix} 3 \\ 3 \\ 0 \end{pmatrix}.$$

Find $\left[T\right]_{\mathcal{B}}^{\mathcal{B}}$.

4. Suppose $T : \mathbb{R}^3 \to \mathbb{R}^3$ with

$$T\begin{pmatrix} 1 \\ 0 \\ 0 \end{pmatrix} = \begin{pmatrix} 3 \\ 5 \\ -1 \end{pmatrix}, \; T\begin{pmatrix} 1 \\ 0 \\ 0 \end{pmatrix} = \begin{pmatrix} 1 \\ 1 \\ 3 \end{pmatrix}, \; T\begin{pmatrix} 0 \\ 0 \\ 1 \end{pmatrix} = \begin{pmatrix} -2 \\ -1 \\ 1 \end{pmatrix}.$$

Consider the basis $\mathcal{B} = \left\{\hat{b}_1, \hat{b}_2, \hat{b}_3\right\}$ with

$$\hat{b}_1 = \begin{pmatrix} 0 \\ 2 \\ 3 \end{pmatrix}, \; \hat{b}_2 = \begin{pmatrix} 1 \\ 1 \\ -2 \end{pmatrix}, \; \hat{b}_3 = \begin{pmatrix} 2 \\ 6 \\ 1 \end{pmatrix}.$$

Find $\left[T\right]_{\mathcal{B}}^{\mathcal{B}}$.

5. If A and B are similar matrices, list the properties that they share; e.g., they have the same determinant.

6. Show that if A and B are similar matrices, then A^2 and B^2 are similar matrices.

7. If A is invertible, show that AB is similar to BA.

Section 5.4 Eigenvalues and Eigenvectors in Systems of Differential Equations

One application of eigenvectors occurs in solving systems of differential equations.

Suppose that a_{ij}, $i,j = 1,\ldots,n$ are constants and $x_1(t),\ldots,x_n(t)$ are continuously differentiable functions and we have the system of n linear differential equations with constant coefficients

$$x_1'(t) = a_{11}x_1(t) + \cdots + a_{1n}x_n(t)$$

$$x_2'(t) = a_{21}x_1(t) + \cdots + a_{2n}x_n(t)$$

$$\vdots$$

$$x_n'(t) = a_{n1}x_1(t) + \cdots + a_{nn}x_n(t)$$

with initial conditions $x_i(0) = c_i$, $i = 1,\ldots,n$.

We can write this system as a vector equation

$$\hat{x}'(t) = A\hat{x}(t) \quad (4)$$

where

$$\hat{x}(t) = \begin{pmatrix} x_1(t) \\ \vdots \\ x_n(t) \end{pmatrix}, \quad \hat{x}'(t) = \begin{pmatrix} x_1'(t) \\ \vdots \\ x_n'(t) \end{pmatrix} \quad \text{and} \quad A = \begin{pmatrix} a_{11} & \cdots & a_{1n} \\ \vdots & & \vdots \\ a_{n1} & \cdots & a_{nn} \end{pmatrix}.$$

We first consider an example where A is diagonal.

Example

Suppose that we have the matrix equation

$$\begin{pmatrix} x_1'(t) \\ x_2'(t) \\ x_3'(t) \end{pmatrix} = \begin{pmatrix} 2 & 0 & 0 \\ 0 & -3 & 0 \\ 0 & 0 & 5 \end{pmatrix} \begin{pmatrix} x_1(t) \\ x_2(t) \\ x_3(t) \end{pmatrix}$$

with initial conditions

$$x_1(0) = -1, x_2(0) = 4, x_3(0) = 9.$$

This yields the equivalent system

$$x_1'(t) = 2x_1(t), \quad x_1(0) = -1$$

$$x_2'(t) = -3x_2(t), \quad x_2(0) = 4$$

$$x_3'(t) = 5x_3(t), \quad x_3(0) = 9.$$

The solutions to these three equations are

$$x_1(t) = -1e^{2t}, \quad x_2(t) = 4e^{-3t}, \quad x_3(t) = 9e^{5t}$$

so we could write the solution as

$$\hat{x}(t) = \begin{pmatrix} x_1(t) \\ x_2(t) \\ x_3(t) \end{pmatrix} = \begin{pmatrix} -1e^{2t} \\ 4e^{-3t} \\ 9e^{5t} \end{pmatrix}.$$

Suppose now that the matrix A in Equation (4) is not diagonal, but can be diagonalized. Recall, this is the same thing as A having a basis of eigenvectors.

Suppose that P is a matrix that diagonalizes A. That is,

$$P^{-1}AP = D$$

where D is diagonal. The hypothesis that A can be diagonalized means the individual equations are independent and there is enough information to ensure a unique solution to the system.

If we have

$$\hat{x}'(t) = A\hat{x}(t)$$

and let

$$\hat{y}(t) = P^{-1}\hat{x}(t)$$

then

$$P\hat{y}(t) = \hat{x}(t)$$

so

$$P\hat{y}'(t) = \hat{x}'(t) = A\hat{x}(t) = AP\hat{y}(t)$$

and we have

$$\hat{y}'(t) = P^{-1}AP\hat{y}(t) = D\hat{y}(t)$$

which we know how to solve.

We must convert the solution to $\hat{x}(t)$, but $\hat{x}(t) = P\widehat{y}(t)$.

An Algorithm to Solve the Matrix Equation $\hat{x}^T(t) = A\hat{x}(t)$ *when A Can Be Diagonalized*

1. The most labor-intensive part is diagonalizing the matrix A, which requires finding the matrix P. Recall that this is done by finding the eigenvalues and eigenvectors of A. Once this is done, the matrix P is the matrix whose columns are the eigenvectors of A and the entries on the diagonal matrix D are the eigenvalues of A.

2. Let

$$\hat{y}(t) = P^{-1}\hat{x}(t)$$

and solve

$$\hat{y}'(t) = P^{-1}AP\hat{y}(t) = D\hat{y}(t).$$

3. The solution is

$$\hat{y}(t) = \begin{pmatrix} c_1 e^{\lambda_1 t} \\ c_2 e^{\lambda_2 t} \\ \vdots \\ c_n e^{\lambda_n t} \end{pmatrix}$$

where λ_i are the eigenvalues of A and c_i are to be determined.

4. Convert the solution back to $\hat{x}(t)$ using $\hat{x}(t) = P\hat{y}(t)$.

It is easy to get lost in the explanation of this step and the next one, but these will be easier to follow after the next example.

The solution will be

$$\begin{pmatrix} x_1(t) \\ x_2(t) \\ \vdots \\ x_4(t) \end{pmatrix} = \begin{pmatrix} P_{11}c_1 e^{\lambda_1 t} + P_{12}c_2 e^{\lambda_2 t} + \cdots + P_{1n}c_n e^{\lambda_n t} \\ P_{21}c_1 e^{\lambda_1 t} + P_{22}c_2 e^{\lambda_2 t} + \cdots + P_{2n}c_n e^{\lambda_n t} \\ \vdots \\ P_{n1}c_1 e^{\lambda_1 t} + P_{n2}c_2 e^{\lambda_2 t} + \cdots + P_{nn}c_n e^{\lambda_n t} \end{pmatrix}$$

so

$$x_i(t) = P_{i1}c_1 e^{\lambda_1 t} + P_{i2}c_2 e^{\lambda_2 t} + \cdots + P_{in}c_n e^{\lambda_n t}$$

5. We now solve for the $c_i's$.

Setting $t=0$ gives the system of equations

$$x_1(0) = P_{11}c_1 + P_{12}c_2 + \cdots + P_{1n}c_n$$

$$x_2(0) = P_{21}c_1 + P_{22}c_2 + \cdots + P_{2n}c_n$$

$$\vdots$$

$$x_n(0) = P_{n1}c_1 + P_{n2}c_2 + \cdots + P_{nn}c_n$$

This can be expressed as the matrix equation

$$\hat{x}(0) = P\hat{c}$$

where

$$\hat{c} = \begin{pmatrix} c_1 \\ c_2 \\ \vdots \\ c_n \end{pmatrix}.$$

Thus

$$\hat{c} = P^{-1}\hat{x}(0).$$

The values $x_i(0)$ are given and the values P_{ij} have been determined. Thus, we can determine the values of c_i. We then know the solution; that is

$$x_i(t) = P_{i1}c_1e^{\lambda_1 t} + P_{i2}c_2e^{\lambda_2 t} + \cdots + P_{in}c_ne^{\lambda_n t}; i = 1, \ldots, n.$$

Example

We solve the system of equations

$$x_1'(t) = -2x_1(t) - 4x_2(t) + 2x_3(t)$$

$$x_2'(t) = -2x_1(t) + x_2(t) + 2x_3(t)$$

$$x_3'(t) = 4x_1(t) + 2x_2(t) + 5x_3(t)$$

with initial conditions

$$x_1(0) = 3, \ x_2(0) = -1, \ x_3(0) = 0.$$

We write this in the form

$$\hat{x}'(t) = A\hat{x}(t)$$

where

$$A = \begin{pmatrix} -2 & -4 & 2 \\ -2 & 1 & 2 \\ 4 & 2 & 5 \end{pmatrix}.$$

1. We first find that the characteristic polynomial for A is

$$P_A(\lambda) = (\lambda - 3)(\lambda + 5)(\lambda - 6)$$

so the eigenvalues are $\lambda = 3, \lambda = -5, \lambda = 6$.
2. For each eigenvalue, we find an eigenvector.
  1. For

$$\lambda_1 = 3, v_1 = \begin{pmatrix} -2 \\ 3 \\ 1 \end{pmatrix}; \; \lambda_2 = -5, v_2 = \begin{pmatrix} -2 \\ -1 \\ 1 \end{pmatrix}; \; \lambda_3 = 6, v_3 = \begin{pmatrix} 1 \\ 6 \\ 16 \end{pmatrix}.$$

3. We then have that

$$P = \begin{pmatrix} -2 & -2 & 1 \\ 3 & -1 & 6 \\ 1 & 1 & 16 \end{pmatrix} \text{ and } D = \begin{pmatrix} 3 & 0 & 0 \\ 0 & -5 & 0 \\ 0 & 0 & 6 \end{pmatrix}.$$

4. The solution to

$$y'(t) = \begin{pmatrix} 3 & 0 & 0 \\ 0 & -5 & 0 \\ 0 & 0 & 6 \end{pmatrix} \begin{pmatrix} y_1(t) \\ y_2(t) \\ y_3(t) \end{pmatrix}$$

is

$$y_1(t) = c_1 e^{3t}$$

$$y_2(t) = c_2 e^{-5t}$$

$$y_3(t) = c_3 e^{6t}$$

So

$$\hat{x}(t) = P\hat{y}(t) = \begin{pmatrix} -2 & -2 & 1 \\ 3 & -1 & 6 \\ 1 & 1 & 16 \end{pmatrix} \begin{pmatrix} c_1 e^{3t} \\ c_2 e^{-5t} \\ c_3 e^{6t} \end{pmatrix} = \begin{pmatrix} -2c_1 e^{3t} - 2c_2 e^{-5t} + c_3 e^{6t} \\ 3c_1 e^{3t} - c_2 e^{-5t} + 6c_3 e^{6t} \\ c_1 e^{3t} + c_2 e^{-5t} + 16c_3 e^{6t} \end{pmatrix}.$$

Now

$$\hat{x}(0) = \begin{pmatrix} 3 \\ -1 \\ 0 \end{pmatrix}$$

So

$$\hat{c} = \begin{pmatrix} c_1 \\ c_2 \\ c_3 \end{pmatrix} = P^{-1}\hat{x}(0) = \begin{pmatrix} -2 & -2 & 1 \\ 3 & -1 & 6 \\ 1 & 1 & 16 \end{pmatrix}^{-1} \begin{pmatrix} 3 \\ -1 \\ 0 \end{pmatrix} = \begin{pmatrix} -3/4 \\ 31/44 \\ 1/11 \end{pmatrix}$$

gives

$$c_1 = -\frac{3}{4}, \quad c_2 = -\frac{31}{44}, \quad c_3 = \frac{1}{11}$$

Thus,

$$x_1(t) = -2\left(-\frac{3}{4}\right)e^{3t} - 2\left(-\frac{31}{44}\right)e^{-5t} + \left(\frac{1}{11}\right)e^{6t} = \frac{3}{2}e^{3t} + \frac{31}{22}e^{-5t} + \frac{1}{11}e^{6t}$$

$$x_2(t) = 3\left(-\frac{3}{4}\right)e^{3t} - \left(-\frac{31}{44}\right)e^{-5t} + 6\left(\frac{1}{11}\right)e^{6t} = -\frac{9}{4}e^{3t} + \frac{31}{44}e^{-5t} + \frac{6}{11}e^{6t}$$

$$x_3(t) = -\frac{3}{4}e^{3t} - \frac{31}{44}e^{-5t} + \frac{16}{11}e^{6t}.$$

We demonstrate that the first equation in the system of equations is satisfied. That is,

$$x_1'(t) = -2x_1(t) - 4x_2(t) + 2x_3(t).$$

Since

$$x_1(t) = \frac{3}{2}e^{3t} + \frac{31}{22}e^{-5t} + \frac{1}{11}e^{6t}$$

we have

$$x_1'(t) = \frac{9}{2}e^{3t} - \frac{155}{22}e^{-5t} + \frac{6}{11}e^{6t}$$

and

$$-2x_1(t) - 4x_2(t) + 2x_3(t)$$

$$= -2\left[\frac{3}{2}e^{3t} + \frac{31}{22}e^{-5t} + \frac{1}{11}e^{6t}\right] - 4\left[-\frac{9}{4}e^{3t} + \frac{31}{44}e^{-5t} + \frac{6}{11}e^{6t}\right]$$

$$+ 2\left[-\frac{3}{4}e^{3t} - \frac{31}{44}e^{-5t} + \frac{16}{11}e^{6t}\right] = \left(-3 + 9 - \frac{3}{2}\right)e^{3t} + \left(-\frac{31}{11} - \frac{31}{11} - \frac{31}{22}\right)e^{-5t}$$

$$+ \left(-\frac{2}{11} - \frac{24}{11} + \frac{32}{11}\right)e^{6t} = \frac{9}{2}e^{3t} - \frac{155}{22}e^{-5t} + \frac{6}{11}e^{6t}.$$

The other two equations can be checked in a similar manner.

Exercises

1. Solve the system of differential equations

$$x_1'(t) = 3x_1(t) + 3x_2(t) - 2x_3(t)$$

$$x_2'(t) = 2x_1(t) - 3x_2(t) + 2x_3(t)$$

$$x_3'(t) = -4x_1(t) + x_2(t) + x_3(t)$$

with initial conditions

$$x_1(0) = 2, x_2(0) = 0, x_3(0) = 1.$$

2. Solve the system of differential equations

$$x_1'(t) = 4x_1(t) + x_3(t)$$

$$x_2'(t) = -2x_1(t) + x_2(t)$$

$$x_3'(t) = -2x_1(t) + x_3(t)$$

with initial conditions

$$x_1(0) = 1, \ x_2(0) = 3, \ x_3(0) = -1.$$

3. Solve the system of differential equations

$$x_1'(t) = x_1(t) + 2x_2(t) + x_3(t)$$

$$x_2'(t) = 6x_1(t) - x_2(t)$$

$$x_3'(t) = -x_1(t) - 2x_2(t) - x_3(t)$$

with initial conditions

$$x_1(0) = 2, x_2(0) = 0, x_3(0) = 1.$$

4. Solve the system of differential equations

$$x_1'(t) = -x_1(t) + 2x_2(t) + 2x_3(t)$$

$$x_2'(t) = 2x_1(t) + 2x_2(t) - x_3(t)$$

$$x_3'(t) = 2x_1(t) - x_2(t) + 2x_3(t)$$

with initial conditions

$$x_1(0) = 3, \ x_2(0) = -2, \ x_3(0) = 1.$$

Chapter 6

Inner Product Spaces

Up to this point, we have made qualitative deductions about vector spaces, but we have not derived quantitative properties (except in Section 3.1). An example of a quantitative property that would almost certainly be of interest is the length of a vector. In this chapter we introduce one way of forming the product of vectors, called the inner product, that associates a scalar with each pair of vectors. Among other things, an inner product enables us to generalize the notion of length that occurs in Euclidean spaces.

The terms "inner product" and "dot product" are often used synonymously, but they are not exactly the same. The term dot product is used only when the scalar field is $\mathbb{R}$. The inner product generalizes the dot product and applies when the scalar field is $\mathbb{R}$ or $\mathbb{C}$.

In Section 3.1 we introduced vector spaces by analyzing $\mathbb{R}^2$ as an example. That vector space is a dot product space and provides some intuition for this chapter.

Note: In this chapter there are some subsections that are marked as optional. The material in these sections is beyond what is normally done in a first linear algebra course. Omitting them will not compromise later material.

Section 6.1 Some Facts About Complex Numbers

In this chapter complex numbers come more to the forefront of our discussion. Here we give enough background in complex numbers to study vector spaces where the scalars are the complex numbers.

Complex numbers, denoted by $\mathbb{C}$, are defined by

$$\mathbb{C} = \left\{ a + bi \,|\, a, b \in \mathbb{R} \right\}$$

where $i = \sqrt{-1}$.

Complex numbers are usually represented as z, where $z = a + bi$, with $a, b \in \mathbb{R}$. The arithmetic operations are what one would expect. For example if

$$z_1 = a_1 + b_1 i \ \text{ and } \ z_2 = a_2 + b_2 i$$

then

$$z_1 + z_2 = (a_1 + b_1 i) + (a_2 + b_2 i) = (a_1 + a_2) + (b_1 + b_2) i$$

and

$$z_1 z_2 = (a_1 + b_1 i)(a_2 + b_2 i) = a_1 a_2 + (a_1 b_2 + b_1 a_2) i + b_1 b_2 i^2$$
$$= (a_1 a_2 - b_1 b_2) + (a_1 b_2 + b_1 a_2) i.$$

If $z = a + bi$, then the real part of z is a and the imaginary part of z is b. These are often denoted *Re z* and *Im z*, respectively.

If $z = a + bi$, then the complex conjugate of z, denoted $\overline{z}$, is defined by

$$\overline{z} = a - bi.$$

To plot complex numbers, one uses a two-dimensional grid (see Figure 6.1.)

The modulus of a complex number z, denoted $|z|$, is the distance of the number z from the origin. For $z = a + bi$, we have

$$|z|^2 = a^2 + b^2 = z\overline{z}.$$

If the vector space is $\mathbb{C}^n$ the formula for the usual dot product breaks down in that it does not necessarily give the length of a vector. For example, with the definition of the Euclidean dot product, the length of the vector $(3i, 2i)$ would be $\sqrt{-13}$. In $\mathbb{C}^n$, the length of the vector $(a_1, \ldots, a_n)$ should be the distance from the origin to the terminal point of the vector, which is

$$\sqrt{a_1 \overline{a_1} + \cdots + a_n \overline{a_n}} = \sqrt{|a_1|^2 + \cdots + |a_n|^2}.$$

The usual inner product in $\mathbb{C}^n$, is if $\hat{u} = (u_1, \ldots, u_n)$ and $\hat{v} = (v_1, \ldots, v_n)$, then

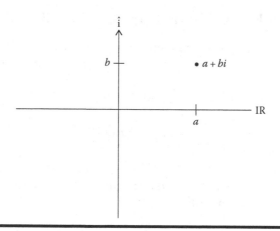

Figure 6.1 **Plotting a complex number in the complex plane.**

$$\langle \hat{u}, \hat{v} \rangle = \sum_{i=1}^{n} u_i \overline{v_i}.$$

Exercises

In Exercises 1–4 perform the arithmetic operations.

1. $(2 + 4i) - 3(-1 + 2i)$

2. $2(5 - 3i)(-6 + i)$

3. $6(2 + i)\overline{(3 - i)}$

4. $|-2 - 3i|$

5. Show that if

$$z = a + bi$$

then

$$z^{-1} = \frac{1}{z} = \frac{1}{a^2 + b^2}(a - bi)$$

if a and b are not both 0.

6. Find the usual inner product, $\langle \hat{u}, \hat{v} \rangle$, for the following pairs of vectors.

   (a.) $\hat{u} = (1 - 2i, 3 + 6i, 2), \hat{v} = (i, -2i, 3 - 2i)$

   (b.) $\hat{u} = (i, 3, 2 - i), \hat{v} = (3, 4 + i, 7i)$

   (c.) $\hat{u} = (2i, 5i, -3i), \hat{v} = (1, 5, 3)$

   (d.) $\hat{u} = (1, 5, 3), \hat{v} = (2i, 5i, -3i)$

7. In this problem we demonstrate a connection between complex numbers and 2×2 matrices. We associate

$$1 \text{ with } \begin{pmatrix} 1 & 0 \\ 0 & 1 \end{pmatrix} \text{ and } i \text{ with } \begin{pmatrix} 0 & -1 \\ 1 & 0 \end{pmatrix}.$$

   (a.) Show that

$$\begin{pmatrix} 0 & -1 \\ 1 & 0 \end{pmatrix}^2 = \begin{pmatrix} -1 & 0 \\ 0 & -1 \end{pmatrix} = -\begin{pmatrix} 1 & 0 \\ 0 & 1 \end{pmatrix},$$

   analogous to $i^2 = -1$.
   Identify

$$a + bi \sim \begin{pmatrix} a & -b \\ b & a \end{pmatrix}.$$

(b.) Identify $z_1 \sim \begin{pmatrix} a & -b \\ b & a \end{pmatrix}$ and $z_2 \sim \begin{pmatrix} c & -d \\ d & c \end{pmatrix}$. Show that

$$\begin{pmatrix} a & -b \\ b & a \end{pmatrix}\begin{pmatrix} c & -d \\ d & c \end{pmatrix} = \begin{pmatrix} c & -d \\ d & c \end{pmatrix}\begin{pmatrix} a & -b \\ b & a \end{pmatrix}$$

as $z_1 z_2 = z_2 z_1$ even though matrices do not typically commute in multiplication.

(c.) Show that if $a + bi \sim \begin{pmatrix} a & -b \\ b & a \end{pmatrix}$, then $\overline{a + bi} \sim \begin{pmatrix} a & -b \\ b & a \end{pmatrix}^T$

(d.) Show that if $z = a + bi = \sim \begin{pmatrix} a & -b \\ b & a \end{pmatrix}$ then

$$\det\begin{pmatrix} a & -b \\ b & a \end{pmatrix} = a^2 + b^2 = |z|^2.$$

(e.) In complex numbers, Euler's theorem states that

$$e^{i\theta} = \cos\theta + i\sin\theta.$$

In the matrix analogy, this says

$$\exp\begin{pmatrix} 0 & -\theta \\ \theta & 0 \end{pmatrix} = \begin{pmatrix} \cos\theta & -\sin\theta \\ \sin\theta & \cos\theta \end{pmatrix}.$$

Recall that $\begin{pmatrix} \cos\theta & -\sin\theta \\ \sin\theta & \cos\theta \end{pmatrix}$ is the rotation matrix in two dimensions.

Taking derivatives gives

$$\frac{d}{d\theta}e^{i\theta} = ie^{i\theta} = i\cos\theta - \sin\theta \text{ whose matrix representation is}$$

$$\begin{pmatrix} -\sin\theta & -\cos\theta \\ \cos\theta & -\sin\theta \end{pmatrix} = \begin{pmatrix} 0 & -1 \\ 1 & 0 \end{pmatrix}\begin{pmatrix} \cos\theta & -\sin\theta \\ \sin\theta & \cos\theta \end{pmatrix} = i\exp\begin{pmatrix} 0 & -\theta \\ \theta & 0 \end{pmatrix}.$$

Section 6.2 Inner Product Spaces

Definition:

Let V be a vector space with scalar field $\mathcal{F}$ ($\mathbb{R}$ or $\mathbb{C}$). An inner product on V is a function

$$\langle,\rangle : V \times V \to \mathcal{F}$$

that satisfies the following conditions for all $\hat{u}, \hat{v}, \hat{w} \in V$ and $\lambda \in \mathcal{F}$:

1. $\langle \hat{u}, \hat{u} \rangle \geq 0$ and $\langle \hat{u}, \hat{u} \rangle = 0$ if and only if $\hat{u} = \hat{0}$
2. $\langle \hat{u} + \hat{v}, \hat{w} \rangle = \langle \hat{u}, \hat{w} \rangle + \langle \hat{v}, \hat{w} \rangle$
3. $\langle \lambda \hat{u}, \hat{v} \rangle = \lambda \langle \hat{u}, \hat{v} \rangle$
4. $\langle \hat{u}, \hat{v} \rangle = \overline{\langle \hat{v}, \hat{u} \rangle}$.

If the scalar field is $\mathbb{R}$, then condition 4 becomes $\langle \hat{u}, \hat{v} \rangle = \langle \hat{v}, \hat{u} \rangle$ so the definition is the same as that for the dot product.

Definition:

A vector space together with an inner product is called an inner product space.

As a consequence of Property 4,

$$\langle \hat{u}, \hat{v} \rangle \langle \hat{v}, \hat{u} \rangle = \langle \hat{u}, \hat{v} \rangle \overline{\langle \hat{u}, \hat{v} \rangle} = \left| \langle \hat{u}, \hat{v} \rangle \right|^2$$

and, as a consequence of Properties 3 and 4,

$$\langle \hat{u}, \lambda \hat{v} \rangle = \overline{\lambda} \langle \hat{u}, \hat{v} \rangle.$$

At the end of this equation, the vertical brackets go outside the pointed brackets and the exponent outside of all: $\left| \langle u, v \rangle \right|^2$

Also, $\langle \hat{v}, \hat{0} \rangle = \langle \hat{0}, \hat{v} \rangle = 0$ for every vector $\hat{v}$ since

$$\langle \hat{0}, \hat{v} \rangle = \langle \hat{0} + \hat{0}, \hat{v} \rangle = \langle \hat{0}, \hat{v} \rangle + \langle \hat{0}, \hat{v} \rangle = 2 \langle \hat{0}, \hat{v} \rangle.$$

Theorem 6.1: (Cauchy–Schwarz Inequality)

If $\hat{u}$ and $\hat{v}$ are vectors in an inner product space V, then

$$\left| \langle \hat{u}, \hat{v} \rangle \right| \leq \|\hat{u}\| \|\hat{v}\|.$$

Proof:

The result is clear if $\hat{u} = \hat{0}$, $\hat{v} = \hat{0}$ or $\left| \langle \hat{u}, \hat{v} \rangle \right| = 0$ so suppose that none of these is the case.

For any scalar α we have

$$0 \leq \|\hat{u} - \alpha \hat{v}\|^2 = \langle \hat{u} - \alpha \hat{v}, \hat{u} - \alpha \hat{v} \rangle = \langle \hat{u}, \hat{u} \rangle - \overline{\alpha} \langle \hat{u}, \hat{v} \rangle - \alpha \langle \hat{v}, \hat{u} \rangle + \alpha \overline{\alpha} \langle \hat{v}, \hat{v} \rangle. \quad (1)$$

The rest of the proof amounts to making a judicious choice for α. Let

$$\alpha = \frac{\langle \hat{u}, \hat{v} \rangle}{\langle \hat{v}, \hat{v} \rangle}.$$

Then

$$\bar{\alpha} = \frac{\overline{\langle \hat{u}, \hat{v} \rangle}}{\langle \hat{v}, \hat{v} \rangle} \quad \text{and} \quad \alpha\bar{\alpha} = \frac{\left|\langle \hat{u}, \hat{v} \rangle\right|^2}{\langle \hat{v}, \hat{v} \rangle \langle \hat{v}, \hat{v} \rangle}$$

so we can rewrite Equation (1) to get

$$0 \le \|\hat{u}\|^2 - \frac{\overline{\langle \hat{u}, \hat{v} \rangle}}{\langle \hat{v}, \hat{v} \rangle} \langle \hat{u}, \hat{v} \rangle - \frac{\langle \hat{u}, \hat{v} \rangle}{\langle \hat{v}, \hat{v} \rangle} \langle \hat{v}, \hat{u} \rangle + \frac{\left|\langle \hat{u}, \hat{v} \rangle\right|^2}{\langle \hat{v}, \hat{v} \rangle \langle \hat{v}, \hat{v} \rangle} \langle \hat{v}, \hat{v} \rangle$$

$$= \|\hat{u}\|^2 - \frac{\left|\langle \hat{u}, \hat{v} \rangle\right|^2}{\|\hat{v}\|^2} - \frac{\left|\langle \hat{u}, \hat{v} \rangle\right|^2}{\|\hat{v}\|^2} + \frac{\left|\langle \hat{u}, \hat{v} \rangle\right|^2}{\|\hat{v}\|^4} \|\hat{v}\|^2.$$

Thus,

$$\|\hat{u}\|^2 + \frac{\left|\langle \hat{u}, \hat{v} \rangle\right|^2}{\|\hat{v}\|^2} \ge 2 \frac{\left|\langle \hat{u}, \hat{v} \rangle\right|^2}{\|\hat{v}\|^2}$$

so

$$\|\hat{u}\|^2 \ge \frac{\left|\langle \hat{u}, \hat{v} \rangle\right|^2}{\|\hat{v}\|^2}$$

and

$$\|\hat{u}\| \, \|\hat{v}\| \ge \left|\langle \hat{u}, \hat{v} \rangle\right|.$$

Historical Note

Hermann Schwarz (born 25 January 1843, died 30 November 1921)

Schwarz began his study in chemistry at Berlin but it was not long before Kummer and Weierstrass had influenced him to change to mathematics. Through them, Schwarz became interested in geometry. Schwarz attended Weierstrass's lectures on *the integral calculus* in 1861 and the notes that Schwarz took at these lectures still exist. His interest in geometry was soon combined with Weierstrass's ideas of analysis.

After Schwarz succeeded Weierstrass in accepting a professorship in Berlin in 1892, the reputation of the most eminent university in Germany for mathematics, which had been Berlin, began to shift toward Göttingen.

The Cauchy–Schwarz inequality appears in arithmetic, geometric, and function-theoretic formulations. The standard modern proof seems to have been first given by Weyl in 1918.

Exercises

1. (a.) Show that for the vector space of real-valued continuous functions on the unit interval [0,1] the function $\langle, \rangle$ given by

$$\langle f, g \rangle = \int_0^1 f(x) g(x) \, dx$$

is an inner product.

(a.) With the inner product above find $\|f\|$ if $f(x) = x^2 + x$.

(b.) With the inner product above find $\|f - g\|$ if

$$f(x) = x^2 + x \text{ and } g(x) = x - 1.$$

2. Show that the function

$$\left\langle \begin{pmatrix} x_1 \\ y_1 \end{pmatrix}, \begin{pmatrix} x_2 \\ y_2 \end{pmatrix} \right\rangle = 2x_1 x_2 + y_1 y_2$$

is an inner product on $\mathbb{R}^2$.

(a.) With the inner product above find $\|\hat{u}\|$ if

$$\hat{u} = \begin{pmatrix} 3 \\ -4 \end{pmatrix}.$$

(b.) With the inner product above find $\|\hat{u} - \hat{v}\|$ if

$$\hat{u} = \begin{pmatrix} 3 \\ -4 \end{pmatrix} \text{ and } \hat{v} = \begin{pmatrix} 2 \\ 2 \end{pmatrix}.$$

3. Show why

$$\left\langle \begin{pmatrix} x_1 \\ y_1 \end{pmatrix}, \begin{pmatrix} x_2 \\ y_2 \end{pmatrix} \right\rangle = x_1^2 x_2 + y_1 y_2$$

is not an inner product on $\mathbb{R}^2$.

4. Show that for any inner product space

$$\|\hat{u} + \hat{v}\|^2 + \|\hat{u} - \hat{v}\|^2 = 2\|\hat{u}\|^2 + 2\|\hat{v}\|^2.$$

5. Show that for any inner product space

$$\|\hat{u} + \hat{v}\|^2 - \|\hat{u} - \hat{v}\|^2 = 4\langle \hat{u}, \hat{v} \rangle.$$

6. Show that in $\mathbb{R}^2$ with the usual inner product and A a 2×2 matrix

$$\langle \hat{u}, A\hat{v} \rangle = \langle A^T \hat{u}, \hat{v} \rangle.$$

Section 6.3 Orthogonality

With the Euclidean dot product in $\mathbb{R}^2$ and $\mathbb{R}^3$, we had a method of defining the angle between two vectors, but with an inner product this cannot be done in $\mathbb{C}^n$ (with the usual interpretation of angle) because $\langle \hat{u}, \hat{v} \rangle$ can be complex. However, we will define what it means for vectors to be orthogonal.

Definition:

A set of non-zero vectors $\{\hat{v}_1, \ldots, \hat{v}_n\}$ in the inner product space V is an orthogonal set if $\langle \hat{v}_i, \hat{v}_j \rangle = 0$ whenever $i \neq j$. An orthonormal set is an orthogonal set that has the additional property that $\langle \hat{v}_i, \hat{v}_i \rangle = 1$ for $i = 1, \ldots, n$.

If we have an orthogonal set $\{\hat{v}_1, \ldots, \hat{v}_n\}$, it can be converted to the orthonormal set

$$\left\{ \frac{\hat{v}_1}{\|\hat{v}_1\|}, \ldots, \frac{\hat{v}_n}{\|\hat{v}_n\|} \right\}.$$

Orthogonality generalizes the idea of being perpendicular to settings other than $\mathbb{R}^n$.

Theorem 6.2:

An orthogonal set of vectors on an inner product space is a linearly independent set.

Proof:

Let $\{\hat{v}_1, \ldots, \hat{v}_n\}$ be an orthogonal set of vectors in the inner product space V, and suppose that

$$a_1 \hat{v}_1 + \cdots + a_n \hat{v}_n = \hat{0}.$$

Now

$$\langle a_1 \hat{v}_1 + \cdots + a_n \hat{v}_n, \hat{v}_1 \rangle = a_1 \langle \hat{v}_1, \hat{v}_1 \rangle + a_2 \langle \hat{v}_2, \hat{v}_1 \rangle + \cdots + a_n \langle \hat{v}_n, \hat{v}_1 \rangle = a_1 \langle \hat{v}_1, \hat{v}_1 \rangle$$

but

$$\langle a_1 \hat{v}_1 + \cdots + a_n \hat{v}_n, \hat{v}_1 \rangle = \langle \hat{0}, \hat{v}_1 \rangle = 0$$

so

$$a_1 \langle \hat{v}_1, \hat{v}_1 \rangle = 0.$$

Since $\langle \hat{v}_1, \hat{v}_1 \rangle \neq 0$, then $a_1 = 0$. Similarly, $a_i = 0$; $i = 2, \ldots, n$.

Corollary:

An orthogonal spanning set of vectors for an inner product space is a basis.

The following example shows why orthogonal bases – and why, even more so, orthonormal bases – are convenient.

Example

Suppose that $\{\hat{v}_1,\ldots,\hat{v}_n\}$ is an orthogonal basis for the inner product space V. If $\hat{v}$ is a vector in V for which

$$\hat{v} = a_1\hat{v}_1 + \cdots + a_n\hat{v}_n,$$

we can find a formula for a_i.

We find the formula for a_1. We have

$$\langle \hat{v},\hat{v}_1 \rangle = \langle a_1\hat{v}_1 + \cdots + a_n\hat{v}_n, \hat{v}_1 \rangle = a_1 \langle \hat{v}_1,\hat{v}_1 \rangle + a_2 \langle \hat{v}_2,\hat{v}_1 \rangle + \cdots + a_n \langle \hat{v}_n,\hat{v}_1 \rangle = a_1 \langle \hat{v}_1,\hat{v}_1 \rangle$$
$$a_1 = \frac{\langle \hat{v},\hat{v}_1 \rangle}{\langle \hat{v}_1,\hat{v}_1 \rangle}.$$

Likewise,

$$a_i = \frac{\langle \hat{v},\hat{v}_i \rangle}{\langle \hat{v}_i,\hat{v}_i \rangle}; \quad i = 1,\ldots,n.$$

If $\{\hat{v}_1,\ldots,\hat{v}_n\}$ is an orthonormal basis for the inner product space V, then we have the simpler formula

$$a_i = \langle \hat{v},\hat{v}_i \rangle; \quad i = 1,\ldots,n.$$

From this we get the following important result.

Theorem 6.3:

If $\{\hat{v}_1,\ldots,\hat{v}_n\}$ is an orthonormal basis for the inner product space V, then for $\hat{v}$ a vector in V we have

$$\hat{v} = \langle \hat{v},\hat{v}_1 \rangle \hat{v}_1 + \cdots + \langle \hat{v},\hat{v}_n \rangle \hat{v}_n.$$

Theorem 6.4:

If $\{\hat{v}_1,\ldots,\hat{v}_n\}$ is an orthonormal basis for the inner product space V, then for $\hat{v}$ a vector in V we have

$$\|\hat{v}\|^2 = \sum_{i=1}^{n} |\langle \hat{v},\hat{v}_i \rangle|^2.$$

Proof:

By Theorem 6.3 we have

$$\|\hat{v}\|^2 = \langle \hat{v}, \hat{v} \rangle = \left\langle \left(\langle \hat{v}, \hat{v}_1 \rangle \hat{v}_1 + \cdots + \langle \hat{v}, \hat{v}_n \rangle \hat{v}_n, \langle \hat{v}, \hat{v}_1 \rangle \hat{v}_1 + \cdots + \langle \hat{v}, \hat{v}_n \rangle \hat{v}_n \right) \right\rangle. \tag{1}$$

Since $\{\hat{v}_1, \ldots, \hat{v}_n\}$ is an orthogonal set expression, Equation (1) is equal to

$$\sum_{i=1}^{n} \left\langle \langle \hat{v}, \hat{v}_i \rangle \hat{v}_i, \langle \hat{v}, \hat{v}_i \rangle \hat{v}_i \right\rangle = \sum_{i=1}^{n} \langle \hat{v}, \hat{v}_i \rangle \overline{\langle \hat{v}, \hat{v}_i \rangle} \langle \hat{v}_i, \hat{v}_i \rangle = \sum_{i=1}^{n} \langle \hat{v}, \hat{v}_i \rangle \overline{\langle v, \hat{v}_i \rangle} \|\hat{v}_i\|^2. \tag{2}$$

Since $\{\hat{v}_1, \ldots, \hat{v}_n\}$ is an orthonormal set expression, Equation (2) is equal to

$$\sum_{i=1}^{n} \langle \hat{v}, \hat{v}_i \rangle \overline{\langle v, \hat{v}_i \rangle} = \sum_{i=1}^{n} |\langle \hat{v}, \hat{v}_i \rangle|^2.$$

Example

The set of vectors

$$\left\{ \begin{pmatrix} 1 \\ 0 \\ 0 \end{pmatrix}, \begin{pmatrix} 0 \\ 1/\sqrt{2} \\ 1/\sqrt{2} \end{pmatrix}, \begin{pmatrix} 0 \\ -1/\sqrt{2} \\ 1/\sqrt{2} \end{pmatrix} \right\}$$

is an orthonormal basis of $\mathbb{R}^3$. We express the vector

$$\hat{v} = \begin{pmatrix} -3 \\ 4 \\ 7 \end{pmatrix}$$

in terms of this basis. We have

$$\langle \hat{v}, \hat{v}_1 \rangle = -3, \quad \langle \hat{v}, \hat{v}_2 \rangle = \frac{4}{\sqrt{2}} + \frac{7}{\sqrt{2}} = \frac{11}{\sqrt{2}}, \quad \langle \hat{v}, \hat{v}_3 \rangle = -\frac{4}{\sqrt{2}} + \frac{7}{\sqrt{2}} = \frac{3}{\sqrt{2}}$$

so

$$\hat{v} = \langle \hat{v}, \hat{v}_1 \rangle \hat{v}_1 + \langle \hat{v}, \hat{v}_2 \rangle \hat{v}_2 + \langle \hat{v}, \hat{v}_3 \rangle \hat{v}_3 = -3 \begin{pmatrix} 1 \\ 0 \\ 0 \end{pmatrix} + \frac{11}{\sqrt{2}} \begin{pmatrix} 0 \\ 1/\sqrt{2} \\ 1/\sqrt{2} \end{pmatrix} + \frac{3}{\sqrt{2}} \begin{pmatrix} 0 \\ -1/\sqrt{2} \\ 1/\sqrt{2} \end{pmatrix}$$

Exercises

1. For the sets of vectors below (i) tell whether the set is orthogonal, orthonormal, or neither; (ii) for those that are orthogonal or orthonormal find the coordinates of the given vector $\hat{v}$ with respect to that basis.

(a.) $\hat{v}_1 = \begin{pmatrix} 1 \\ -4 \end{pmatrix}$, $\hat{v}_2 = \begin{pmatrix} 4 \\ 1 \end{pmatrix}$, $\hat{v} = \begin{pmatrix} 7 \\ -3 \end{pmatrix}$, (b.) $\hat{v}_1 = \begin{pmatrix} \dfrac{1}{\sqrt{2}} \\ \dfrac{1}{\sqrt{2}} \end{pmatrix}$, $\hat{v}_2 = \begin{pmatrix} -\dfrac{1}{\sqrt{2}} \\ \dfrac{1}{\sqrt{2}} \end{pmatrix}$, $\hat{v} = \begin{pmatrix} 2 \\ 5 \end{pmatrix}$

(c.) $\hat{v}_1 = \begin{pmatrix} \dfrac{1}{\sqrt{5}} \\ -\dfrac{2}{\sqrt{5}} \end{pmatrix}$, $\hat{v}_2 = \begin{pmatrix} -\dfrac{2}{\sqrt{5}} \\ -\dfrac{1}{\sqrt{5}} \end{pmatrix}$, $\hat{v} = \begin{pmatrix} 0 \\ 3 \end{pmatrix}$

(d.) $\hat{v}_1 = \begin{pmatrix} \dfrac{1}{\sqrt{3}} \\ \dfrac{1}{\sqrt{3}} \\ \dfrac{1}{\sqrt{3}} \end{pmatrix}$, $\hat{v}_2 = \begin{pmatrix} \dfrac{-2}{\sqrt{6}} \\ \dfrac{1}{\sqrt{6}} \\ \dfrac{1}{\sqrt{6}} \end{pmatrix}$, $\hat{v}_3 = \begin{pmatrix} 0 \\ \dfrac{-1}{\sqrt{2}} \\ \dfrac{1}{\sqrt{2}} \end{pmatrix}$, $\hat{v} = \begin{pmatrix} 3 \\ 0 \\ -8 \end{pmatrix}$

(e.) $\hat{v}_1 = \begin{pmatrix} -3 \\ -1 \\ 1 \end{pmatrix}$, $\hat{v}_2 = \begin{pmatrix} \dfrac{1}{\sqrt{6}} \\ \dfrac{-2}{\sqrt{6}} \\ \dfrac{-1}{\sqrt{6}} \end{pmatrix}$, $\hat{v}_3 = \begin{pmatrix} 1 \\ 4 \\ -7 \end{pmatrix}$, $\hat{v} = \begin{pmatrix} 5 \\ -1 \\ 3 \end{pmatrix}$.

Section 6.4 The Gram–Schmidt Process

Our goal in this section is to show that if V is an inner product space, then there is an orthonormal basis for V and to determine an algorithm to construct such a basis.

The process we will use to convert a given set of vectors to an orthonormal set of vectors is called the Gram–Schmidt process. There are two approaches that can be used to apply the Gram–Schmidt process. One is to construct an orthogonal set of vectors and when that task is completed, convert the orthogonal set to an orthonormal set by normalizing the vectors. The other approach is to normalize the vectors at each step of the process. We will use the former method.

Historical Note

Erhardt Schmidt (born 13 January 1876, died 6 December 1959)

Schmidt obtained his doctorate from the University of Göttingen in 1905 under Hilbert's supervision.

Schmidt's main interest was in integral equations and Hilbert space. He took various ideas of Hilbert on integral equations and combined these into the concept of a Hilbert space around 1905. He expanded functions related to the integral of the kernel function as an infinite series in a set of orthonormal eigenfunctions.

Schmidt published a two-part paper on integral equations in 1907 in which he re-proved Hilbert's results in a simpler fashion, and also with fewer restrictions. In this paper he gave what is now called the Gram–Schmidt orthonormalization process for constructing an orthonormal set of functions from a linearly independent set. We note that Laplace presented the Gram–Schmidt process before either Gram or Schmidt.

Jurgen Gram (born 27 June 1850, died 29 April 1916)

Gram is best remembered for the Gram–Schmidt orthogonalization process which constructs an orthogonal set of vectors from an independent one. He was not, however, the first to use this method. The process seems to be a result of Laplace and it was essentially used by Cauchy in 1836.

Gram's mathematical career was a balance between pure mathematics and practical applications of the subject. The practical applications to forestry, which he continued to study at this time, and his work on probability and numerical analysis involved both the theory and its application to practical situations.

Gram's work on probability and numerical analysis led him to study abstract problems in number theory. In 1884 he won the Gold Medal of the Royal Danish Academy of Sciences for his paper *Investigations of the number of primes less than a given number* which he published in the *Journal of the Society*. Gram also worked on the Riemann zeta function.

Gram died at age 65 while he was on his way to a meeting of the Danish Academy when he was struck and killed by a bicycle.

Algorithm for the Gram–Schmidt Process

We first give an algorithm that converts a basis to an orthogonal basis, demonstrate the use of the algorithm, and then prove a theorem that validates the ideas.

Let $\{\hat{x}_1,\ldots,\hat{x}_k\}$ be a basis for the subspace X of the vector space V. Define

$$\hat{v}_1 = \hat{x}_1$$

$$\hat{v}_2 = \hat{x}_2 - \frac{\langle \hat{x}_2,\hat{v}_1 \rangle}{\langle \hat{v}_1,\hat{v}_1 \rangle}\hat{v}_1$$

$$\hat{v}_3 = \hat{x}_3 - \frac{\langle \hat{x}_3,\hat{v}_1 \rangle}{\langle \hat{v}_1,\hat{v}_1 \rangle}\hat{v}_1 - \frac{\langle \hat{x}_3,\hat{v}_2 \rangle}{\langle \hat{v}_2,\hat{v}_2 \rangle}\hat{v}_2$$

$$\vdots$$

$$\hat{v}_k = \hat{x}_k - \frac{\langle \hat{x}_k,\hat{v}_1 \rangle}{\langle \hat{v}_1,\hat{v}_1 \rangle}\hat{v}_1 - \frac{\langle \hat{x}_k,\hat{v} \rangle_2}{\langle \hat{v}_2,\hat{v}_2 \rangle}\hat{v}_2 - \cdots - \frac{\langle \hat{x}_k,\hat{v}_{k-1} \rangle}{\langle \hat{v}_{k-1},\hat{v}_{k-1} \rangle}\hat{v}_{k-1}.$$

After the next example, we will show that $\{\hat{v}_1,\ldots,\hat{v}_k\}$ is an orthogonal set and

$$\text{span}\{\hat{v}_1,\ldots,\hat{v}_k\} = \text{span}\{\hat{x}_1,\ldots,\hat{x}_k\}; \quad k=1,\ldots,n.$$

Example

Find the orthogonal basis that is derived from the basis

$$\left\{ \begin{pmatrix} 1 \\ 1 \\ 1 \end{pmatrix}, \begin{pmatrix} 2 \\ 0 \\ 4 \end{pmatrix}, \begin{pmatrix} -3 \\ 1 \\ 2 \end{pmatrix} \right\}$$

using the Gram–Schmidt process.

We have

$$\hat{x}_1 = \begin{pmatrix} 1 \\ 1 \\ 1 \end{pmatrix}, \quad \hat{x}_2 = \begin{pmatrix} 2 \\ 0 \\ 4 \end{pmatrix}, \quad \hat{x}_3 = \begin{pmatrix} -3 \\ 1 \\ 2 \end{pmatrix}$$

so

$$\hat{v}_1 = \hat{x}_1 = \begin{pmatrix} 1 \\ 1 \\ 1 \end{pmatrix}.$$

Also,

$$\hat{v}_2 = \hat{x}_2 - \frac{\langle \hat{x}_2, \hat{v}_1 \rangle}{\langle \hat{v}_1, \hat{v}_1 \rangle} \hat{v}_1 = \begin{pmatrix} 2 \\ 0 \\ 4 \end{pmatrix} - \frac{\begin{pmatrix} 2 \\ 0 \\ 4 \end{pmatrix} \cdot \begin{pmatrix} 1 \\ 1 \\ 1 \end{pmatrix}}{\begin{pmatrix} 1 \\ 1 \\ 1 \end{pmatrix} \cdot \begin{pmatrix} 1 \\ 1 \\ 1 \end{pmatrix}} \begin{pmatrix} 1 \\ 1 \\ 1 \end{pmatrix} = \begin{pmatrix} 2 \\ 0 \\ 4 \end{pmatrix} - \frac{6}{3} \begin{pmatrix} 1 \\ 1 \\ 1 \end{pmatrix} = \begin{pmatrix} 0 \\ -2 \\ 2 \end{pmatrix}$$

and

$$\hat{v}_3 = \hat{x}_3 - \frac{\langle \hat{x}_3, \hat{v}_1 \rangle}{\langle \hat{v}_1, \hat{v}_1 \rangle} \hat{v}_1 - \frac{\langle \hat{x}_3, \hat{v}_2 \rangle}{\langle \hat{v}_2, \hat{v}_2 \rangle} \hat{v}_2$$

$$= \begin{pmatrix} -3 \\ 1 \\ 2 \end{pmatrix} - \frac{\begin{pmatrix} -3 \\ 1 \\ 2 \end{pmatrix} \cdot \begin{pmatrix} 1 \\ 1 \\ 1 \end{pmatrix}}{\begin{pmatrix} 1 \\ 1 \\ 1 \end{pmatrix} \cdot \begin{pmatrix} 1 \\ 1 \\ 1 \end{pmatrix}} \begin{pmatrix} 1 \\ 1 \\ 1 \end{pmatrix} - \frac{\begin{pmatrix} -3 \\ 1 \\ 2 \end{pmatrix} \cdot \begin{pmatrix} 0 \\ -2 \\ 2 \end{pmatrix}}{\begin{pmatrix} 0 \\ -2 \\ 2 \end{pmatrix} \cdot \begin{pmatrix} 0 \\ -2 \\ 2 \end{pmatrix}} \begin{pmatrix} 0 \\ -2 \\ 2 \end{pmatrix}$$

$$= \begin{pmatrix} -3 \\ 1 \\ 2 \end{pmatrix} - \frac{0}{3} \begin{pmatrix} 1 \\ 1 \\ 1 \end{pmatrix} - \frac{2}{8} \begin{pmatrix} 0 \\ -2 \\ 2 \end{pmatrix} = \begin{pmatrix} -3 \\ 1 \\ 2 \end{pmatrix} - \begin{pmatrix} 0 \\ -\frac{1}{2} \\ \frac{1}{2} \end{pmatrix} = \begin{pmatrix} -3 \\ \frac{3}{2} \\ \frac{3}{2} \end{pmatrix}.$$

Thus, the orthogonal basis generated by the Gram–Schmidt process is

$$\{\hat{v}_1, \hat{v}_2, \hat{v}_3\} = \left\{ \begin{pmatrix} 1 \\ 1 \\ 1 \end{pmatrix}, \begin{pmatrix} 0 \\ -2 \\ 2 \end{pmatrix}, \begin{pmatrix} -3 \\ \frac{3}{2} \\ \frac{3}{2} \end{pmatrix} \right\}.$$

It is often the case that one would prefer an orthonormal basis, which in this case is

$$\left\{ \frac{\hat{v}_1}{\|\hat{v}_1\|}, \frac{\hat{v}_2}{\|\hat{v}_2\|}, \frac{\hat{v}_3}{\|\hat{v}_3\|} \right\} = \left\{ \begin{pmatrix} \frac{1}{\sqrt{3}} \\ \frac{1}{\sqrt{3}} \\ \frac{1}{\sqrt{3}} \end{pmatrix}, \begin{pmatrix} 0 \\ \frac{-2}{\sqrt{8}} \\ \frac{2}{\sqrt{8}} \end{pmatrix}, \begin{pmatrix} \frac{-3}{\sqrt{27/2}} \\ \frac{3/2}{\sqrt{27/2}} \\ \frac{3/2}{\sqrt{27/2}} \end{pmatrix} \right\}.$$

The formulas in the Gram–Schmidt process may appear daunting, but the intuition is not. The process is based on the following fact from Fourier analysis.

Let $\{\hat{v}_1,\ldots,\hat{v}_n\}$ be an orthogonal basis for the inner product space V and suppose that $\hat{v} \in V$. Then for each $k = 1, \ldots, n$

Fact:

$$\left\| \hat{v} - \sum_{j=1}^{k} a_i \hat{v}_i \right\|$$

is minimized when

$$a_i = \frac{\langle \hat{v}, \hat{v}_i \rangle}{\langle \hat{v}_i, \hat{v}_i \rangle}; \quad i = 1,\ldots,n.$$

For a proof of this fact see Kirkwood [*An Introduction to Analysis*].

Later, we will call

$$\sum_{i=1}^{k} \frac{\langle \hat{v}, \hat{v}_i \rangle}{\langle \hat{v}_i, \hat{v}_i \rangle} \hat{v}_i$$

the projection of $\hat{v}$ onto the subspace generated by $\{\hat{v}_1,\ldots,\hat{v}_n\}$. In terms of distance, this is the closest vector in the subspace to the vector $\hat{v}$.

What we are doing in the Gram–Schmidt process is this:

We begin by letting $\hat{v}_1 = \hat{x}_1$. Because of linear independence

$$\hat{x}_k \notin \operatorname{span}\{\hat{x}_1,\ldots,\hat{x}_{k-1}\}; \quad k = 2,\ldots,n.$$

So $\hat{x}_2 \notin \operatorname{span}\{\hat{x}_1\} = \operatorname{span}\{\hat{v}_1\}$ and

$$\frac{\langle \hat{x}_2, \hat{v}_1 \rangle}{\langle \hat{v}_1, \hat{v}_1 \rangle} \hat{v}_1$$

is the best approximation that $\operatorname{span}\{\hat{v}_1\}$ can provide to $\hat{x}_2$. The vector

$$\hat{v}_2 = \hat{x}_2 - \frac{\langle \hat{x}_2, \hat{v}_1 \rangle}{\langle \hat{v}_1, \hat{v}_1 \rangle} \hat{v}_1$$

is the residual between $\hat{x}_2$ and the best approximation that span$\{\hat{v}_1\}$ can provide to $\hat{x}_2$.

Similarly, $\hat{x}_3 \notin$ span$\{\hat{x}_1, \hat{x}_2\} =$ span$\{\hat{v}_1, \hat{v}_2\}$ and

$$\frac{\langle \hat{x}_3, \hat{v}_1 \rangle}{\langle \hat{v}_1, \hat{v}_1 \rangle} \hat{v}_1 + \frac{\langle \hat{x}_3, \hat{v}_2 \rangle}{\langle \hat{v}_2, \hat{v}_2 \rangle} \hat{v}_2$$

is the best approximation that span$\{\hat{v}_1, \hat{v}_2\}$ can provide to $\hat{x}_3$. The vector

$$\hat{v}_3 = \hat{x}_3 - \frac{\langle \hat{x}_3, \hat{v}_1 \rangle}{\langle \hat{v}_1, \hat{v}_1 \rangle} \hat{v}_1 - \frac{\langle \hat{x}_3, \hat{v}_2 \rangle}{\langle \hat{v}_2, \hat{v}_2 \rangle} \hat{v}_2$$

is the residual between $\hat{x}_2$ and the best approximation that span$\{\hat{v}_1, \hat{v}_2\}$ can provide to $\hat{x}_3$.

Figure 6.2 gives a visual image of why $\hat{v}_3$ is orthogonal to any vector in span$\{\hat{v}_1, \hat{v}_2\}$. The vector $\hat{x}_3$ is not in span$\{\hat{v}_1, \hat{v}_2\}$ and $\hat{v}_3$ is the shortest distance from $\hat{x}_3$ to the plane spanned by $\{\hat{v}_1, \hat{v}_2\}$, which is the perpendicular distance from $\hat{x}_3$ to the plane spanned by $\{\hat{v}_1, \hat{v}_2\}$.

The next result highlights the importance of the Gram–Schmidt process.

Theorem 6.5:

Let V be a finite dimensional inner product space with basis $\{\hat{x}_1, \ldots, \hat{x}_n\}$ and let $\{\hat{v}_1, \ldots, \hat{v}_n\}$ be the vectors generated by the Gram–Schmidt process from $\{\hat{x}_1, \ldots, \hat{x}_n\}$. Then

(a.) $\{\hat{v}_1, \ldots, \hat{v}_n\}$ is an orthogonal set.

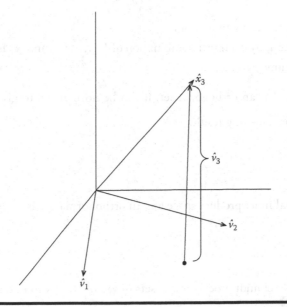

Figure 6.2   The shortest distance from vector $\hat{x}_3$ to the span of $\hat{v}_1$ and $\hat{v}_2$.

(b.) For any $k = 1, \ldots, n$ we have

$$\text{span}\{\hat{v}_1, \ldots, \hat{v}_k\} = \text{span}\{\hat{x}_1, \ldots, \hat{x}_k\}.$$

Proof:

(a.) We proceed by induction. We have

$$\langle \hat{v}_2, \hat{v}_1 \rangle = \left\langle \hat{x}_2 - \frac{\langle \hat{x}_2, \hat{v}_1 \rangle}{\langle \hat{v}_1, \hat{v}_1 \rangle} \hat{v}_1, \hat{v}_1 \right\rangle = \langle \hat{x}_2, \hat{v}_1 \rangle - \frac{\langle \hat{x}_2, \hat{v}_1 \rangle}{\langle \hat{v}_1, \hat{v}_1 \rangle} \langle \hat{v}_1, \hat{v}_1 \rangle = 0.$$

Suppose that $\{\hat{v}_1, \ldots, \hat{v}_{k-1}\}$ is an orthogonal set and

$$\hat{v}_k = \hat{x}_k - \frac{\langle \hat{x}_k, \hat{v}_1 \rangle}{\langle \hat{v}_1, \hat{v}_1 \rangle} \hat{v}_1 - \frac{\langle \hat{x}_k, \hat{v}_2 \rangle}{\langle \hat{v}_2, \hat{v}_2 \rangle} \hat{v}_2 - \cdots - \frac{\langle \hat{x}_k, \hat{v}_{k-1} \rangle}{\langle \hat{v}_{k-1}, \hat{v}_{k-1} \rangle} \hat{v}_{k-1}.$$

Then if $j < k$ we have

$$\langle \hat{v}_k, \hat{v}_j \rangle = \left\langle \hat{x}_k - \frac{\langle \hat{x}_k, \hat{v}_1 \rangle}{\langle \hat{v}_1, \hat{v}_1 \rangle} \hat{v}_1 - \frac{\langle \hat{x}_k, \hat{v}_2 \rangle}{\langle \hat{v}_2, \hat{v}_2 \rangle} \hat{v}_2 - \cdots - \frac{\langle \hat{x}_k, \hat{v}_{k-1} \rangle}{\langle \hat{v}_{k-1}, \hat{v}_{k-1} \rangle} \hat{v}_{k-1}, \hat{v}_j \right\rangle$$

$$= \langle \hat{x}_k, \hat{v}_j \rangle - \frac{\langle \hat{x}_k, \hat{v}_1 \rangle}{\langle \hat{v}_1, \hat{v}_1 \rangle} \langle \hat{v}_1, \hat{v}_j \rangle - \cdots - \frac{\langle \hat{x}_k, \hat{v}_{k-1} \rangle}{\langle \hat{v}_{k-1}, \hat{v}_{k-1} \rangle} \langle \hat{v}_{k-1}, \hat{v}_j \rangle.$$

Since $\{\hat{v}_1, \ldots, \hat{v}_{k-1}\}$ is an orthogonal set, each term $\langle \hat{v}_i, \hat{v}_j \rangle = 0$ except $\langle \hat{v}_j, \hat{v}_j \rangle$. Thus,

$$\langle \hat{v}_k, \hat{v}_j \rangle = \langle \hat{x}_k, \hat{v}_j \rangle - \frac{\langle \hat{x}_k, \hat{v}_j \rangle}{\langle \hat{v}_j, \hat{v}_j \rangle} \langle \hat{v}_j, \hat{v}_j \rangle = 0$$

and the result is true by induction.

(b.) By the construction, $\hat{v}_j$ is a linear combination of $\{\hat{x}_1, \ldots, \hat{x}_j\}$ and $\hat{x}_j$ is a linear combination of $\{\hat{v}_1, \ldots, \hat{v}_j\}$ for any $j = 1, \ldots, n$.

Since $\{\hat{v}_1, \ldots, \hat{v}_n\}$ is an orthogonal set, it can be normalized to give an orthonormal set.

Thus, we have the following result.

Corollary:

Every finite dimensional inner product space has an orthonormal basis.

Exercises

1. Apply the Gram–Schmidt process to the sets of vectors below to find an orthonormal basis.

(a.) $\left\{ \begin{pmatrix} 1 \\ 0 \end{pmatrix} \begin{pmatrix} 2 \\ 7 \end{pmatrix} \right\}$

(b.) $\left\{ \begin{pmatrix} 3 \\ -5 \end{pmatrix} \begin{pmatrix} 4 \\ 11 \end{pmatrix} \right\}$

(c.) $\left\{ \begin{pmatrix} 2 \\ 7 \end{pmatrix} \begin{pmatrix} -5 \\ -9 \end{pmatrix} \right\}$

(d.) $\left\{ \begin{pmatrix} 1 \\ 0 \\ 1 \end{pmatrix} \begin{pmatrix} 2 \\ -5 \\ 4 \end{pmatrix} \begin{pmatrix} 1 \\ 1 \\ 1 \end{pmatrix} \right\}$

(e.) $\left\{ \begin{pmatrix} -2 \\ 5 \\ -5 \end{pmatrix} \begin{pmatrix} 1 \\ 3 \\ 1 \end{pmatrix} \begin{pmatrix} 0 \\ 4 \\ 3 \end{pmatrix} \right\}$

(f.) $\left\{ \begin{pmatrix} 3 \\ 1 \\ 0 \end{pmatrix} \begin{pmatrix} 3 \\ 3 \\ 2 \end{pmatrix} \begin{pmatrix} -1 \\ -2 \\ 5 \end{pmatrix} \right\}$

(g.) $\left\{ \begin{pmatrix} 1 \\ 1 \\ 1 \end{pmatrix} \begin{pmatrix} 0 \\ 2 \\ -9 \end{pmatrix} \begin{pmatrix} 4 \\ 2 \\ 10 \end{pmatrix} \right\}$

(h.) $\left\{ \begin{pmatrix} 1 \\ 0 \\ 1 \\ -2 \end{pmatrix} \begin{pmatrix} 1 \\ 1 \\ -3 \\ 4 \end{pmatrix} \begin{pmatrix} 3 \\ 4 \\ 0 \\ 1 \end{pmatrix} \begin{pmatrix} 1 \\ 1 \\ 1 \\ 1 \end{pmatrix} \right\}$

(i.) $\left\{ \begin{pmatrix} 2 \\ 3 \\ 0 \\ 0 \end{pmatrix} \begin{pmatrix} 1 \\ -4 \\ 6 \\ 3 \end{pmatrix} \begin{pmatrix} 5 \\ 1 \\ 1 \\ 2 \end{pmatrix} \begin{pmatrix} 1 \\ 0 \\ 0 \\ 0 \end{pmatrix} \right\}$

2. Use the Gram–Schmidt process to find an orthonormal basis for $P_1(x)$ from $\{1-x, 4+3x\}$ where the inner product is

$$\langle f, g \rangle = \int_0^1 f(x)g(x)\, dx$$

3. Use the Gram–Schmidt process to find an orthonormal basis for $P_2(x)$ from $\{3+x^2, 1+2x, 3x\}$ where the inner product is

$$\langle f, g \rangle = \int_{-1}^1 f(x)g(x)\, dx$$

4. Find an orthonormal basis of $\mathbb{R}^2$ where the inner product is

$$\langle (x_1, y_1), (x_2, y_2) \rangle = x_1 x_2 + 2 y_1 y_2$$

5. Use the Gram–Schmidt process to derive an orthonormal basis of $\mathbb{R}^2$ from $\{(-1,3),(2,8)\}$ where the inner product is

$$\langle (x_1, y_1), (x_2, y_2) \rangle = x_1 x_2 + 3 y_1 y_2$$

Section 6.5 Representation of a Linear Transformation on Inner Product Spaces (Optional)

For inner product spaces finding the representation of a linear transformation with respect to given bases is easier if the basis of the image vector space is orthonormal. Recall the following facts:

Suppose that U and V are inner product spaces and

$$T : U \to V$$

is a linear transformation. We want to find the matrix representation of T with respect to the basis $B_1 = \{\hat{u}_1, \ldots, \hat{u}_m\}$ for U the basis $B_2 = \{\hat{v}_1, \ldots, \hat{v}_n\}$ for V.

Recall that the ith column of the matrix we seek is $\left[T(\hat{u}_i) \right]_{B_2}$. That is, if

$$T(\hat{u}_i) = a_{1i}\hat{v}_1 + a_{2i}\hat{v}_2 + \cdots + a_{ni}\hat{v}_n$$

then the ith column of the matrix we seek is

$$\begin{pmatrix} a_{1i} \\ a_{2i} \\ \vdots \\ a_{ni} \end{pmatrix}.$$

Further recall that in an inner product space with *orthonormal* basis $\{\hat{v}_1, \ldots, \hat{v}_n\}$ we have

$$\hat{v} = a_1\hat{v}_1 + a_1\hat{v}_1 + \cdots + a_n\hat{v}_{n1} = \langle \hat{v}, \hat{v}_1 \rangle \hat{v}_1 + \langle \hat{v}, \hat{v}_2 \rangle \hat{v}_2 + \cdots + \langle \hat{v}, \hat{v}_n \rangle \hat{v}_n$$

so that

$$T(\hat{u}_i) = a_{1i}\hat{v}_1 + a_{2i}\hat{v}_2 + \cdots + a_{ni}\hat{v}_n$$
$$= \langle T(\hat{u}_i), \hat{v}_1 \rangle \hat{v}_1 + \langle T(\hat{u}_i), \hat{v}_2 \rangle \hat{v}_2 + \cdots + \langle T(\hat{u}_i), \hat{v}_n \rangle \hat{v}_n.$$

So under these conditions the matrix representation of T with respect to the basis $B_1 = \{\hat{u}_1, \ldots, \hat{u}_m\}$ for U and the basis $B_2 = \{\hat{v}_1, \ldots, \hat{v}_n\}$ for V is

$$\begin{pmatrix} \langle T(\hat{u}_1),\hat{v}_1\rangle & \langle T(\hat{u}_2),\hat{v}_1\rangle & \cdots & \langle T(\hat{u}_m),\hat{v}_1\rangle \\ \langle T(\hat{u}_1),\hat{v}_2\rangle & \langle T(\hat{u}_2),\hat{v}_2\rangle & \cdots & \langle T(\hat{u}_m),\hat{v}_2\rangle \\ \vdots & \vdots & \vdots & \vdots \\ \langle T(\hat{u}_1),\hat{v}_n\rangle & \langle T(\hat{u}_2),\hat{v}_n\rangle & \cdots & \langle T(\hat{u}_m),\hat{v}_n\rangle \end{pmatrix}.$$

Example

Let

$$T:\mathbb{R}^3 \to \mathbb{R}^2$$

have the matrix representation

$$\begin{pmatrix} 2 & -1 & 1 \\ 0 & 3 & 4 \end{pmatrix}$$

in the usual basis. Find the matrix representation of T with respect to the bases

$$\left\{\begin{pmatrix}1\\0\\0\end{pmatrix},\begin{pmatrix}1\\1\\0\end{pmatrix},\begin{pmatrix}1\\1\\1\end{pmatrix}\right\} \text{ and } \left\{\begin{pmatrix}1/\sqrt{2}\\1/\sqrt{2}\end{pmatrix}\begin{pmatrix}1/\sqrt{2}\\-1/\sqrt{2}\end{pmatrix}\right\}$$

Note that the basis for $\mathbb{R}^2$ is an orthonormal basis.

We have

$$T\begin{pmatrix}1\\0\\0\end{pmatrix}=\begin{pmatrix}2\\0\end{pmatrix},T\begin{pmatrix}1\\1\\0\end{pmatrix}=\begin{pmatrix}1\\3\end{pmatrix},T\begin{pmatrix}1\\1\\1\end{pmatrix}=\begin{pmatrix}2\\7\end{pmatrix}$$

$$\left\langle T\begin{pmatrix}1\\0\\0\end{pmatrix},\begin{pmatrix}\frac{1}{\sqrt{2}}\\\frac{1}{\sqrt{2}}\end{pmatrix}\right\rangle=\left\langle\begin{pmatrix}2\\0\end{pmatrix},\begin{pmatrix}\frac{1}{\sqrt{2}}\\\frac{1}{\sqrt{2}}\end{pmatrix}\right\rangle=\frac{2}{\sqrt{2}};$$

$$\left\langle T\begin{pmatrix}1\\0\\0\end{pmatrix},\begin{pmatrix}1/\sqrt{2}\\-1/\sqrt{2}\end{pmatrix}\right\rangle=\left\langle\begin{pmatrix}2\\0\end{pmatrix},\begin{pmatrix}1/\sqrt{2}\\-1/\sqrt{2}\end{pmatrix}\right\rangle=\frac{2}{\sqrt{2}}$$

$$\left\langle T\begin{pmatrix}1\\1\\0\end{pmatrix},\begin{pmatrix}\frac{1}{\sqrt{2}}\\\frac{1}{\sqrt{2}}\end{pmatrix}\right\rangle=\left\langle\begin{pmatrix}1\\3\end{pmatrix},\begin{pmatrix}\frac{1}{\sqrt{2}}\\\frac{1}{\sqrt{2}}\end{pmatrix}\right\rangle=\frac{4}{\sqrt{2}};$$

$$\left\langle T\begin{pmatrix}1\\1\\0\end{pmatrix},\begin{pmatrix}1/\sqrt{2}\\-1/\sqrt{2}\end{pmatrix}\right\rangle=\left\langle\begin{pmatrix}1\\3\end{pmatrix},\begin{pmatrix}1/\sqrt{2}\\-1/\sqrt{2}\end{pmatrix}\right\rangle=\frac{-2}{\sqrt{2}}$$

$$\left\langle T\begin{pmatrix}1\\1\\1\end{pmatrix}, \begin{pmatrix}\frac{1}{\sqrt{2}}\\\frac{1}{\sqrt{2}}\end{pmatrix}\right\rangle = \left\langle \begin{pmatrix}2\\7\end{pmatrix}, \begin{pmatrix}\frac{1}{\sqrt{2}}\\\frac{1}{\sqrt{2}}\end{pmatrix}\right\rangle = \frac{9}{\sqrt{2}};$$

$$\left\langle T\begin{pmatrix}1\\1\\1\end{pmatrix}, \begin{pmatrix}1/\sqrt{2}\\-1/\sqrt{2}\end{pmatrix}\right\rangle = \left\langle \begin{pmatrix}2\\7\end{pmatrix}, \begin{pmatrix}1/\sqrt{2}\\-1/\sqrt{2}\end{pmatrix}\right\rangle = \frac{-5}{\sqrt{2}}$$

so the matrix representation of T with respect to these bases is

$$\begin{pmatrix} \dfrac{2}{\sqrt{2}} & \dfrac{4}{\sqrt{2}} & \dfrac{9}{\sqrt{2}} \\ \dfrac{2}{\sqrt{2}} & \dfrac{-2}{\sqrt{2}} & \dfrac{-5}{\sqrt{2}} \end{pmatrix}.$$

If we use the earlier method to compute the matrix with the formula $P_{B_2}^{-1} T P_{B_1}$ with

$$P_{B_1} = \begin{pmatrix} 1 & 1 & 1 \\ 0 & 1 & 1 \\ 0 & 0 & 1 \end{pmatrix}, T = \begin{pmatrix} 2 & -1 & 1 \\ 0 & 3 & 4 \end{pmatrix}, P_{B_2} = \begin{pmatrix} 1/\sqrt{2} & 1/\sqrt{2} \\ 1/\sqrt{2} & -1/\sqrt{2} \end{pmatrix}$$

we of course get the same result.

Exercises

In Exercises 1–2 find the matrix representation of the linear transformation of T with the usual (standard) dot product.

1. Let

   $$T : \mathbb{R}^2 \to \mathbb{R}^2$$

   have the matrix representation

   $$\begin{pmatrix} -2 & 0 \\ 4 & 3 \end{pmatrix}$$

   in the usual basis. Find the matrix representation of T with respect to the bases

   $$\left\{ \begin{pmatrix} 1 \\ -2 \end{pmatrix} \begin{pmatrix} -5 \\ 6 \end{pmatrix} \right\} \text{ and } \left\{ \begin{pmatrix} 1/\sqrt{2} \\ 1/\sqrt{2} \end{pmatrix} \begin{pmatrix} 1/\sqrt{2} \\ -1/\sqrt{2} \end{pmatrix} \right\}.$$

2. Let

   $$T : \mathbb{R}^3 \to \mathbb{R}^2$$

   have the matrix representation

$$\begin{pmatrix} 0 & 3 & -1 \\ 2 & -6 & 2 \end{pmatrix}$$

in the usual basis. Find the matrix representation of T with respect to the bases

$$\left\{ \begin{pmatrix} -2 \\ 0 \\ 1 \end{pmatrix}, \begin{pmatrix} 4 \\ 5 \\ 0 \end{pmatrix}, \begin{pmatrix} 0 \\ 0 \\ 3 \end{pmatrix} \right\} \text{ and } \left\{ \begin{pmatrix} 1/\sqrt{2} \\ 1/\sqrt{2} \end{pmatrix} \begin{pmatrix} 1/\sqrt{2} \\ -1/\sqrt{2} \end{pmatrix} \right\}.$$

Section 6.6 Orthogonal Complement

Definition:

Let V be an inner product space and let W be a non-empty subset of V. The orthogonal complement of W, denote $W^\perp$, is the set of vectors

$$W^\perp = \left\{ \hat{v} \in V | \langle \hat{v}, \hat{w} \rangle = 0 \text{ for every } \hat{w} \in W \right\}.$$

For any set W, the set $W^\perp$ is non-empty because $\hat{0} \in W^\perp$ for any set W.

The major goal of this section is to derive an algorithm to determine $W^\perp$ from any set W. This algorithm is demonstrated by an example at the end of this section. The next two theorems give the justification for the algorithm.

Theorem 6.6:

If V is an inner product space and W is a non-empty subset of V, then $W^\perp$ is a subspace of V.

The proof is left as an exercise. Note that $W^\perp$ in the theorem is a subspace of V even if W is only a subset of V.

Theorem 6.7:

Suppose that V is an inner product space and let W be a subspace of V. Then

$$W \cap W^\perp = \{\hat{0}\}$$

$$W + W^\perp = V$$

Proof:

1. Suppose that $\hat{v} \in W \cap W^\perp$. Then $\langle \hat{v}, \hat{v} \rangle = 0$, so $\hat{v} = \hat{0}$.
2. Let $U = W + W^\perp$. We show that $U = V$. Suppose $U \neq V$. We have $U \subset V$ since $W \subset V$ and $W^\perp \subset V$. We suppose $U \neq V$ and will get a contradiction.

Form an orthonormal basis of U and extend this basis to be an orthonormal basis of V. Since $U \neq V$, there is a vector $\hat{v}$ in the orthonormal basis of V with $\hat{v} \notin U$. Since $\hat{v}$ is orthogonal to U, then $\hat{v}$ is orthogonal to W which means $\hat{v} \in W^{\perp}$ which is a contradiction.

Corollary:

If V is an inner product space and W is a subspace of V then

1. $V = W \oplus W^{\perp}$.
2. Any vector $\hat{v} \in V$ is uniquely expressible as

$$\hat{v} = \hat{w}_1 + \hat{w}_2$$

where $\hat{w}_1 \in W, \hat{w}_2 \in W^{\perp}$.
3. $\dim(V) = \dim(W) + \dim(W^{\perp})$.

Example

Find the orthogonal complement of

$$W = \left\{ \begin{pmatrix} 1 \\ 2 \\ 0 \\ 1 \end{pmatrix} \begin{pmatrix} 1 \\ 1 \\ 3 \\ 3 \end{pmatrix} \right\}.$$

A vector

$$\hat{v} = \begin{pmatrix} v_1 \\ v_2 \\ v_3 \\ v_4 \end{pmatrix}$$

is in $W^{\perp}$ if and only if

$$\left\langle \begin{pmatrix} v_1 \\ v_2 \\ v_3 \\ v_4 \end{pmatrix}, \begin{pmatrix} 1 \\ 2 \\ 0 \\ 1 \end{pmatrix} \right\rangle = 0 \text{ and } \left\langle \begin{pmatrix} v_1 \\ v_2 \\ v_3 \\ v_4 \end{pmatrix}, \begin{pmatrix} 1 \\ 1 \\ 3 \\ 3 \end{pmatrix} \right\rangle = 0.$$

This yields the system of equations

$$1v_1 + 2v_2 + 0v_3 + 1v_4 = 0$$

$$1v_1 + 1v_2 + 3v_3 + 3v_4 = 0$$

The augmented matrix for this system is

$$\begin{pmatrix} 1 & 2 & 0 & 1 & 0 \\ 1 & 1 & 3 & 3 & 0 \end{pmatrix}$$

which, when row reduced is

$$\begin{pmatrix} 1 & 0 & 6 & 5 & 0 \\ 0 & 1 & -3 & -2 & 0 \end{pmatrix}.$$

The free variables are v_3 and v_4.

$$v_1 = -6v_4 - 5v_4$$

$$v_2 = 3v_3 + 2v_4$$

So, letting $v_3 = s$, $v_4 = t$, we have

$$\begin{pmatrix} v_1 \\ v_2 \\ v_3 \\ v_4 \end{pmatrix} = \begin{pmatrix} -6v_3 - 5v_4 \\ 3v_3 + 2v_4 \\ v_3 \\ v_4 \end{pmatrix} = \begin{pmatrix} -6s - 5t \\ 3s + 2t \\ s \\ t \end{pmatrix} = s\begin{pmatrix} -6 \\ 3 \\ 1 \\ 0 \end{pmatrix} + t\begin{pmatrix} -5 \\ 2 \\ 0 \\ 1 \end{pmatrix}.$$

Thus,

$$W^\perp = \text{span}\left\{ \begin{pmatrix} -6 \\ 3 \\ 1 \\ 0 \end{pmatrix}, \begin{pmatrix} -5 \\ 2 \\ 0 \\ 1 \end{pmatrix} \right\}.$$

Exercises

1. Find $W^\perp$ when W is the span of the sets of vectors below.

(a.) $\left\{ \begin{pmatrix} 1 \\ 0 \\ -4 \end{pmatrix}, \begin{pmatrix} 2 \\ 2 \\ 1 \end{pmatrix} \right\}$

(b.) $\left\{ \begin{pmatrix} 3 \\ -2 \\ 5 \end{pmatrix} \right\}$

(c.) $\left\{ \begin{pmatrix} 2 \\ 2 \\ 1 \end{pmatrix}, \begin{pmatrix} 6 \\ 0 \\ 3 \end{pmatrix}, \begin{pmatrix} 10 \\ 4 \\ 5 \end{pmatrix} \right\}$

(d.) $\left\{ \begin{pmatrix} 1 \\ 0 \\ 0 \\ 4 \end{pmatrix}, \begin{pmatrix} 2 \\ 1 \\ 3 \\ 5 \end{pmatrix}, \begin{pmatrix} 5 \\ 2 \\ 0 \\ -1 \end{pmatrix} \right\}$

(e.) $\left\{ \begin{pmatrix} 3 \\ 1 \\ 1 \\ 2 \end{pmatrix}, \begin{pmatrix} 0 \\ 1 \\ 4 \\ 0 \end{pmatrix} \right\}$

2. Show that a vector $\hat{u}$ is in the orthogonal complement of the span of $\{\hat{v}_1 \ldots, \hat{v}_n\}$ if and only if $u, v_i = 0$, for $i = 1, \ldots, n$.

Section 6.7 Four Subspaces Associated with a Matrix (Optional)

Let A be an $m \times n$ matrix. There are four fundamental subspaces associated with A.

1. The null space of A, $\mathcal{N}(A) = \left\{ \hat{x} \in \mathbb{R}^n \middle| A\hat{x} = \hat{0} \right\}$
2. The row space of A, which is the set of linear combinations of the rows of A.
3. The column space of A which is the set of linear combinations of the columns of A, which is also the row space of A^T. Recall from Theorem 3.19, the column space of A is the range of A, which we denote here as $\mathcal{R}(A)$.
4. The null space of A^T, $\mathcal{N}(A^T) = \left\{ x \in \mathbb{R}^m \middle| A^T x = 0 \right\}$.

Theorem 6.8:

If A is a linear transformation on a finite dimensional inner product space, then $\mathcal{N}(A) = \left(\mathcal{R}(A^T) \right)^{\perp}$.

Proof:

We first give the idea of the proof when A is a 2×3 matrix. Let

$$A = \begin{pmatrix} a_{11} & a_{12} & a_{13} \\ a_{21} & a_{22} & a_{23} \end{pmatrix}, \quad x \in \mathcal{N}(A), \quad y \in \mathcal{R}(A^T).$$

Since $\hat{y} \in \mathcal{R}(A^T)$, there is a $\hat{z}$ with $\hat{y} = A^T \hat{z}$.
    Then

$$A\hat{x} = \begin{pmatrix} a_{11} & a_{12} & a_{13} \\ a_{21} & a_{22} & a_{23} \end{pmatrix} \begin{pmatrix} x_1 \\ x_2 \\ x_3 \end{pmatrix} = \begin{pmatrix} a_{11}x_1 + a_{12}x_2 + a_{13}x_3 \\ a_{21}x_1 + a_{22}x_2 + a_{23}x_3 \end{pmatrix} = \begin{pmatrix} 0 \\ 0 \end{pmatrix}$$

$$\hat{y} = A^T \hat{z} = \begin{pmatrix} a_{11} & a_{21} \\ a_{12} & a_{22} \\ a_{13} & a_{23} \end{pmatrix} \begin{pmatrix} z_1 \\ z_2 \end{pmatrix} = \begin{pmatrix} a_{11}z_1 + a_{21}z_2 \\ a_{12}z_1 + a_{22}z_2 \\ a_{13}z_1 + a_{23}z_2 \end{pmatrix}.$$

So

$$\begin{aligned} \langle \hat{x}, \hat{y} \rangle &= x_1 \left(a_{11}z_1 + a_{21}z_2 \right) + x_2 \left(a_{12}z_1 + a_{22}z_2 \right) + x_3 \left(a_{13}z_1 + a_{23}z_2 \right) \\ &= z_1 \left(a_{11}x_1 + a_{12}x_2 + a_{13}x_3 \right) + z_2 \left(a_{21}x_1 + a_{22}x_2 + a_{23}x_3 \right) \\ &= z_1(0) + z_2(0) = 0. \end{aligned}$$

This proof can be generalized to the case of an $m \times n$ matrix.

A more "elegant" proof is the following.

We have for $\hat{x} \in \mathcal{N}(A)$, $\hat{y} \in \mathcal{R}(A^T)$ with $\hat{y} = A^T \hat{z}$

$$\langle \hat{x}, \hat{y} \rangle = \langle \hat{x}, A^T \hat{z} \rangle = \langle A\hat{x}, \hat{z} \rangle = \langle \hat{0}, \hat{z} \rangle = 0.$$

Corollary:

(a.) $\left(\mathcal{N}(A) \right)^{\perp} = \mathcal{R}(A^T)$

(b.) $\mathcal{N}(A^T) = \left(\mathcal{R}(A) \right)^{\perp}$

(c.) $\mathcal{R}(A) = \left(\mathcal{N}(A^T) \right)^{\perp}$

Proof:

Part (a.) is true because in finite dimensions $S = \left(S^{\perp} \right)^{\perp}$ for any subspace S. Note that this is not true if S is a subset that is not a subspace.

Parts (b.) and (c.) are true by replacing A with A^T.

Section 6.8 Projections

Definition:

Suppose that V is a finite dimensional vector space. A linear transformation

$$P : V \to V$$

is a projection if

$$P^2 = P.$$

The word projection suggests a shadow. If an object makes a shadow, then observing the shadow can tell us something about the object, but not everything. If V is a vector space with basis $\{\hat{v}_1, \dots, \hat{v}_n\}$ and $\hat{v}$ is a vector in V, then $\hat{v}$ has a unique representation

$$\hat{v} = a_1 \hat{v}_1 + \cdots + a_n \hat{v}_n.$$

Knowing some of the a_is tells us something about $\hat{v}$. A central topic of the section is, given a vector and a subspace, find "how much" of the vector is in the subspace.

Example

Suppose that V is a vector space with basis $\{\hat{v}_1, \dots, \hat{v}_n\}$ and U is the subspace of V whose basis is $\{\hat{v}_1, \dots, \hat{v}_k\}$. If

$$\hat{v} = a_1\hat{v}_1 + \cdots + a_k\hat{v}_k + a_{k+1}\hat{v}_{k+1} + \cdots + a_n\hat{v}_n$$

then the function

$$\text{Proj}_U\hat{v} = a_1\hat{v}_1 + \cdots + a_k\hat{v}_k$$

is a projection of $\hat{v}$ onto U.

The proof of the following theorem is left as an exercise.

Theorem 6.9:

If U and W are the range and null spaces respectively of a projection P, then

$$P\hat{x} = \hat{x} \text{ for all } \hat{x} \in U, \quad P\hat{y} = \hat{0} \text{ for all } \hat{y} \in W$$

and

$$V = U \oplus W.$$

Also, $I-P$ is a projection onto W.

We will restrict our attention to inner product spaces for the remainder of this section.

Definition:

In a real (complex) inner product space, the projection P is orthogonal if the matrix representation of P is symmetric (Hermitian) with respect to some basis.

Example

If $V = \mathbb{R}^3$ with the standard basis, and if U is the subspace generated by $\{\hat{e}_1, \hat{e}_2\}$ then

$$\text{Proj}_U(x,y,z) = (x,y,0).$$

The matrix for this projection with respect to the standard basis is

$$\begin{pmatrix} 1 & 0 & 0 \\ 0 & 1 & 0 \\ 0 & 0 & 0 \end{pmatrix}$$

which is symmetric.

The Projection Matrix

Our next goal is, given an inner product space V and subspace W, to find the projection matrix P of V onto W. This will give us a method of finding the distance from a vector not in a subspace to the subspace and also the best approximation for a vector in a subspace to a vector not in the subspace. Both of these questions have important applications. If one wishes to ignore the theory, the formula that you need is contained in Theorem 6.11. Here are the major facts that enable us to accomplish that theorem.

1. If A is a matrix, then by Corollary (c.) of Theorem 6.8

$$\mathcal{R}(A) = \left(\mathcal{N}(A^T)\right)^{\perp}$$

      so that

$$V = \mathcal{R}(A) \oplus \left(\mathcal{N}(A^T)\right)$$

2. We have shown that if A is a matrix whose columns are linearly independent, then A is a one-to-one function as a linear transformation.
3. We show in Theorem 6.10 that if A is a matrix whose columns are linearly independent, then $A^T A$ is an invertible matrix.
4. We show in Theorem 6.11 that if W is a subspace of V with basis $\{\hat{a}_1, \ldots, \hat{a}_k\}$ then the projection of V onto W is given by the matrix

$$P = A\left(A^T A\right)^{-1} A^T$$

Where A is the matrix whose columns are $\hat{a}_1, \ldots, \hat{a}_k$.

Theorem 6.10:

Suppose A is a matrix whose columns are linearly independent. Then $A^T A$ is invertible.

Proof:

We have $A^T A$ is a square matrix. Suppose there is a non-zero vector $\hat{x}$ for which

$$\left(A^T A\right)\hat{x} = \hat{0}.$$

Then

$$0 = \left\langle \left(A^T\ A\right)\hat{x}, \hat{x} \right\rangle = \langle A\hat{x}, A\hat{x} \rangle = \|A\hat{x}\|^2$$

so

$$A\hat{x} = \hat{0}.$$

If

$$\hat{x} = \begin{pmatrix} x_1 \\ \vdots \\ x_n \end{pmatrix}$$

then

$$A\hat{x} = x_1\hat{a}_1 + \cdots + x_n\hat{a}_n = 0,$$

where $\hat{a}_i$ is the ith column of A. But since the columns of A are linearly independent, the only way this can occur is if every $x_i = 0$; i.e., $\hat{x} = \hat{0}$.

Theorem 6.11:

Let W be the subspace spanned by the linearly independent vectors $\{\hat{a}_1,\ldots,\hat{a}_k\}$ and let A be the matrix whose columns are the vectors $\hat{a}_1,\ldots,\hat{a}_k$.

Then the projection matrix of V onto W is

$$P = A\left(A^T A\right)^{-1} A^T.$$

Proof:

Let $\hat{b} \in V$. The projection of $\hat{b}$ onto the column space of A is $P\hat{b}$. Since $P\hat{b}$ is in the column space of A, we have $P\hat{b} = A\hat{x}$ for some $\hat{x} \in V$. Set

$$\hat{e} = \hat{b} - P\hat{b} = \hat{b} - A\hat{x}.$$

Now $P\hat{e} = P(\hat{b} - P\hat{b}) = P\hat{b} - P^2\hat{b} = P\hat{b} - P\hat{b} = \hat{0}$

Therefore, $\hat{e}$ is in $\mathcal{N}(A) = \mathcal{R}(A)^\perp = \mathcal{N}(A^T)$.

So $\hat{0} = A^T\hat{e} = A^T\left(\hat{b} - A\hat{x}\right)$

which gives $A^T\hat{b} = A^T A\hat{x}$.

Since $A^T A$ is invertible, we have

$$\hat{x} = \left(A^T A\right)^{-1} A^T\hat{b}.$$

Now $P\hat{b} = A\hat{x}$ so

$$P\hat{b} = A\left(A^T A\right)^{-1} A^T \hat{b}$$

and we have

$$P = A\left(A^T A\right)^{-1} A^T.$$

Example

Let

$$\hat{v}_1 = \begin{pmatrix} 1 \\ 2 \\ 1 \\ 3 \end{pmatrix}, \ \hat{v}_2 = \begin{pmatrix} 1 \\ 1 \\ 1 \\ 1 \end{pmatrix} \text{ and so } A = \begin{pmatrix} 1 & 1 \\ 2 & 1 \\ 1 & 1 \\ 3 & 1 \end{pmatrix}.$$

Then

$$A^T A = \begin{pmatrix} 15 & 7 \\ 7 & 4 \end{pmatrix}$$

$$\left(A^T A\right)^{-1} = \begin{pmatrix} \dfrac{4}{11} & \dfrac{-7}{11} \\ \dfrac{-7}{11} & \dfrac{15}{11} \end{pmatrix}$$

$$A\left(A^T A\right)^{-1} A^T = \begin{pmatrix} \dfrac{5}{11} & \dfrac{2}{11} & \dfrac{5}{11} & -\dfrac{1}{11} \\ \dfrac{2}{11} & \dfrac{3}{11} & \dfrac{2}{11} & \dfrac{4}{11} \\ \dfrac{5}{11} & \dfrac{2}{11} & \dfrac{5}{11} & -\dfrac{1}{11} \\ -\dfrac{1}{11} & \dfrac{4}{11} & -\dfrac{1}{11} & \dfrac{9}{11} \end{pmatrix}.$$

Let

$$\hat{x} = \begin{pmatrix} 6 \\ 3 \\ 2 \\ 9 \end{pmatrix}.$$

Then

$$\left[A\left(A^T A\right)^{-1} A^T\right]\hat{x} = \begin{pmatrix} \dfrac{5}{11} & \dfrac{2}{11} & \dfrac{5}{11} & -\dfrac{1}{11} \\[2mm] \dfrac{2}{11} & \dfrac{3}{11} & \dfrac{2}{11} & \dfrac{4}{11} \\[2mm] \dfrac{5}{11} & \dfrac{2}{11} & \dfrac{5}{11} & -\dfrac{1}{11} \\[2mm] -\dfrac{1}{11} & \dfrac{4}{11} & -\dfrac{1}{11} & \dfrac{9}{11} \end{pmatrix}\begin{pmatrix} 6 \\ 3 \\ 2 \\ 9 \end{pmatrix}$$

$$= \begin{pmatrix} \dfrac{37}{11} \\[2mm] \dfrac{61}{11} \\[2mm] \dfrac{37}{11} \\[2mm] \dfrac{85}{11} \end{pmatrix} = \dfrac{24}{11}\begin{pmatrix} 1 \\ 2 \\ 1 \\ 3 \end{pmatrix} + \dfrac{13}{11}\begin{pmatrix} 1 \\ 1 \\ 1 \\ 1 \end{pmatrix}.$$

Let V be a finite dimensional inner product space and let W be a subspace of V. Using the Gram–Schmidt process, we can find an orthonormal basis for W, say $\left\{\hat{u}_1,\ldots,\hat{u}_k\right\}$. For $\hat{v} \in V$, we have

$$\text{proj}_W\hat{v} = \left\langle \hat{v},\hat{u}_1 \right\rangle \hat{u}_1 + \cdots + \left\langle \hat{v},\hat{u}_n \right\rangle \hat{u}_n.$$

This formulation will be used to show $\text{proj}_W\hat{v}$ is the vector in W that is closest to the vector $\hat{v}$.

Theorem 6.12:

Let V be a finite dimensional inner product space and with W a subspace of V. Let $\left\{\hat{u}_1,\ldots,\hat{u}_k\right\}$ be an orthonormal basis for W. Let $\hat{v} \in V$. Then

$$\hat{w}_1 = \text{proj}_W\hat{v}$$

is the vector in W that is closest to $\hat{v}$ in the sense that
if $\hat{x} \in W$, $\hat{x} \neq \hat{w}_1$, then $\left\|\hat{v} - \hat{w}_1\right\| < \left\|\hat{v} - \hat{x}\right\|$.

Proof:

Add the vectors $\hat{u}_{k+1},\ldots,\hat{u}_n$ to the set $\left\{\hat{u}_1,\ldots,\hat{u}_k\right\}$ so that $\left\{\hat{u}_1,\ldots,\hat{u}_k,\hat{u}_{k+1},\ldots,\hat{u}_n\right\}$ is an orthonormal basis of V. Let $\hat{v} \in V$. Then

$$\hat{v} = \left\langle \hat{v},\hat{u}_1 \right\rangle \hat{u}_1 + \cdots + \left\langle \hat{v},\hat{u}_k \right\rangle \hat{u}_k + \left\langle \hat{v},\hat{u}_{k+1} \right\rangle \hat{u}_{k+1} + + \cdots + \left\langle \hat{v},\hat{u}_n \right\rangle \hat{u}_n.$$

Let $\hat{x} \in W$, $\hat{x} \neq \hat{w}_1$. Then

$$\hat{x} = \left\langle \hat{x},\hat{u}_1 \right\rangle \hat{u}_1 + \cdots + \left\langle \hat{x},\hat{u}_k \right\rangle \hat{u}_k.$$

Now

$$\|\hat{v} - \hat{x}\| = \left\|\left(\langle\hat{v},\hat{u}_1\rangle\hat{u}_1 + \cdots + \langle\hat{v},\hat{u}_k\rangle\hat{u}_k + \langle\hat{v},\hat{u}_{k+1}\rangle\hat{u}_{k+1} + \cdots + \langle\hat{v},\hat{u}_n\rangle\hat{u}_n\right)\right.$$
$$\left. - \left(\langle\hat{x},\hat{u}_1\rangle\hat{u}_1 + \cdots + \langle\hat{x},\hat{u}_k\rangle\hat{u}_k\right)\right\|$$
$$= \left\|\left(\langle\hat{v},\hat{u}_1\rangle - \langle\hat{x},\hat{u}_1\rangle\right)\hat{u}_1 + \cdots + \left(\langle\hat{v},\hat{u}_k\rangle - \langle\hat{x},\hat{u}_k\rangle\right)\hat{u}_k\right\|$$
$$+ \left\|\langle\hat{v},\hat{u}_{k+1}\rangle\hat{u}_{k+1} + \cdots + \langle\hat{v},\hat{u}_n\rangle\hat{u}_n\right\|$$

and

$$\|\hat{v} - \hat{w}_1\| = \left\|\left(\langle\hat{v},\hat{u}_1\rangle\hat{u}_1 + \cdots + \langle\hat{v},\hat{u}_k\rangle\hat{u}_k + \langle\hat{v},\hat{u}_{k+1}\rangle\hat{u}_{k+1} + \cdots + \langle\hat{v},\hat{u}_n\rangle\hat{u}_n\right)\right.$$
$$\left. - \left(\langle\hat{v},\hat{u}_1\rangle\hat{u}_1 + \cdots + \langle\hat{v},\hat{u}_k\rangle\hat{u}_k\right)\right\|$$
$$= \left\|\|\hat{v},\hat{u}_{k+1}\|\hat{u}_{k+1} + \cdots + \langle\hat{v},\hat{u}_n\rangle\hat{u}_n\right\|.$$

Thus

$$\|\hat{v} - \hat{x}\| - \|\hat{v} - \hat{w}_1\| = \left\|\left(\langle\hat{v},\hat{u}_1\rangle - \langle\hat{x},\hat{u}_1\rangle\right)\hat{u}_1 + \cdots + \left(\langle\hat{v},\hat{u}_k\rangle - \langle\hat{x},\hat{u}_k\rangle\right)\hat{u}_k\right\| \geq 0$$

and

$$\left\|\left(\langle\hat{v},\hat{u}_1\rangle - \langle\hat{x},\hat{u}_1\rangle\right)\hat{u}_1 + \cdots + \left(\langle\hat{v},\hat{u}_k\rangle - \langle\hat{x},\hat{u}_k\rangle\right)\hat{u}_k\right\| = 0$$

if and only if

$$\langle\hat{v},\hat{u}_i\rangle - \langle\hat{x},\hat{u}_i\rangle = 0, \quad i = 1,\ldots,k$$

which is true if and only if $\hat{w}_1 = \hat{x}$.

Example

(a.) Find the point in the plane

$$x + y - 2z = 0$$

that is nearest to the point

$$\begin{pmatrix} 5 \\ 6 \\ 7 \end{pmatrix}.$$

We have

$$x = -y + 2z$$

so y and z are free variables. We set

$$y = s, z = t$$

so a vector in the plane is

$$\begin{pmatrix} -s+2t \\ s \\ t \end{pmatrix} = \begin{pmatrix} -s \\ s \\ 0 \end{pmatrix} + \begin{pmatrix} 2t \\ 0 \\ t \end{pmatrix} = s\begin{pmatrix} -1 \\ 1 \\ 0 \end{pmatrix} + t\begin{pmatrix} 2 \\ 0 \\ 1 \end{pmatrix}$$

and a basis for the plane is

$$\left\{ \begin{pmatrix} -1 \\ 1 \\ 0 \end{pmatrix}, \begin{pmatrix} 2 \\ 0 \\ 1 \end{pmatrix} \right\}.$$

Thus, we take

$$A = \begin{pmatrix} -1 & 2 \\ 1 & 0 \\ 0 & 1 \end{pmatrix}$$

and

$$A\left(A^T A\right)^{-1} A^T = \begin{pmatrix} \dfrac{5}{6} & -\dfrac{1}{6} & \dfrac{1}{3} \\[2mm] -\dfrac{1}{6} & \dfrac{5}{6} & \dfrac{1}{3} \\[2mm] \dfrac{1}{3} & \dfrac{1}{3} & \dfrac{1}{3} \end{pmatrix}.$$

The projection of $\begin{pmatrix} 5 \\ 6 \\ 7 \end{pmatrix}$ onto the plane is

$$\begin{pmatrix} \dfrac{5}{6} & -\dfrac{1}{6} & \dfrac{1}{3} \\[2mm] -\dfrac{1}{6} & \dfrac{5}{6} & \dfrac{1}{3} \\[2mm] \dfrac{1}{3} & \dfrac{1}{3} & \dfrac{1}{3} \end{pmatrix} \begin{pmatrix} 5 \\ 6 \\ 7 \end{pmatrix} = \begin{pmatrix} \dfrac{11}{2} \\[2mm] \dfrac{13}{2} \\[2mm] 6 \end{pmatrix}.$$

(b.) The distance between $\begin{pmatrix} 5 \\ 6 \\ 7 \end{pmatrix}$ and the plane is

$$\begin{pmatrix} 5 \\ 6 \\ 7 \end{pmatrix} - \begin{pmatrix} \dfrac{11}{2} \\[2mm] \dfrac{13}{2} \\[2mm] 6 \end{pmatrix} = \begin{pmatrix} -1/2 \\ -1/2 \\ 1 \end{pmatrix} = \sqrt{\dfrac{3}{2}}$$

Exercises

1. Find the projection of the vector

$$\hat{v} = \begin{pmatrix} 3 \\ 2 \end{pmatrix}$$

onto the subspace $W = \text{span}\left\{ \begin{pmatrix} 4 \\ -5 \end{pmatrix} \right\}$ of $\mathbb{R}^2$.

In Exercises 2–7, W is the subspace spanned by the set of vectors S.
(a.) Find the projection of the vector $\hat{y}$ onto W.
(b.) Find the distance from the point to the subspace.

2. $S = \left\{ \begin{pmatrix} 1 \\ 2 \\ 2 \end{pmatrix}, \begin{pmatrix} -1 \\ 0 \\ 2 \end{pmatrix} \right\} \hat{y} = \begin{pmatrix} 3 \\ 2 \\ -4 \end{pmatrix}$

3. $S = \left\{ \begin{pmatrix} 1 \\ -1 \\ 1 \end{pmatrix}, \begin{pmatrix} 1 \\ 0 \\ 1 \end{pmatrix} \right\} \hat{y} = \begin{pmatrix} 0 \\ 6 \\ 5 \end{pmatrix}$

4. $S = \left\{ \begin{pmatrix} 0 \\ 1 \\ 1 \end{pmatrix}, \begin{pmatrix} 1 \\ 0 \\ 1 \end{pmatrix} \right\} \hat{y} = \begin{pmatrix} 8 \\ 1 \\ 3 \end{pmatrix}$

5. $S = \left\{ \begin{pmatrix} 1 \\ 1 \\ 1 \\ 1 \end{pmatrix}, \begin{pmatrix} 4 \\ -1 \\ 4 \\ 1 \end{pmatrix} \right\} \hat{y} = \begin{pmatrix} 2 \\ -3 \\ 0 \\ 0 \end{pmatrix}$

6. $S = \left\{ \begin{pmatrix} 1 \\ 0 \\ 0 \\ 1 \end{pmatrix}, \begin{pmatrix} 0 \\ 1 \\ 1 \\ 0 \end{pmatrix} \right\} \hat{y} = \begin{pmatrix} 4 \\ 2 \\ 0 \\ 7 \end{pmatrix}$

7. $S = \left\{ \begin{pmatrix} -1 \\ 1 \\ -1 \\ 1 \end{pmatrix}, \begin{pmatrix} 1 \\ 3 \\ 1 \\ -1 \end{pmatrix} \right\} y = \begin{pmatrix} -3 \\ 4 \\ 4 \\ 1 \end{pmatrix}$

8. Show that if P is the projection in problem 6, then $P^2 = P$ and $P^T = P$.
9. Show that if P is any projection, then $I - P$ is a projection.
10. (a.) Find the projection of the point $(4, -1, 2)$ onto the plane

$$3x - 2y + z = 0.$$

(b.) Find the distance between the point $(4, -1, 2)$ and the plane

$$3x - 2y + z = 0.$$

Section 6.9 Least Squares Estimates in Statistics (Optional)

One of the most widely used and versatile methods in statistics is linear regression. The general form of the regression model is

$$Y = \beta_0 + \beta_1 X_1 + \beta_2 X_2 + \ldots + \beta_n X_n + \varepsilon.$$

Y is called the response variable, the X_is are explanatory or predictor variables, and ε is a random error term. The versatility comes from the possibility that some Xs can be functions of others; the model is linear in the β_is but not necessarily in the Xs. Some examples of linear regression models are:

$$Y = \beta_0 + \beta_1 X_1 + \beta_2 X_2 + \varepsilon$$

$$Y = \beta_0 + \beta_1 X_1 + \beta_2 X_2 + \beta_3 X_1 X_2 + \varepsilon$$

$$Y = \beta_0 + \beta_1 X_1 + \beta_2 X_2 + \beta_3 X_2^2 + \varepsilon$$

In application, data is obtained by experiment or observation and used to produce estimates of the coefficients. A linear regression model may be used to predict YY for given values of the explanatory variables, or to describe the relationship of the explanatory variables to Y. Analysis may show which variables are useful in the model and which ones can be dropped without much loss of information.

The purpose of this section is to show how linear algebra provides the estimates of the coefficients in a linear regression model.

To make the notation simple, suppose we have two predictor variables, X_1 and X_2 and we have four vectors of observed data (x_{i1}, x_{i2}, y_i), $i = 1, 2, 3, 4$. Each of the data vectors exemplifies the model

$$y_i = \beta_0 + \beta_1 x_{i1} + \beta_2 x_{i2} + \varepsilon_i$$

where ε_i is not known.

These four equations may be expressed in matrix form as

$$\hat{Y} = X\hat{\beta} + \hat{\varepsilon}$$

where

$$\hat{Y} = \begin{bmatrix} y_1 \\ y_2 \\ y_3 \\ y_4 \end{bmatrix}, \quad X = \begin{bmatrix} 1 & x_{11} & x_{12} \\ 1 & x_{21} & x_{22} \\ 1 & x_{31} & x_{32} \\ 1 & x_{41} & x_{42} \end{bmatrix}, \quad \hat{\beta} = \begin{bmatrix} \beta_0 \\ \beta_1 \\ \beta_2 \end{bmatrix}, \quad \text{and } \hat{\varepsilon} = \begin{bmatrix} \varepsilon_1 \\ \varepsilon_2 \\ \varepsilon_3 \\ \varepsilon_4 \end{bmatrix}.$$

We want to find a vector value of $\hat{\beta}$ that makes the vector $\hat{\varepsilon}$ as small as possible.

We begin by ignoring $\hat{\varepsilon}$ temporarily and working on the rest of the equation, $\hat{Y} = X\hat{\beta}$, which in all likelihood does not have an exact solution. Although the matrix X can't be inverted, $X^t X$ can be, so we left-multiply both sides of the equation by $\left(X^t X \right)^{-1} X^t$:

$$\left(X^t X \right)^{-1} X^t \hat{Y} = \left(X^t X \right)^{-1} X^t X \hat{\beta} = \hat{\beta}$$

In statistics, it is customary to denote this solution by $\hat{\beta}$, but since our vectors already have hats, we will resort to using an asterisk to denote an estimate obtained from data:

$$\widehat{\beta^*} = \left(X^t X \right)^{-1} X^t \widehat{Y}$$

Using the estimated coefficients, the predicted value (also called the fitted value) of $\hat{Y}$, which we denote with an asterisk, is

$$\widehat{Y^*} = X \widehat{\beta^*} = X \left(X^t X \right)^{-1} X^t Y.$$

It is now apparent that the predicted value of $\hat{Y}$ is the projection of $\hat{Y}$ into the column space of the data matrix X. Furthermore, the difference

$$\hat{E} = \hat{Y} - \widehat{Y^*} = \hat{Y} - X\widehat{\beta^*}$$

is the difference between the actual, observed value of $\hat{Y}$ and the value $\hat{Y}^*$ predicted by the linear model; since $\widehat{\beta^*}$ is an estimate of $\hat{\beta}$, $\hat{E}$ is an estimate of $\hat{\varepsilon}$. The components of $\hat{E}$ are called residuals and $\hat{E}$ is the residual vector. The theory of projections guarantees that $\hat{E}$ is in the orthogonal complement of the column space of X, and that $\hat{E}$ is the smallest possible residual among all choices for $\hat{\beta}$.

Note that $\left| \hat{E} \right|^2 = \sum_{i=1}^{n} \left(y_i - y_i^* \right)^2$ is the sum of the squares of the residuals.

This technique for estimating $\hat{\beta}$ is called *the method of least squares,* $\widehat{\beta^*}$ is called the least squares estimate of $\hat{\beta}$, and the resulting equation $\widehat{Y^*} = X \widehat{\beta^*}$ is called the least squares regression equation.

Example

Suppose our four data points are (1, 2, 4), (0, 4, –3), (2, 0, 10), and (3, 3, 9).

Then $Y = \begin{bmatrix} 4 \\ -3 \\ 10 \\ 9 \end{bmatrix}$ and $X = \begin{bmatrix} 1 & 1 & 2 \\ 1 & 0 & 4 \\ 1 & 2 & 0 \\ 1 & 3 & 3 \end{bmatrix}$.

So $X^t X = \begin{bmatrix} 4 & 6 & 9 \\ 6 & 14 & 11 \\ 9 & 11 & 29 \end{bmatrix}$ and $\left(X^t X \right)^{-1} = \dfrac{1}{30} \begin{bmatrix} 57 & -15 & -12 \\ -15 & 7 & 2 \\ -12 & 2 & 4 \end{bmatrix}$.

It follows that

$$\widehat{\beta^*} = \left(X^t X \right)^{-1} X^t Y = \begin{bmatrix} 33/10 \\ 103/30 \\ -23/15 \end{bmatrix} \text{ and } \widehat{Y^*} = X\widehat{\beta^*} = \begin{bmatrix} 11/3 \\ -17/6 \\ 61/6 \\ 9 \end{bmatrix}.$$

$$\hat{E} = \hat{Y} - \widehat{Y^*} = \begin{bmatrix} 1/3 \\ -1/6 \\ -1/6 \\ 0 \end{bmatrix}$$

The least squares regression equation is

$$\widehat{Y^*} = (33/10) + (103/30)X_1 + (-23/15)X_2.$$

The sum of squared residuals is $\left(\dfrac{1}{3} \right)^2 + \left(-\dfrac{1}{6} \right)^2 + \left(-\dfrac{1}{6} \right)^2 + 0^2 = \dfrac{1}{6}$.
Check that $X^t \hat{E} = 0$.

Summary

The general form of the least squares regression has k explanatory variables $X_1, X_2, \ldots, X_k$, and the response variable Y. The linear model is

$$Y = \beta_0 + \beta_1 X_1 + \cdots + \beta_k X_k + \epsilon.$$

Data for estimating the coefficients of the model consists of n vectors in $\mathbb{R}^{k+1}$ of the form $\left(x_{i1}, x_{i2}, \ldots, x_{ik}, y_i \right)$, $1 \le i \le n$, with $n \ge k+1$.

We construct the vector $\hat{Y} = \begin{bmatrix} y_1 \\ y_2 \\ \vdots \\ y_n \end{bmatrix}$ and the matrix $X = \begin{bmatrix} 1 & x_{11} & x_{12} & \cdots & x_{1k} \\ 1 & x_{21} & x_{22} & \cdots & x_{2k} \\ \vdots & \vdots & \vdots & \ddots & \vdots \\ 1 & x_{n1} & x_{n2} & \cdots & x_{nk} \end{bmatrix}$

Provided that the rank of X is $k+1$, so that $X^t X$ is invertible, the least squares estimate of $\hat{\beta}$ is

$$\widehat{\beta^*} = \left(X^t X \right)^{-1} X^t \widehat{Y}.$$

Then $\widehat{Y^*} = X\widehat{\beta^*}$ is the least squares regression equation.

The residual vector is $\hat{E} = \hat{Y} - \widehat{Y^*} = \hat{Y} - X\widehat{\beta^*}$

Because $\widehat{Y^*}$ is the projection of $\hat{Y}$ into the column space of X, the least squares estimate minimizes the sum of the squared residuals .

Exercise

1. The table below gives percent body fat measured by a water immersion method, with weight (in pounds), height (in inches), and neck circumference (in centimeters) for eight men. (Excerpted from a larger data set in Lock, Lock, Lock, Lock, Lock, *Statistics: Unlocking the Power of Data*, page 562.)

Bodyfat	Weight	Height	Neck
32.3	247.25	73.5	42.1
22.5	177.25	71.5	36.2
22	156.25	69	35.5
12.3	154.25	67.75	36.2
20.5	177	70	37.2
22.6	198	72	39.9
28.7	200.5	71.5	37.9
21.3	163	70.25	35.3

(a.) Taking $Y =$ Bodyfat, $X_1 =$ Weight, $X_2 =$ Height, and $X_3 =$ Neck circumference, set up the matrix X for the linear model $\hat{Y} = X\hat{\beta} + \hat{\varepsilon}$. Using matrix operations, calculate the least squares estimate of $\hat{\beta}$ and the vector of fitted values, $\widehat{Y^*}$.

(b.) Calculate the vector $\hat{E}$ of residuals and find $\left|\hat{E}\right|^2$.

(c.) Let $\hat{b} = \left[6, 0.3, 1, -2 \right]$. Compare $\hat{b}$ with the least squares estimate of $\hat{\beta}$. Use $\hat{b}$ to calculate $\hat{e} = \hat{Y} - X \cdot \hat{b}$. Will $\left|\hat{e}\right|^2$ be larger or smaller than $\left|\hat{E}\right|^2$ from part b above? Verify this by calculation.

Section 6.10 Weighted Inner Products (Optional)

While the Euclidean inner product is the most often used inner product, there are applications where other inner products are useful. Among these are weighted inner products, which we now consider. The role of the weights is to distinguish between components if some components have more impact than others.

On $\mathbb{R}^n$, for

$$\hat{x} = \begin{pmatrix} x_1 \\ \vdots \\ x_n \end{pmatrix}, \quad \hat{y} = \begin{pmatrix} y_1 \\ \vdots \\ y_n \end{pmatrix}, \quad \hat{z} = \begin{pmatrix} z_1 \\ \vdots \\ z_n \end{pmatrix}$$

we show that the function

$$\hat{x} \cdot \hat{y} = c_1 x_1 y_1 + \cdots + c_n x_n y_n; \quad c_i > 0, \quad i = 1, \ldots, n$$

is a dot product. The "weights" are the numbers c_i. We verify that the axioms are satisfied.

1. $\hat{x} \cdot \hat{x} = c_1 x_1 x_1 + \cdots + c_n x_n x_n \geq 0$ and $\hat{x} \cdot \hat{x} = 0$ fi and only if $\hat{x} = \hat{0}$.
2. $\hat{x} \cdot \hat{y} = c_1 x_1 y_1 + \cdots + c_n x_n y_n = c_1 y_1 x_1 + \cdots + c_n y_n x_n = \hat{y} \cdot \hat{x}$.
3. $\hat{x} \cdot (\hat{y} + \hat{z}) = c_1 x_1 (y_1 + z_1) \cdots + c_n x_n (y_n + z_n)$

$$= c_1 x_1 y_1 + c_1 x_1 z_1 + \cdots + c_n x_n y_n + c_n x_n z_n$$
$$= c_1 x_1 y_1 + \cdots + c_n x_n y_n + c_1 x_1 z_1 + \cdots + c_n x_n z_n$$
$$= \hat{x} \cdot \hat{y} + \hat{x} \cdot \hat{z}.$$

4. $a(\hat{x} \cdot \hat{y}) = a(c_1 x_1 y_1 + \cdots + c_n x_n y_n)$

$$= \left[c_1 (a x_1) \right] y_1 + \cdots + \left[c_n (a x_n) \right] y_n = (a\hat{x} \cdot \hat{y}).$$

Also,

$$a(\hat{x} \cdot \hat{y}) = a(c_1 x_1 y_1 + \cdots + c_n x_n y_n)$$
$$= c_1 x_1 (a y_1) + \cdots + c_n x_n (a y_n) = (\hat{x} \cdot a\hat{y}).$$

Example

Find the projection of the vector

$$\begin{pmatrix} 2 \\ -1 \\ 3 \end{pmatrix}$$

onto the subspace W of $\mathbb{R}^3$ that has the basis

$$\left\{ \begin{pmatrix} 2 \\ 4 \\ 4 \end{pmatrix}, \begin{pmatrix} -1 \\ 0 \\ 2 \end{pmatrix} \right\}.$$

We first use the Gram–Schmidt process to find an orthonormal basis for W.
    Let

$$\hat{v}_1 = \begin{pmatrix} 2 \\ 4 \\ 4 \end{pmatrix}, \quad \hat{v}_2 = \begin{pmatrix} -1 \\ 0 \\ 2 \end{pmatrix}, \quad \hat{u}_1 = \frac{\hat{v}_1}{\|\hat{v}_1\|} = \begin{pmatrix} 1/3 \\ 2/3 \\ 2/3 \end{pmatrix}$$

$$\hat{w}_2 = \hat{v}_2 - \langle \hat{v}_2, \hat{u}_1 \rangle \hat{u}_1$$

$$= \begin{pmatrix} -1 \\ 0 \\ 2 \end{pmatrix} - \left\langle \begin{pmatrix} -1 \\ 0 \\ 2 \end{pmatrix}, \begin{pmatrix} 1/3 \\ 2/3 \\ 2/3 \end{pmatrix} \right\rangle \begin{pmatrix} 1/3 \\ 2/3 \\ 2/3 \end{pmatrix} = \begin{pmatrix} -1 \\ 0 \\ 2 \end{pmatrix} - 1 \begin{pmatrix} 1/3 \\ 2/3 \\ 2/3 \end{pmatrix} = \begin{pmatrix} -4/3 \\ -2/3 \\ 4/3 \end{pmatrix}$$

and

$$\|\hat{w}_2\| = \sqrt{\left(-\frac{4}{3}\right)^2 + \left(-\frac{2}{3}\right)^2 + \left(\frac{4}{3}\right)^2} = \sqrt{\frac{36}{9}} = 2$$

so

$$\hat{u}_2 = \frac{\hat{w}_2}{\|\hat{w}_2\|} = \begin{pmatrix} -2/3 \\ -1/3 \\ 2/3 \end{pmatrix}.$$

Then

$$\mathrm{proj}_W \hat{v} = \langle \hat{v}, \hat{u}_1 \rangle \hat{u}_1 + \langle \hat{v}, \hat{u}_2 \rangle \hat{u}_2$$

$$= \left\langle \begin{pmatrix} 2 \\ -1 \\ 3 \end{pmatrix}, \begin{pmatrix} 1/3 \\ 2/3 \\ 2/3 \end{pmatrix} \right\rangle \begin{pmatrix} 1/3 \\ 2/3 \\ 2/3 \end{pmatrix} + \left\langle \begin{pmatrix} 2 \\ -1 \\ 3 \end{pmatrix}, \begin{pmatrix} -2/3 \\ -1/3 \\ 2/3 \end{pmatrix} \right\rangle \begin{pmatrix} -2/3 \\ -1/3 \\ 2/3 \end{pmatrix}$$

$$= 2 \begin{pmatrix} 1/3 \\ 2/3 \\ 2/3 \end{pmatrix} + 1 \begin{pmatrix} -2/3 \\ -1/3 \\ 2/3 \end{pmatrix} = \begin{pmatrix} 0 \\ 1 \\ 2 \end{pmatrix}.$$

Also,

$$\hat{v} - \mathrm{proj}_W \hat{v} = \begin{pmatrix} 2 \\ -1 \\ 3 \end{pmatrix} - \begin{pmatrix} 0 \\ 1 \\ 2 \end{pmatrix} = \begin{pmatrix} 2 \\ -2 \\ 1 \end{pmatrix}$$

so

$$\|\hat{v} - \mathrm{proj}_W \hat{v}\|^2 = 9 \quad \text{and} \quad \|\hat{v} - \mathrm{proj}_W \hat{v}\| = 3.$$

Chapter 7

Linear Functionals, Dual Spaces, and Adjoint Operators

Note: In this chapter there are some subsections that are marked as optional. The material in these sections is beyond what is normally done in a first linear algebra course. Omitting them will not compromise later material.

Section 7.1 Linear Functionals

In previous chapters we studied linear transformations between vector spaces. We denoted the class of linear transformations from the vector space V to the vector space W by $L(V, W)$. Since the scalar field $\mathcal{F}$ is a vector space (the simplest of all non-trivial vector spaces), one class of linear transformations from V is $L(V, \mathcal{F})$. This class of linear transformations is the subject of this section.

Definition:

If V is a vector space with scalar field $\mathcal{F}$, then a linear transformation

$$T : V \to \mathcal{F}$$

is called a linear functional on V.

We denote the set of linear functionals on V by V'. Thus, in our earlier notation we are now describing $L(V, \mathcal{F})$ as V'. These linear transformations take vectors from V as inputs and give scalars as outputs.

Example

1. Let $T : \mathbb{R}^2 \to \mathbb{R}$ be defined by $T(x, y) = 4x - 3y$.
2. Let $T : C[0,1] \to \mathbb{R}$ be defined

$$T(f) = \int_0^1 f(x)\,dx.$$

3. If V is an inner product space we can define a linear functional on V as follows: Fix a vector $\hat{w} \in V$. Define $T : V \to \mathcal{F}$ by

$$T(\hat{v}) = \langle \hat{v}, \hat{w} \rangle.$$

Note that if the scalar field were the complex numbers, then

$$T(\hat{v}) = \langle \hat{w}, \hat{v} \rangle$$

would not be a linear functional but it would be if the scalar field were the real numbers. (Why?) For $v_1', v_2' \in V'$ and $\alpha \in \mathcal{F}$ we define

$$v_1' + v_2' \in V' \text{ and } \alpha v_1' \in V'$$

by

$$\left(v_1' + v_2'\right)(\hat{x}) = v_1'(\hat{x}) + v_2'(\hat{x}) \text{ for } \hat{x} \in V$$

$$\left(\alpha v_1'\right)(\hat{x}) = \alpha\left(v_1'(\hat{x})\right) \text{ for } \hat{x} \in V.$$

We know V' is a vector space because $L(V, W)$ is a vector space for any vector space W.

Definition:

With these operations, V' is a vector space, called the dual of the vector space V.

There are actually two duals of a vector space – the topological dual and the algebraic dual. The algebraic dual is what we have just defined. The topological dual consists of continuous linear functionals (a term that we will not define). For finite dimensional vector spaces these coincide, but for infinite dimensional vector spaces the topological dual space is a proper subset of the algebraic dual space. Since we are dealing only with finite dimensions, we will simply use the term "dual space".

Theorem 7.1:

Suppose that V is a finite dimensional vector space with basis $\mathcal{B} = \left\{\hat{v}_1, \ldots, \hat{v}_n\right\}$. Let

$$v_i' : V \to \mathcal{F}; i = 1, \ldots, n$$

be the linear functional defined by

$$v_i'(\hat{v}_j) = \delta_{ij} = \begin{cases} 1 & \text{if } i = j \\ 0 & \text{if } i \neq j \end{cases}.$$

Then $\mathcal{B}' = \{v_1', \ldots, v_n'\}$ is a basis for V' called the basis of V' dual to the basis $\mathcal{B}$.

Proof:

We first show that $\mathcal{B}'$ is a linearly independent set.

Let $0'$ denote the zero linear functional on V; that is,

$$0'(v) = 0 \text{ for all } v \in V$$

and suppose that

$$0' = \alpha_1 v_1' + \cdots + \alpha_n v_n'.$$

Choose $\hat{v}_j$ from the basis. Then

$$
\begin{aligned}
0 = 0'(\hat{v}_j) &= (\alpha_1 v_1' + \cdots + \alpha_n v_n')\hat{v}_j \\
&= \alpha_1 v_1'(\hat{v}_j) + \cdots + \alpha_j v_j'(\hat{v}_j) + \cdots + \alpha_n v_n'(\hat{v}_j) \\
&= \alpha_1 0 + \cdots + \alpha_j v_j'(\hat{v}_j) + \cdots + \alpha_n 0 = \alpha_j.
\end{aligned}
$$

Thus $\alpha_j = 0$. Since j was arbitrary, $\mathcal{B}'$ is linearly independent.

We next show that $\mathcal{B}'$ spans V'. Choose $v' \in V'$. Now v' is determined by its effect on the elements of $\mathcal{B}$. Suppose that $v'(\hat{v}_j) = \alpha_j$, $j = 1, \ldots, n$. Then we claim

$$v' = \sum_{i=1}^{n} \alpha_i v_i'.$$

We demonstrate the claim by showing each has the same effect on a basis element. We have

$$\left(\sum_{i=1}^{n} \alpha_i v_i'\right)(\hat{v}_j) = \sum_{i=1}^{n} \alpha_i v_i'(\hat{v}_j) = \alpha_j$$

and $v'(\hat{v}_j) = \alpha_j$ so the claim follows.

Corollary:

If the dimension of the vector space V is n, then the dimension of V' is n.

The vector spaces in this chapter will be finite dimensional inner product spaces. In that setting, the most important result is the next theorem

Theorem 7.2: (Riesz Representation Theorem)

If U is a finite dimensional inner product space, then for y' a linear functional on U there is a unique vector $\hat{y} \in U$ for which

$$y'(\hat{x}) = \langle \hat{x}, \hat{y} \rangle \text{ for all } \hat{x} \in U.$$

Proof:

Let $y' \in U'$ and let $\{\hat{u}_1, \ldots, \hat{u}_n\}$ be an orthonormal basis of U (which is possible by the Gram–Schmidt process). Suppose that

$$y'(\hat{u}_i) = \alpha_i. \text{ Let } \hat{y} = \overline{\alpha_1}\hat{u}_1 + \cdots + \overline{\alpha_n}\hat{u}_n. \text{ If } \hat{x} = \beta_1\hat{u}_1 + \cdots + \beta_n\hat{u}_n, \text{ then}$$

$$y'(\hat{x}) = y'(\beta_1\hat{u}_1 + \cdots + \beta_n\hat{u}_n) = \beta_1 y'(\hat{u}_1) + \cdots$$
$$+ \beta_n y'(\hat{u}_n) = \beta_1\alpha_1 + \cdots + \beta_n\alpha_n.$$

But

$$\langle \hat{x}, \hat{y} \rangle = \langle \beta_1\hat{u}_1 + \cdots + \beta_n\hat{u}_n, \overline{\alpha_1}\hat{u}_1 + \cdots + \overline{\alpha_n}\hat{u}_n \rangle$$

which, by the orthonormality of $\{\hat{u}_1, \ldots, \hat{u}_n\}$, is equal to

$$\langle \beta_1\hat{u}_1, \overline{\alpha_1}\hat{u}_1 \rangle + \cdots + \langle \beta_1\hat{u}_1, \overline{\alpha_n}\hat{u}_n \rangle + \langle \beta_2\hat{u}_2, \overline{\alpha_1}\hat{u}_1 \rangle + \cdots + \langle \beta_2\hat{u}_2, \overline{\alpha_n}\hat{u}_n \rangle$$
$$+ \langle \beta_n\hat{u}_n, \overline{\alpha_1}\hat{u}_1 \rangle + \cdots + \langle \beta_n\hat{u}_n, \overline{\alpha_n}\hat{u}_n \rangle = \beta_1\alpha_1 + \cdots + \beta_n\alpha_n.$$

To show uniqueness, suppose that there are $\hat{y}_1, \hat{y}_2 \in U$ for which

$$y'(\hat{x}) = \langle \hat{x}, \hat{y}_1 \rangle = \langle \hat{x}, \hat{y}_2 \rangle$$

for all $\hat{x} \in U$.

Then

$$0 = \langle \hat{x}, \hat{y}_1 \rangle - \langle \hat{x}, \hat{y}_2 \rangle = \langle \hat{x}, \hat{y}_1 - \hat{y}_2 \rangle$$

for all $\hat{x} \in U$. Taking $\hat{x} = \hat{y}_1 - \hat{y}_2$, we have

$$0 = \langle \hat{x}, \hat{y}_1 - \hat{y}_2 \rangle = \langle \hat{y}_1 - \hat{y}_2, \hat{y}_1 - \hat{y}_2 \rangle = \|\hat{y}_1 - \hat{y}_2\|^2$$

so $\hat{y}_1 = \hat{y}_2$.

In the proof of the theorem, it was demonstrated how to find the vector $\hat{y}$ from the linear functional y'. The corollary that follows highlights this connection.

Corollary:

Let U be a vector space with orthonormal basis $\{\hat{u}_1, \ldots, \hat{u}_n\}$. Let $y' \in U'$ with

$$y'\left(\hat{u}_i\right) = \alpha_i.$$

Then the vector

$$\hat{y} = \overline{\alpha_1}\hat{u}_1 + \cdots + \overline{\alpha_n}\hat{u}_n$$

satisfies

$$y'\left(\hat{x}\right) = \langle \hat{x}, \hat{y} \rangle \text{ for all } \hat{x} \in U.$$

Example

Let $T : \mathbb{C}^3 \to \mathbb{C}$ be defined by

$$T\left(x_1, x_2, x_3\right) = \left(2 + 3i\right)x_1 + 4x_2 + 2ix_3.$$

So T is y' in the theorem.

We want to find the vector $\hat{y}$ for which

$$T\left(x_1, x_2, x_3\right) = \langle \hat{x}, \hat{y} \rangle = x_1\overline{y_1} + x_2\overline{y_2} + x_3\overline{y_3}.$$

This will be true if

$$\overline{y_1} = \left(2 + 3i\right), \quad \overline{y_2} = 4, \quad \overline{y_3} = 2i$$

so

$$y_1 = 2 - 3i, \quad y_2 = 4, \quad y_3 = -2i$$

or

$$\hat{y} = \begin{pmatrix} 2 - 3i \\ 4 \\ -2i \end{pmatrix}.$$

What occurs in this example highlights what the proof of Theorem 7.2 says. Namely, the components of $\hat{y}$ are the complex conjugates of the coefficients of x_i in the definition of $T\left(\hat{x}\right)$.

The Second Dual of a Vector Space (Optional)

Since V' is a vector space, it has a dual space that is denoted V''. The elements of V'' are linear functionals on V'. In this section we describe the relationship between V and V''. We can use the elements of V to build the elements of V'', as we now describe.

Let V be a vector space. Fix $x_0 \in V$. For each $y \in V'$ we have $y\left(x_0\right) \in \mathcal{F}$. Also for each $z \in V''$ and each $y \in V'$, we have $z\left(y\right) \in \mathcal{F}$. We demonstrate that if V is a finite dimensional vector space, then there is a special relationship between V and V''. To that end, for the fixed $x_0 \in V$ we seek a unique $z_0 \in V''$ for which $y(x_0) = z_0(y)$ for every $y \in V'$.

Let

$\{e_1, \ldots, e_n\}$ be a basis for V,

$\{e_1', \ldots, e_n'\}$ be the basis for V' where $e_i'(e_j) = \delta_{ij}$,

and

$\{e_1'', \ldots, e_n''\}$ be the basis for V'' where $e_i''(e_j') = \delta_{ij}$.

Let the fixed $x_0 \in V$ that we know be given by

$$x_0 = a_1 e_1 + \cdots + a_n e_n$$

and the $z_0 \in V''$ to be determined be given by

$$z_0 = c_1 e_1'' + \cdots + c_n e_n''.$$

An element $y \in V'$ can be represented as

$$y = b_1 e_1' + \cdots + b_n e_n'.$$

Thus,

$$y(x_0) = a_1 b_1 + \cdots + a_n b_n$$

and

$$z_0(y) = b_1 c_1 + \cdots + b_n c_n$$

so the condition that

$$y(x_0) = z_0(y) \text{ for every } y \in V'$$

can be expressed as

$$a_1 b_1 + \cdots + a_n b_n = b_1 c_1 + \cdots + b_n c_n$$

for every choice of b_i, $i = 1, \ldots, n$.

This means that $a_i = c_i$, $i = 1, \ldots, n$.

The function

$$\Phi : V \to V''$$

defined by

$$\Phi(a_1 e_1 + \cdots + a_n e_n) = a_1 e_1'' + \cdots + a_n e_n''$$

is one-to-one and onto and satisfies

$$\Phi\left(a_1 x_1 + a_2 x_2\right) = a_1 \Phi\left(x_1\right) + a_2 \Phi\left(x_2\right).$$

This says that V and V'' are (naturally) isomorphic.

Exercises

1. For each of the linear functionals $T : V \to \mathcal{F}$ given below, find the vector $\hat{w}$ for which $T\left(\hat{v}\right) = \langle \hat{v}, \hat{w} \rangle$.

   (a.) $T : \mathbb{R}^2 \to \mathbb{R}$; $T\left(x, y\right) = 4x - 2y$,

   (b.) $T : \mathbb{R}^3 \to \mathbb{R}$; $T\left(x, y, z\right) = 3x - 7z$

   (c.) $T : \mathbb{R}^3 \to \mathbb{R}$; $T\left(x, y, z\right) = x$

   (d.) $T : \mathbb{C}^2 \to \mathbb{C}$; $T\left(x, y\right) = \left(2 - 4i\right)x - 8y$

   (e.) $T : \mathbb{C}^3 \to \mathbb{C}$; $T\left(x, y, z\right) = ix + \left(3 - 2i\right)y + \left(5 + 4i\right)z$

   (f.) $T : \mathbb{C}^3 \to \mathbb{C}$; $T\left(x, y, z\right) = \left(3 - i\right)x + \left(7 + 5i\right)y + \left(1 - 2i\right)z$.

2. Suppose $T : \mathbb{R}^2 \to \mathbb{R}$; $T\left(x, y\right) = 5x + 3y$. Find the vector $\hat{w}$ for which $T\left(\hat{v}\right) = \langle \hat{v}, \hat{w} \rangle$.

   (a.) In the standard basis.

   (b.) In the basis B where

   $$B = \left\{ \begin{pmatrix} 3 \\ 1 \end{pmatrix} \begin{pmatrix} -2 \\ 5 \end{pmatrix} \right\}.$$

   (c.) Find $T\begin{pmatrix} 6 \\ 1 \end{pmatrix}$ in the standard basis.

   (d.) Find $T\begin{pmatrix} 6 \\ 1 \end{pmatrix}$ in the basis $\mathcal{B}$.

3. Suppose $T : \mathbb{R}^3 \to \mathbb{R}$; $T\left(x, y, z\right) = 5x + 3y - 4z$. find the vector $\hat{w}$ for which $T\left(\hat{v}\right) = \langle \hat{v}, \hat{w} \rangle$

   (a.) In the basis $\mathcal{B} = \left\{ \begin{pmatrix} 1 \\ 1 \\ 1 \end{pmatrix} \begin{pmatrix} 1 \\ 1 \\ 0 \end{pmatrix} \begin{pmatrix} 2 \\ 0 \\ 0 \end{pmatrix} \right\}$

   (b.) Find $T\begin{pmatrix} a \\ b \\ c \end{pmatrix}$ in the basis $\mathcal{B}$.

Section 7.2 The Adjoint of a Linear Operator

The adjoint of a linear operator has classically been defined for all vector spaces (see Halmos (*Finite Dimensional Vector Spaces*)), but the most common applications occur with inner product spaces.

It has become common in the literature to either differentiate between the adjugate of a linear transformation on a vector space and reserve the term "adjoint" for inner product spaces, or to only deal with inner product spaces. We will only consider inner product spaces.

The Adjoint Operator

In this section V and W are finite dimensional inner product spaces and $T : V \to W$ is a linear transformation. We want to find a linear transformation

$$T^* : W \to V$$

that satisfies

$$\langle T\hat{v}, \hat{w} \rangle_W = \langle \hat{v}, T^*\hat{w} \rangle_V$$

for every $\hat{v} \in V$, $\hat{w} \in W$.

The notation $\langle, \rangle_W$ and $\langle, \rangle_V$ will be used to emphasize that V and W may have different inner products.

The major tool in this construction will be the Riesz Representation Theorem.

To establish some intuition, we consider an example that has the features of the abstract theory.

Let

$$T : \mathbb{R}^3 \to \mathbb{R}^2$$

be defined by

$$T \begin{pmatrix} x \\ y \\ z \end{pmatrix} = \begin{pmatrix} 3x + 2y - z \\ -4x + 3y \end{pmatrix}.$$

For a given $\hat{w} \in \mathbb{R}^2$ we can define a linear functional

$$T_{\hat{w}} : \mathbb{R}^3 \to \mathbb{R}$$

by

$$T_{\hat{w}} \begin{pmatrix} x \\ y \\ z \end{pmatrix} = \langle T\hat{v}, \hat{w} \rangle_{\mathbb{R}^2}.$$

If we take $\hat{w} = \begin{pmatrix} -1 \\ 4 \end{pmatrix}$ in our example, we get

$$\langle T\hat{v},\hat{w}\rangle_{\mathbb{R}^2} = \left\langle T\begin{pmatrix}x\\y\\z\end{pmatrix},\begin{pmatrix}-1\\4\end{pmatrix}\right\rangle_{\mathbb{R}^2} = \left\langle\begin{pmatrix}3x+2y-z\\-4x+3y\end{pmatrix},\begin{pmatrix}-1\\4\end{pmatrix}\right\rangle_{\mathbb{R}^2}$$

$$= -3x-2y+z-16x+12y = -19x+10y+z = T_{\hat{w}}\begin{pmatrix}x\\y\\z\end{pmatrix}.$$

Thus,

$$T_{\hat{w}}:\mathbb{R}^3\to\mathbb{R}.$$

By the Riesz Representation Theorem, associated with $T_{\hat{w}}$ is a unique vector in $\mathbb{R}^3$ that we denote $T^*(\hat{w})$ for which

$$T_{\hat{w}}(\hat{v}) = \left\langle\hat{v},T^*(\hat{w})\right\rangle_{\mathbb{R}^3}.$$

Thus,

$$\left\langle T(\hat{v}),\hat{w}\right\rangle_{\mathbb{R}^2} = \left\langle\hat{v},T^*(\hat{w})\right\rangle_{\mathbb{R}^3}$$

Definition:

The function $T^*:W\to V$ defined above is called the adjoint of T.
   We now

 (i) show T^* is a linear transformation and
 (ii) show how to find the representation of T^* given the representation of T.

Theorem 7.3:

Let V and W be finite dimensional inner product spaces and

$$T:V\to W$$

be a linear transformation. The function $T^*:W\to V$ defined by

$$\left\langle T(\hat{v}),\hat{w}\right\rangle_W = \left\langle\hat{v},T^*(\hat{w})\right\rangle_V$$

is a linear transformation.

Proof:

Let $\hat{w}_1,\hat{w}_2\in W$ and $\alpha\in\mathcal{F}$. We show

$$T^*(\hat{w}_1+\hat{w}_2) = T^*(\hat{w}_1)+T^*(\hat{w}_2).$$

For $\hat{v} \in V$ we have

$$\left\langle \hat{v}, T^* \left(\hat{w}_1 + \hat{w}_2 \right) \right\rangle_V = \left\langle T\hat{v}, \hat{w}_1 + \hat{w}_2 \right\rangle_W = \left\langle T\hat{v}, \hat{w}_1 \right\rangle_W + \left\langle T\hat{v}, \hat{w}_2 \right\rangle_W$$
$$= \left\langle \hat{v}, T^*(\hat{w}_1) \right\rangle_V + \left\langle \hat{v}, T^* \left(\hat{w}_2 \right) \right\rangle_V$$

and so, by the uniqueness of the Riesz Representation, it follows that

$$T^* \left(\hat{w}_1 + \hat{w}_2 \right) = T^* \left(\hat{w}_1 \right) + T^* \left(\hat{w}_2 \right).$$

We next show $T^* \left(\alpha \hat{w}_1 \right) = \alpha T^* \left(\hat{w}_1 \right)$.

We have

$$\left\langle \hat{v}, T^* \left(\alpha \hat{w}_1 \right) \right\rangle_V = \left\langle T\hat{v}, \alpha \hat{w}_1 \right\rangle_W = \bar{\alpha} \left\langle T\hat{v}, \hat{w}_1 \right\rangle_W = \bar{\alpha} \left\langle \hat{v}, T^* \left(\hat{w}_1 \right) \right\rangle_V$$
$$= \left\langle \hat{v}, \alpha T^* \left(\hat{w}_1 \right) \right\rangle_V$$

and again, by the uniqueness of the Riesz Representation, it follows that

$$T^* \left(\alpha \hat{w}_1 \right) = \alpha T^* \left(\hat{w}_1 \right).$$

We want to determine a formula for T^* in terms of T. The following example shows how to do this. In our subsequent discussion we assume that the inner product is the usual inner product for the different vector spaces.

Example

Suppose $T : \mathbb{R}^3 \to \mathbb{R}^2$ is defined by

$$T\left(x, y, z\right) = \left(2x - 5y + 2z, 3x + 4y - 5z\right)$$

so that the matrix representation of T is

$$\begin{pmatrix} 2 & -5 & 2 \\ 3 & 4 & -5 \end{pmatrix}$$

We find $T^*\hat{w}$ by solving $\langle T\hat{v}, \hat{w} \rangle = \langle \hat{v}, T^*\hat{w} \rangle$.

Let $\hat{v} = \left(x, y, z\right)$ and $\hat{w} = \left(a, b\right)$.

We have

$$\langle T\hat{v}, \hat{w} \rangle = \left\langle \left(2x - 5y + 2z, 3x + 4y - 5z\right), \left(a, b\right) \right\rangle$$
$$= a\left(2x - 5y + 2z\right) + b\left(3x + 4y - 5z\right).$$

We want to express

$$a\left(2x - 5y + 2z\right) + b\left(3x + 4y - 5z\right)$$

as

$$xf(a,b) + yg(a,b) + zh(a,b).$$

Since $T^*\hat{w} \in V$ this will allow us to express our answer in the proper form which is

$$T^*\hat{w} = \big(f(a,b), g(a,b), h(a,b)\big)$$

Now

$$a(2x - 5y + 2z) + b(3x + 4y - 5z)$$
$$= x(2a + 3b) + y(-5a + 4b) + z(2a - 5b)$$

so

$$T^*\hat{w} = T^*(a,b) = (2a + 3b, -5a + 4b, 2a - 5b).$$

Notice that the matrix representation of T^* is

$$\begin{pmatrix} 2 & 3 \\ -5 & 4 \\ 2 & -5 \end{pmatrix}$$

which is the transpose of the matrix T.

We show later that if the scalar field is $\mathbb{R}$ then $T^* = T^T$ and if the scalar field is $\mathbb{C}$ then the representation of the adjoint is the matrix whose entries are the complex conjugate of the entries of the transpose of T.

We next formally address the question of finding the matrix representation of T^* with respect to a particular basis. We reiterate a point we have made several times earlier that while a linear transformation does not change with a change in basis, the matrix representation of the transformation does. The following "equation" might be helpful in keeping this straight:

$$\text{linear transformation} + \text{basis} \Rightarrow \text{matrix representation.}$$

We first relate the matrix representation of T^* to that of T if we have the usual basis for both vector spaces.

Theorem 7.4:

Let U and V be finite dimensional inner product spaces over $\mathbb{C}$ with the usual inner product. Let

$$A : U \to V$$

be a linear transformation with matrix

$$A = \begin{pmatrix} a_{11} & a_{12} & \cdots & a_{1n} \\ a_{21} & a_{22} & \cdots & a_{2n} \\ \vdots & \vdots & & \vdots \\ a_{m1} & a_{m2} & \cdots & a_{mn} \end{pmatrix}.$$

If the i, j entry of A is a_{ij}, then the i, j entry of the adjoint of A is $\overline{a_{ji}}$.

Proof:

Let

$$A = \begin{pmatrix} a_{11} & a_{12} & \cdots & a_{1n} \\ a_{21} & a_{22} & \cdots & a_{2n} \\ \vdots & \vdots & & \vdots \\ a_{m1} & a_{m2} & \cdots & a_{mn} \end{pmatrix} \qquad A^* = \begin{pmatrix} b_{11} & b_{12} & \cdots & b_{1m} \\ b_{21} & b_{22} & \cdots & b_{2n} \\ \vdots & \vdots & & \vdots \\ b_{n1} & b_{n2} & \cdots & b_{nm} \end{pmatrix}$$

$$\hat{x} = \begin{pmatrix} x_1 \\ \vdots \\ x_n \end{pmatrix} \text{ and } \hat{y} = \begin{pmatrix} y_1 \\ \vdots \\ y_m \end{pmatrix}.$$

Then

$$A\hat{x} = \begin{pmatrix} a_{11}x_1 + \cdots + a_{1n}x_n \\ \vdots \\ a_{m1}x_1 + \cdots + a_{mn}x_n \end{pmatrix}$$

and

$$\langle A\hat{x}, \hat{y} \rangle = \left(a_{11}x_1 + \cdots + a_{1n}x_n \right)\overline{y_1} + \cdots + \left(a_{m1}x_1 + \cdots + a_{mn}x_n \right)\overline{y_m}.$$

Also

$$A^*y = \begin{pmatrix} b_{11}y_1 + \cdots + b_{1m}y_m \\ \vdots \\ b_{1n}y_1 + \cdots + b_{nm}y_m \end{pmatrix}$$

and

$$\langle \hat{x}, A^*\hat{y} \rangle = x_1 \left(\overline{b_{11}y_1 + \cdots + b_{1m}y_m} \right) + \cdots + x_n \left(\overline{b_{n1}y_1 + \cdots + b_{nm}y_m} \right)$$

$$= \left(x_1\overline{b_{11}} + x_2\overline{b_{21}} + \cdots + x_n\overline{b_{n1}} \right)\overline{y_1} + \cdots + \left(x_1\overline{b_{1m}} + \cdots + x_n\overline{b_{nm}} \right)\overline{y_m}$$

for all x_i, y_j; $i = 1,\ldots,n$, $j = 1,\ldots,m$. Thus,

$$b_{ij} = \overline{a}_{ji}; \ i, j = 1,\ldots,n.$$

Notation: The matrix A^* denotes the conjugate transpose of the matrix A, and it is obtained by first taking the transpose of A and then taking the complex conjugate of each entry of the resulting matrix.

Example

Suppose $T : \mathbb{C}^3 \to \mathbb{C}^2$ is defined by

$$T(x,y,z) = \left(3x + (2+5i)y + iz, (2-4i)x + (1-3i)y + 9z\right).$$

The matrix of T in the standard basis is

$$\begin{pmatrix} 3 & 2+5i & i \\ 2-4i & 1-3i & 9 \end{pmatrix}$$

so the matrix representation of T^* in the standard basis is

$$A^* = \begin{pmatrix} 3 & 2+4i \\ 2-5i & 1+3i \\ -i & 9 \end{pmatrix}.$$

Not surprisingly, the situation when other than the usual bases are used is not as simple. However, if orthonormal bases are used, it is still manageable. This is because if $\{\hat{v}_1,\ldots,\hat{v}_n\}$ is an orthonormal basis for V and $\hat{v} \in V$ then

$$\hat{v} = \langle \hat{v}, \hat{v}_1 \rangle \hat{v}_1 + \cdots + \langle \hat{v}, \hat{v}_n \rangle \hat{v}_n$$

so that the representation of $\hat{v}$ as a column vector in the basis $\{\hat{v}_1 \ldots, \hat{v}_n\}$ is

$$\begin{pmatrix} \langle \hat{v}, \hat{v}_1 \rangle \\ \vdots \\ \langle \hat{v}, \hat{v}_n \rangle \end{pmatrix}.$$

This fact, together with the work we did on representing a linear transformation in other than the usual bases in Section 6.5 gives the following result.

Theorem 7.5:

If V and W are finite dimensional inner product spaces with orthonormal bases $\mathcal{B} = \{\hat{v}_1 \ldots, \hat{v}_n\}$ and $\mathcal{C} = \{\hat{w}_1 \ldots, \hat{w}_m\}$, respectively, and $T : V \to W$ is a linear transformation, then the matrix representation of T with respect to these bases is

$$A = P_C^{-1} T P_B.$$

where P_B and P_C are the transition matrices for $\mathcal{B}$ and $\mathcal{C}$, respectively.

With an argument identical to that of Theorem 7.3 we get the following.

Corollary:

The matrix for T^* under the hypotheses of Theorem 7.4 is the conjugate transpose of the matrix A. That is, if

$$T^* : W \to V$$

with orthonormal bases $\mathcal{B}$ of V and $\mathcal{C}$ of W, then the matrix of T^* with respect to these bases is A^* where

$$\left(A^*\right)_{ij} = \bar{A}_{ji} \,.$$

Adjoint on Weighted Inner Product Spaces (Optional)

The adjoint of a linear operator depends on the inner product as well as the operator. Consider the following example:

Define an inner product on $\mathbb{R}^2$ by

$$\langle \hat{x}, \hat{y} \rangle = 2x_1 y_1 + x_2 y_2$$

and let

$$A = \begin{pmatrix} a_{11} & a_{12} \\ a_{21} & a_{22} \end{pmatrix}$$

So

$$\langle A\hat{x}, \hat{y} \rangle = \left\langle \begin{pmatrix} a_{11}x_1 + a_{12}x_2 \\ a_{21}x_1 + a_{22}x_2 \end{pmatrix}, \left(\begin{pmatrix} y_1 \\ y_2 \end{pmatrix} \right) \right\rangle \cdot$$

$$= 2\left(a_{11}x_1 + a_{12}x_2\right)y_1 + \left(a_{21}x_1 + a_{22}x_2\right)y_2$$

Suppose that

$$B = \begin{pmatrix} b_{11} & b_{12} \\ b_{21} & b_{22} \end{pmatrix}$$

is the adjoint of A. Then

$$\langle \hat{x}, B\,\hat{y} \rangle = \left\langle \begin{pmatrix} x_1 \\ x_2 \end{pmatrix}, \begin{pmatrix} b_{11}y_1 + b_{12}y_2 \\ b_{21}y_1 + b_{22}y_2 \end{pmatrix} \right\rangle$$

$$= 2x_1\left(b_{11}y_1 + b_{12}y_2\right) + x_2\left(b_{21}y_1 + b_{22}y_2\right)$$

This gives

$$2a_{11}x_1 y_1 = 2b_{11}x_1 y_1 \quad \text{so} \quad a_{11} = b_{11}$$

$$2a_{12}x_2 y_1 = b_{21}x_2 y_1 \quad \text{so} \quad 2a_{12} = b_{21}$$

$$a_{21}x_1 y_2 = 2b_{12}x_1 y_2 \quad \text{so} \quad a_{21} = 2b_{12} \quad \text{or} \quad b_{12} = \frac{a_{21}}{2}$$

$$a_{22}x_2y_2 = b_{22}x_2y_2 \quad \text{so} \quad a_{22} = b_{22}$$

Thus, the adjoint of A is

$$A' = \begin{pmatrix} b_{11} & b_{12} \\ b_{21} & b_{22} \end{pmatrix} = \begin{pmatrix} a_{11} & \dfrac{a_{21}}{2} \\ 2a_{12} & a_{22} \end{pmatrix}.$$

Exercises

1. For the linear transformations given below find

   (i) T^*w, (ii) The matrix representation of T in the usual basis, and (iii) The matrix representation of T^* in the usual basis.

   (a.) $T : \mathbb{R}^3 \to \mathbb{R}^2$ is defined by

   $$T(x,y,z) = (x + 2y - 3z, 3x - 5z).$$

   (b.) $T : \mathbb{R}^2 \to \mathbb{R}^4$ is defined by

   $$T(x,y) = (4y, 3x + 2z, x + 2y - 6z, 2x).$$

   (c.) $T : \mathbb{R} \to \mathbb{R}^2$ is defined by

   $$T(x) = (4x, -3x).$$

   (d.) $T : \mathbb{C}^3 \to \mathbb{C}^2$ is defined by

   $$T(x,y,z) = \left((2 - 3i)x + (2 - 4i)y + (1 - 3i)y + iz, 4x - 3iy + z \right)$$

   (e.) $T : \mathbb{C}^2 \to \mathbb{C}^3$ is defined by

   $$T(x,y) = \left(3x + (4 + 5i)y, 6ix - 2iy, (6 - 2i)x + (1 - i)y \right)$$

   (f.) $T : \mathbb{C}^3 \to \mathbb{C}^3$ is defined by

   $$T(x,y,z) = \left(ix + y, 4iy - 2z, x + (3 + i)z \right)$$

2. (a.) If
$T : \mathbb{R}^2 \to \mathbb{R}^3$ is defined by

   $$T(x,y) = (x - 2y, 2x + 4y, 3x)$$

find $T_{\hat{w}}$ for

    (i) $\hat{w} = (1,-2,0)$

    (ii) $\hat{w} = (-3,5,-2)$

(b). If

$$T : \mathbb{R}^3 \to \mathbb{R}^2 \text{ is defined by}$$

$$T(x,y,z) = (2x - 2y + 3z, x + 4y + z)$$

    find $T_{\hat{w}}$ for
    (i) $\hat{w} = (-4,-2)$

    (ii) $\hat{w} = (1,5)$

3. Show if $T : U \to V$ is a linear transformation then

  (a.) $\left(T^*\right)^* = T$

  (b.) $\left(aT\right)^* = \bar{a}\left(T^*\right)$

4. Recall that if $T : U \to V$ is a linear transformation then

$$\mathcal{N}(T) = \left\{\hat{u} \in U \,|\, T\hat{u} = \hat{0}\right\}, \ \mathcal{R}(T) = \left\{\hat{v} \in V \,|\, \hat{v} = T\hat{u} \text{ for some } \hat{u} \in U\right\}$$

Show that

  (a.) $\mathcal{N}\left(T^*\right) = \left(\mathcal{R}(T)\right)^{\perp}$

  (b.) $\mathcal{R}\left(T^*\right) = \left(\mathcal{N}(T)\right)^{\perp}$

  (c.) $\mathcal{N}(T) = \left(\mathcal{R}(T)^*\right)^{\perp}$

  (d.) $\mathcal{R}(T) = \left(\mathcal{N}\left(T^*\right)\right)^{\perp}$

Compare with Theorem 6.9 (Chapter 6).

5. Let $T : U \to V$ be a linear transformation. Show
  (a.) T is one-to-one if and only if T^* is onto.
  (b.) T is onto if and only if T^* is one-to-one.

Section 7.3 The Spectral Theorem

In some areas of science, including mathematical physics, eigenvectors on inner product spaces are of central importance.

Of particular interest are symmetric matrices when the vector space is $\mathbb{R}^n$ and Hermitian matrices when the vector space is $\mathbb{C}^n$.

We repeat a definition.

Definition:

A symmetric matrix A is an $n \times n$ matrix for which $A^T = A$;
that is, $A_{ij} = A_{ji}$ $i, j = 1, \ldots, n$.

A Hermitian matrix A is an $n \times n$ matrix for which $\overline{A}_{ji} = A_{ij}$ $i, j = 1, \ldots, n$.

Note that if the scalar field is $\mathbb{R}$, then a symmetric matrix has the property that $A = A^T$ and if the scalar field is $\mathbb{C}$ then a Hermitian matrix has the property that $A = A^*$.

Also, note that the main diagonal entries of a Hermitian matrix are real.

Definition:

A linear operator A on the inner product space V is self-adjoint if $A = A^*$; that is, if

$$\langle A\hat{v}, \hat{w} \rangle = \langle \hat{w}, A\hat{v} \rangle \text{ for all } \hat{v}, \hat{w} \in V.$$

Theorem 7.6:

Let $V = \mathcal{F}^n$ be an inner product space and let A be a symmetric matrix. Then

1. The eigenvalues of A are real.
2. Eigenvectors corresponding to different eigenvalues are orthogonal.

Proof:

1. Suppose that λ is an eigenvalue of A with eigenvector $\hat{v}$. Then

$$\lambda \|\hat{v}\|^2 = \lambda \langle \hat{v}, \hat{v} \rangle = \lambda \langle \hat{v}, \hat{v} \rangle = \langle A\hat{v}, \hat{v} \rangle$$
$$= \langle \hat{v}, A^T \hat{v} \rangle = \langle \hat{v}, A\hat{v} \rangle = \langle \hat{v}, \lambda\hat{v} \rangle = \overline{\lambda} \langle \hat{v}, \hat{v} \rangle = \overline{\lambda} \|\hat{v}\|^2$$

   and since $\|\hat{v}\|^2 \neq 0$, we have $\lambda = \overline{\lambda}$.
2. Suppose that λ is an eigenvalue of A with eigenvector $\hat{v}$ and μ is an eigenvalue of A with eigenvector $\hat{u}$. (So λ and μ are real.) Then

$$\lambda \langle \hat{v}, \hat{u} \rangle = \langle \lambda\hat{v}, \hat{u} \rangle = \langle A\hat{v}, \hat{u} \rangle = \langle \hat{v}, A\hat{u} \rangle = \langle \hat{v}, \mu\hat{u} \rangle = \mu \langle \hat{v}, \hat{u} \rangle.$$

Thus,

$$\lambda \langle \hat{v}, \hat{u} \rangle - \mu \langle \hat{v}, \hat{u} \rangle = (\lambda - \mu) \langle \hat{v}, \hat{u} \rangle = 0,$$

but

$$(\lambda - \mu) \neq 0 \text{ so } \langle \hat{v}, \hat{u} \rangle = 0.$$

Corollary:

If A is a self-adjoint matrix, then

1. The eigenvalues of A are real.
2. Eigenvectors corresponding to different eigenvalues are orthogonal.

The spectral theorem, which we now study, is arguably the most important theorem in linear algebra. The proof of the spectral theorem in the real case is (perhaps surprisingly) more difficult than in the complex case. The reason is that the proof requires that we know that a symmetric matrix has an eigenvalue. We know this is true in the complex case because, by the fundamental theorem of algebra, the characteristic polynomial factors into linear factors over the complex numbers. To prove the spectral theorem in the real case, we need the next two theorems.

Theorem 7.7:

Suppose that A is a symmetric matrix. If α and β are real numbers with $\alpha^2 < 4\beta$ then $A^2 + \alpha A + \beta I$ is invertible.

Proof:

Let $\hat{v}$ be a nonzero vector. Now

$$\left\langle \left(A^2 + \alpha A + \beta I \right)\hat{v}, \hat{v} \right\rangle = \left\langle A^2 v, \hat{v} \right\rangle + \alpha \left\langle A\hat{v}, \hat{v} \right\rangle + \beta \|\hat{v}\|^2$$

$$= \left\langle A\hat{v}, A^T \hat{v} \right\rangle + \alpha \left\langle A\hat{v}, \hat{v} \right\rangle + \beta \|\hat{v}\|^2$$

$$= \left\langle A\hat{v}, A\hat{v} \right\rangle + \alpha \left\langle A\hat{v}, \hat{v} \right\rangle + \beta \|\hat{v}\|^2$$

$$= \|A\hat{v}\|^2 + \alpha \left\langle A\hat{v}, \hat{v} \right\rangle + \beta \|\hat{v}\|^2 .$$

since A is symmetric.

By the Cauchy–Schwartz inequality

$$|\alpha| \|A\hat{v}\| \|\hat{v}\| \ge \alpha \left\langle A\hat{v}, \hat{v} \right\rangle$$

so

$$\|A\hat{v}\|^2 + \alpha \left\langle A\hat{v}, \hat{v} \right\rangle + \beta \|\hat{v}\|^2 \ge \|A\hat{v}\|^2 - |\alpha| \|A\hat{v}\| \|\hat{v}\| + \beta \|\hat{v}\|^2 .$$

Now

$$\|A\hat{v}\|^2 - |\alpha| \|A\hat{v}\| \|\hat{v}\| + \beta \|\hat{v}\|^2$$

$$= \left(\|A\hat{v}\|^2 - |\alpha| \|A\hat{v}\| \|\hat{v}\| + \frac{\alpha^2}{4} \|\hat{v}\|^2 \right) + \left(\beta \|\hat{v}\|^2 - \frac{\alpha^2}{4} \|\hat{v}\|^2 \right)$$

$$= \left(\|A\hat{v}\| - \frac{|\alpha|}{2} \|\hat{v}\| \right)^2 + \left(\beta - \frac{\alpha^2}{4} \right) \|\hat{v}\|^2$$

and

$$\left(\|A\hat{v}\|-\frac{|\alpha|}{2}\|\hat{v}\|\right)^2+\left(\beta-\frac{\alpha^2}{4}\right)\|\hat{v}\|^2>0$$

if $\alpha^2<4\beta$.

Thus, $\left(A^2+\alpha A+\beta I\right)\hat{v}\neq\hat{0}$ so $\mathcal{N}\left(A^2+\alpha A+\beta I\right)=\{\hat{0}\}$ and so $\left(A^2+\alpha A+\beta I\right)$ is invertible.

Theorem 7.8:

If A is a real $n\times n$ symmetric matrix, then A has a real eigenvalue.

Proof:

Suppose that
$\hat{v}\neq\hat{0}$. The set $\{\hat{v},A\hat{v},A^2\hat{v},\ldots,A^n\hat{v}\}$ has $n+1$ vectors, and so is linearly dependent. Thus, there are real numbers $c_0,c_1,\ldots,c_n$ not all zero for which

$$c_0\hat{v}+c_1A\hat{v}+\cdots+c_nA^n\hat{v}=\hat{0}.$$

Consider the polynomial

$$p(x)=c_0+c_1x+c_2x^2+\cdots+c_nx^n.$$

This is a polynomial with real coefficients and may be factored over the real numbers so that each factor is a linear factor or an irreducible quadratic factor. Thus, we have

$$p(x)=\gamma(x-\lambda_1)\cdots(x-\lambda_k)\left(x^2+\alpha_1x+\beta_1\right)\cdots\left(x^2+\alpha_jx+\beta_j\right).$$

According to the quadratic formula, there are no real roots to
$x^2+\alpha x+\beta=0$ exactly when $\alpha^2<4b$. According to Theorem 7.7, this means

$$A^2+\alpha A+\beta I$$

is invertible.
   Now consider

$$p(A)=c_0I+c_1A+c_2A^2+\cdots+c_nA^n$$
$$=\gamma(A-\lambda_1I)\cdots(A-\lambda_kI)\left(A^2+\alpha_1A+\beta_1\right)\cdots\left(A^2+\alpha_jA+\beta_j\right).$$

Now,

$$\gamma(A-\lambda_1I)\cdots(A-\lambda_kI)\left(A^2+\alpha_1A+\beta_1\right)\cdots\left(A^2+\alpha_jA+\beta_j\right)\hat{v}=\hat{0}$$

and each $\left(A^2 + \alpha_i A + \beta_i\right)$ is invertible, so

$$\left(A - \lambda_1 I\right)\cdots\left(A - \lambda_k I\right)\hat{v} = \hat{0}.$$

Thus, there is a k for which $\left(A - \lambda_k I\right)\hat{v} = \hat{0}$ and so λ_k is an eigenvalue for A.

There are different ways to state the spectral theorem. One of the more intuitive ways is given here.

Theorem 7.9: (Spectral Theorem)

Let V be an inner product space. A linear operator T on V is self-adjoint if and only if there is an orthonormal basis of V consisting of eigenvectors of T.

Proof:

Suppose that T is a self-adjoint operator on V. We show there is an orthonormal basis of eigenvectors.

We prove the result by induction on the dimension of V. Let n denote the dimension of V.

For $n = 1$, let $\{\hat{u}\}$ be an orthonormal basis for V. Then $T\hat{u} = c\hat{u}$ for some $c \in \mathcal{F}$. Since T is self-adjoint, c is real.

Assume the result holds for $n = k$.

Suppose that $n = k + 1$. If the field is the complex numbers, then the characteristic polynomial factors into linear factors, and there is an eigenvalue. If the field is the real numbers, there is an eigenvalue by Theorem 7.8. Since T is self-adjoint, the eigenvalues are real. Let $\hat{u}$ be a normalized eigenvector of T with eigenvalue λ. Let W be the subspace of T spanned by $\hat{u}$.

Now $T\left(W\right) \subset W$. We show

$$T\left(W^{\perp}\right) \subset W^{\perp}.$$

Let $c\hat{u} \in U$ and $\hat{y} \in W^{\perp}$. We show that $T\left(W^{\perp}\right) \subset W^{\perp}$ by showing $\left\langle c\hat{u}, T\hat{y} \right\rangle = 0$.

We have

$$\left\langle c\hat{u}, T\hat{y} \right\rangle = c\left\langle \hat{u}, T\hat{y} \right\rangle = c\left\langle T\hat{u}, \hat{y} \right\rangle = c\left\langle \lambda\hat{u}, \hat{y} \right\rangle = c\lambda\left\langle \hat{u}, \hat{y} \right\rangle = 0.$$

Consider the matrix for T that we have constructed thus far.

Let the basis to be constructed be $\mathcal{B} = \{\hat{u}, *, *, \ldots, *\}$ where the vectors other than $\hat{u}$ are not yet determined.

Let A be the matrix representation of T in the basis $\mathcal{B}$. Because $T\hat{u} = \lambda\hat{u}$, the first column of A is

$$\begin{pmatrix} \lambda \\ 0 \\ \vdots \\ 0 \end{pmatrix}$$

and because $T\left(W^\perp\right) \subset W^\perp$ the first row of A is

$$\begin{pmatrix} \lambda & 0 & \cdots & 0 \end{pmatrix}.$$

Thus, the matrix for T to this point is

$$\begin{pmatrix} \lambda & 0 & \cdots & 0 \\ 0 & * & \cdots & * \\ \vdots & \vdots & & \vdots \\ 0 & * & \cdots & * \end{pmatrix}.$$

Thus,

$$V = W \oplus W^\perp$$

and the dimension of $W^\perp$ is k.

There is a somewhat subtle point that needs to be addressed before we can claim to use the induction hypothesis; namely, that T restricted to $W^\perp$ is a self-adjoint operator on $W^\perp$. We denote T restricted to $W^\perp$ by $T_{W^\perp}$.

Suppose that $\hat{u}, \hat{v} \in W^\perp$. We have

$$\left\langle \left(T_{W^\perp}\hat{u}\right), \hat{v}\right\rangle = \left\langle T\hat{u}, \hat{v}\right\rangle$$

because if $\hat{u} \in W^\perp \subset V$ then $T_{W^\perp}\left(\hat{u}\right) = T\left(\hat{u}\right)$.

Also,

$$\left\langle T\hat{u}, \hat{v}\right\rangle = \left\langle \hat{u}, T\hat{v}\right\rangle$$

because T is self-adjoint on V and

$$\left\langle \hat{u}, T\hat{v}\right\rangle = \left\langle \hat{u}, T_{W^\perp}\left(\hat{v}\right)\right\rangle$$

because $T\left(W^\perp\right) \subset W^\perp$ so if $\hat{v} \in W^\perp$ then $T_{W^\perp}\left(\hat{v}\right) = T\left(\hat{v}\right)$.

Thus, we may apply the induction principle to conclude that the result holds for all n.

Conversely, suppose there is an orthonormal basis of eigenvectors of T. We show that T is self-adjoint.

In the orthonormal basis, T is diagonal so $T = T^T$ or $T = T^*$

We collect the results of the spectral theorem:

If A is a self-adjoint matrix then

(i) The eigenvalues of A are real.
(ii) The algebraic dimension of each eigenvalue is equal to the geometric dimension.
(iii) Eigenvectors corresponding to different eigenvalues are orthogonal.

(iv) The matrix P whose columns are the normalized eigenvectors of A, has the properties

$$P = P^T = P^{-1}$$

and

$$A = PDP^{-1}$$

where D is a diagonal matrix whose diagonal elements are the eigenvalues of A.

Example

We construct a spectral diagonalization of the matrix

$$A = \begin{pmatrix} 4 & 1 & 1 \\ 1 & 4 & 1 \\ 1 & 1 & 4 \end{pmatrix}.$$

The characteristic polynomial of A is

$$-\lambda^3 + 12\lambda^2 - 45\lambda + 54 = -(\lambda - 3)^2(\lambda - 6).$$

For $\lambda = 3$ we have

$$A - 3I = \begin{pmatrix} 1 & 1 & 1 \\ 1 & 1 & 1 \\ 1 & 1 & 1 \end{pmatrix}$$

which row reduces to

$$\begin{pmatrix} 1 & 1 & 1 \\ 0 & 0 & 0 \\ 0 & 0 & 0 \end{pmatrix}.$$

Two linearly independent eigenvectors are

$$\hat{v}_1 = \begin{pmatrix} -1 \\ 1 \\ 0 \end{pmatrix} \text{ and } \hat{v}_2 = \begin{pmatrix} -1 \\ 0 \\ 1 \end{pmatrix}.$$

For $\lambda = 6$ we have

$$A - 6I = \begin{pmatrix} -2 & 1 & 1 \\ 1 & -2 & 1 \\ 1 & 1 & -2 \end{pmatrix}$$

which row reduces to

$$\begin{pmatrix} 1 & 0 & -1 \\ 0 & 1 & -1 \\ 0 & 0 & 0 \end{pmatrix} \text{ and } \hat{v}_3 = \begin{pmatrix} 1 \\ 1 \\ 1 \end{pmatrix}$$

is an eigenvector.

When we normalize the eigenvectors we have

$$\frac{\hat{v}_1}{\|\hat{v}_1\|} = \hat{w}_1 = \begin{pmatrix} -\dfrac{1}{\sqrt{2}} \\ \dfrac{1}{\sqrt{2}} \\ 0 \end{pmatrix}, \quad \frac{\hat{v}_2}{\|\hat{v}_2\|} = \hat{w}_2 = \begin{pmatrix} -\dfrac{1}{\sqrt{2}} \\ 0 \\ \dfrac{1}{\sqrt{2}} \end{pmatrix},$$

$$\frac{\hat{v}_3}{\|\hat{v}_3\|} = \hat{w}_3 = \begin{pmatrix} \dfrac{1}{\sqrt{3}} \\ \dfrac{1}{\sqrt{3}} \\ \dfrac{1}{\sqrt{3}} \end{pmatrix}.$$

One choice of P is

$$P = \begin{pmatrix} -\dfrac{1}{\sqrt{2}} & -\dfrac{1}{\sqrt{2}} & \dfrac{1}{\sqrt{3}} \\ \dfrac{1}{\sqrt{2}} & 0 & \dfrac{1}{\sqrt{3}} \\ 0 & \dfrac{1}{\sqrt{2}} & \dfrac{1}{\sqrt{3}} \end{pmatrix}$$

and for this choice of P we have

$$D = \begin{pmatrix} 3 & 0 & 0 \\ 0 & 3 & 0 \\ 0 & 0 & 6 \end{pmatrix}.$$

Exercises

Give a spectral diagonalization for the matrices below.

1. $A = \begin{pmatrix} 2 & 1 \\ 1 & 2 \end{pmatrix}$

2. $A = \begin{pmatrix} 0 & 3 \\ 3 & 0 \end{pmatrix}$

3. $A = \begin{pmatrix} 3 & 2 \\ 2 & 3 \end{pmatrix}$

4. $A = \begin{pmatrix} 1 & 2 & 2 \\ 2 & 1 & 2 \\ 2 & 2 & 1 \end{pmatrix}$

5. $A = \begin{pmatrix} 3 & 1 & -1 \\ 1 & 3 & -1 \\ -1 & -1 & 5 \end{pmatrix}$

6. $A = \begin{pmatrix} 0 & 1 & 1 \\ 1 & 0 & 1 \\ 1 & 1 & 0 \end{pmatrix}$

Chapter 8

Two Decompositions of a Matrix

In the remaining chapters, we discuss topics that are not normally covered in a first linear algebra course but are nonetheless important in the study of the subject. The style, in some places, will be more succinct than in the earlier chapters.

Section 8.1 Polar Decomposition of a Matrix

We recall some definitions.

Definition:

The square matrix U is a unitary matrix if $UU^* = U^*U = I$, where U^* is the conjugate transpose of U.

Definition:

A self-adjoint linear operator B on an inner product space V is positive semi-definite if $\langle \hat{v}, B\hat{v} \rangle \geq 0$ for all vectors $\hat{v} \ \varepsilon \ V$.

Given a square matrix A, we want to find matrices U and P with $A = UP$ where U is unitary and P is positive semi-definite. This is similar in spirit to converting a complex number $z = x + iy$ to the polar form

$$z = re^{i\theta}$$

where

$$r = \sqrt{\bar{z}z} = |z|$$

and θ is real so that

$$\left| e^{i\theta} \right| = 1.$$

The matrix U plays a role analogous to $e^{i\theta}$ and P a role analogous to r.

We first present the theoretical foundation of the construction and then give an algorithm for determining the decomposition.

Let

$$T : \mathbb{R}^n \to \mathbb{R}^m$$

and let A be the $m \times n$ matrix for T with respect to some basis.

Theorem 8.1:

If A is an $m \times n$ matrix, then $A^* A$ is an $n \times n$ matrix that is symmetric and positive semi-definite.

Proof:

$A^* A$ is symmetric because $\left(A^* A \right)^* = \left(A^* \right)\left(A^{**} \right) = A^* A$ and $A^* A$ is positive semi-definite because for any vector $\hat{v} \in V$

$$\left\langle \hat{v}, A^* A \hat{v} \right\rangle = \left\langle A\hat{v}, A\hat{v} \right\rangle = \left\| A\hat{v}^2 \right\|$$

Thus, $A^* A$ can be diagonalized where the diagonal entries are non-negative. That is, there is a unitary matrix S for which

$$A^* A = S \begin{pmatrix} \lambda_1^{\ 2} & 0 & 0 \\ 0 & \ddots & 0 \\ 0 & \cdots & \lambda_n^{\ 2} \end{pmatrix} S^{-1}$$

where the $\left(\lambda_i^{\ 2} \right)$'s are the eigenvalues of $A^* A$.

Now

$$A^* A = S \begin{pmatrix} \lambda_1^{\ 2} & 0 & 0 \\ 0 & \ddots & 0 \\ 0 & \cdots & \lambda_n^{\ 2} \end{pmatrix} S^{-1} = \left[S \begin{pmatrix} \lambda_1 & 0 & 0 \\ 0 & \ddots & 0 \\ 0 & \cdots & \lambda_n \end{pmatrix} S^{-1} \right]\left[S \begin{pmatrix} \lambda_1 & 0 & 0 \\ 0 & \ddots & 0 \\ 0 & \cdots & \lambda_n \end{pmatrix} S^{-1} \right]$$

so it is appropriate to say

$$\sqrt{A^* A} = S \begin{pmatrix} \lambda_1 & 0 & 0 \\ 0 & \ddots & 0 \\ 0 & \cdots & \lambda_n \end{pmatrix} S^{-1}.$$

Note that

$$\left(\sqrt{A^* A}\right)^* = \left(S \begin{pmatrix} \lambda_1 & 0 & 0 \\ 0 & \ddots & 0 \\ 0 & \cdots & \lambda_n \end{pmatrix} S^{-1} \right)^* = \left(S \begin{pmatrix} \lambda_1 & 0 & 0 \\ 0 & \ddots & 0 \\ 0 & \cdots & \lambda_n \end{pmatrix} S^* \right)^*$$

$$= S^{**} \begin{pmatrix} \lambda_1 & 0 & 0 \\ 0 & \ddots & 0 \\ 0 & \cdots & \lambda_n \end{pmatrix} S^* = S \begin{pmatrix} \lambda_1 & 0 & 0 \\ 0 & \ddots & 0 \\ 0 & \cdots & \lambda_n \end{pmatrix} S^*$$

$$= S \begin{pmatrix} \lambda_1 & 0 & 0 \\ 0 & \ddots & 0 \\ 0 & \cdots & \lambda_n \end{pmatrix} S^{-1}$$

so $\sqrt{A^* A}$ is self-adjoint.

Theorem 8.2:

For $\hat{v}$ a vector in $\mathbb{R}^n$, we have $\|A\hat{v}\| = \left\| \sqrt{A^* A} \, \hat{v} \right\|$. That is, A and $\sqrt{A^* A}$ have the same effect on the length of a vector.

Proof:

We have

$$\|A\hat{v}^2\| = \langle A\hat{v}, A\hat{v} \rangle = \langle \hat{v}, A^* A\hat{v} \rangle = \left\langle \hat{v}, \sqrt{A^* A} \sqrt{A^* A} \, \hat{v} \right\rangle$$

$$= \left\langle \sqrt{A^* A} \, \hat{v}, \sqrt{A^* A} \, \hat{v} \right\rangle = \left\| \sqrt{A^* A} \, \hat{v} \right\|^2$$

where the second to last equality is valid because $\sqrt{A^* A}$ is self-adjoint.

Corollary:

We have

(a.) $\ker A = \ker \sqrt{A^* A}$

(b.) $\operatorname{rank} A = \operatorname{rank} \sqrt{A^* A}$.

Part (b.) is true because $\dim(\ker A) + \operatorname{rank} A = n$; or, said another way,

$$\operatorname{nullity}(A) + \operatorname{rank} A = n$$

Theorem 8.3:

Let A be a matrix. There are matrices U and P where U is unitary, P is positive semi-definite and $A = UP$.

Proof:

Since $\sqrt{A^*A}$ is self-adjoint, there is an orthonormal basis of eigenvectors, $\hat{v}_1,\ldots,\hat{v}_n$. Suppose that

$$\sqrt{A^*A}\,\hat{v}_i = \lambda_i \hat{v}_i.$$

Since $\sqrt{A^*A}$ is positive semi-definite, $\lambda_i \geq 0$. Enumerate the eigenvectors so that

$$\lambda_i > 0 \ \text{ for } \ i = 1,\ldots,k \text{ and } \lambda_i \doteq 0 \text{ for } i = k+1,\ldots,n.$$

We show that

$$\left\{ \frac{1}{\lambda_i} A\hat{v}_i, i = 1,\ldots,k \right\}$$

is an orthonormal set of vectors. We have

$$\left\langle \frac{1}{\lambda_i} A\hat{v}_i, \frac{1}{\lambda_j} A\hat{v}_j \right\rangle = \frac{1}{\lambda_i \lambda_j} \left\langle A\hat{v}_i, A\hat{v}_j \right\rangle = \frac{1}{\lambda_i \lambda_j} \left\langle \hat{v}_i, A^* A\hat{v}_j \right\rangle = \frac{1}{\lambda_i \lambda_j} \left\langle \hat{v}_i, \sqrt{A^*A}\sqrt{A^*A}\,\hat{v}_j \right\rangle$$

$$= \frac{1}{\lambda_i \lambda_j} \left\langle \hat{v}_i, \sqrt{A^*A}\,\lambda_j \hat{v}_j \right\rangle = \frac{1}{\lambda_i \lambda_j} \left\langle \hat{v}_i, \lambda_j^2 \hat{v}_j \right\rangle = \frac{\lambda_j}{\lambda_i} \left\langle \hat{v}_i, \hat{v}_j \right\rangle = \delta_{i,j}.$$

Thus, we have

$$\left\{ \frac{1}{\lambda_i} A\hat{v}_i, i = 1,\ldots,k \right\}$$

is a basis for the image of A. We extend this by completing an orthonormal basis for A by adding the orthonormal vectors $\{\hat{u}_i, i = k+1,\ldots,n\}$. It follows that $\{\hat{u}_i, i = k+1,\ldots,n\}$ is a basis for Ker A.

We form an orthonormal basis for $\mathbb{R}^n$ by extending

$$\left\{ \frac{1}{\lambda_i} A\hat{v}_i, i = 1,\ldots,k \right\}$$

to include $\hat{u}_{k+1},\ldots,\hat{u}_n$. *Note that extending the basis is necessary if and only if 0 is an eigenvalue.*

Define the linear transformation $U : \mathbb{R}^n \to \mathbb{R}^n$ by

$$U\hat{v}_i = \begin{cases} \dfrac{1}{\lambda_i} A\hat{v}_i, i = 1,\ldots,k \\[2ex] \hat{u}_i, i = k+1,\ldots,n \end{cases}.$$

Thus, with respect to the basis

$$\{\hat{v}_1,\ldots,\hat{v}_k,\hat{v}_{k+1},\ldots,\hat{v}_n\}$$

the matrix of U is

$$\left(\begin{array}{ccc|c} \dfrac{1}{\lambda_1}A\hat{v}_1 & \cdots & \dfrac{1}{\lambda_k}A\hat{v}_k & \mathcal{O} \\ & & & I \end{array}\right)$$

where the ith column of U, $1\le i\le k$, is the column vector

$$\frac{1}{\lambda_i}A\hat{v}_i$$

I is the $(n-k)\times(n-k)$ identity matrix and $\mathcal{O}$ is the $k\times(n-k)$ zero matrix.

By the construction, U is a unitary matrix. We now show that

$$A=U\sqrt{A^*A}.$$

Choose a basis vector $\hat{v}_i$. If $1\le i\le k$, then

$$U\sqrt{A^*A}\hat{v}_i=U\left(\sqrt{A^*A}\hat{v}_i\right)=U\left(\lambda_i\hat{v}_i\right)=\lambda_i\left(U\hat{v}_i\right)=\lambda_i\left(\frac{1}{\lambda_i}A\hat{v}_i\right)=A\hat{v}_i.$$

If $k+1\le i\le n$, then $\hat{v}_i\in\ker A$, so $A\hat{v}_i=\hat{0}$. But $\ker A=\ker\sqrt{A^*A}$, so

$$\sqrt{A^*A}\hat{v}_i=\hat{0}.$$

We give an example of how to apply the construction. The construction will depend on whether 0 is an eigenvalue for A^*A.

Example

Case 1. 0 is not an eigenvalue for A^*A.

Let

$$A=\begin{pmatrix}1 & 2\\ -1 & 2\end{pmatrix}.$$

Step 1. We compute A^*A and $\sqrt{A^*A}$.

$$A=\begin{pmatrix}1 & 2\\ -1 & 2\end{pmatrix},A^*=\begin{pmatrix}1 & -1\\ 2 & 2\end{pmatrix};\;A^*A=\begin{pmatrix}2 & 0\\ 0 & 8\end{pmatrix},\sqrt{A^*A}=\begin{pmatrix}\sqrt{2} & 0\\ 0 & \sqrt{8}\end{pmatrix}.$$

In this case, computing $\sqrt{A^*A}$ was simple because of the form of A^*A. At the end of this example we discuss methods to compute $\sqrt{A^*A}$ if A^*A is in a less convenient form.

Step 2. Find an orthonormal basis of eigenvectors and their eigenvalues for $\sqrt{A^*A}$.
 In this example,

$$\hat{v}_1 = \begin{pmatrix} 1 \\ 0 \end{pmatrix}, \lambda_1 = \sqrt{2}; \quad \hat{v}_2 = \begin{pmatrix} 0 \\ 1 \end{pmatrix}, \lambda_2 = \sqrt{8}.$$

Step 3. For the eigenvectors whose eigenvalues are non-zero, compute

$$\frac{1}{\lambda_i} A \hat{v}_i.$$

These will be columns for the matrix U.
 In this example,

$$\frac{1}{\lambda_1} A \hat{v}_1 = \frac{1}{\sqrt{2}} \begin{pmatrix} 1 & 2 \\ -1 & 2 \end{pmatrix} \begin{pmatrix} 1 \\ 0 \end{pmatrix} = \frac{1}{\sqrt{2}} \begin{pmatrix} 1 \\ -1 \end{pmatrix}; \quad \frac{1}{\lambda_2} A \hat{v}_2 = \frac{1}{\sqrt{8}} \begin{pmatrix} 1 & 2 \\ -1 & 2 \end{pmatrix} \begin{pmatrix} 0 \\ 1 \end{pmatrix} = \frac{1}{\sqrt{8}} \begin{pmatrix} 2 \\ 2 \end{pmatrix}.$$

Step 4. If there are no eigenvectors whose eigenvalue is 0, then we have sufficient information to compute the matrix U.
 In this example

$$U = \begin{pmatrix} \dfrac{1}{\sqrt{2}} & \dfrac{2}{\sqrt{8}} \\ -\dfrac{1}{\sqrt{2}} & \dfrac{2}{\sqrt{8}} \end{pmatrix}.$$

Step 5. We have $A = U\sqrt{A^*A}$.
 In this example, we can check that

$$U\sqrt{A^*A} = \begin{pmatrix} \dfrac{1}{\sqrt{2}} & \dfrac{2}{\sqrt{8}} \\ -\dfrac{1}{\sqrt{2}} & \dfrac{2}{\sqrt{8}} \end{pmatrix} \begin{pmatrix} \sqrt{2} & 0 \\ 0 & \sqrt{8} \end{pmatrix} = \begin{pmatrix} 1 & 2 \\ -1 & 2 \end{pmatrix}.$$

Note:
 Because of the form of A^*A in this particular example, we can immediately say

$$\sqrt{A^*A} = \begin{pmatrix} \sqrt{2} & 0 \\ 0 & \sqrt{8} \end{pmatrix}.$$

In the case that A^*A is more complicated, we could use the Jordan canonical form, which is a topic of a later chapter, to find $\sqrt{A^*A}$.

 We present a third way to find $\sqrt{A^*A}$ that is sometimes useful.

 If $\hat{u}_i$ is an eigenvector for $A^* A$ with eigenvalue λ_i^2, then,

$$A^*A\, \hat{u}_i = \lambda_i^2\, \hat{u}_i$$

and so

$$\sqrt{A^*A}\, \hat{u}_i = \lambda_i\, \hat{v}_i.$$

Thus

$$\sqrt{A^*A}\left(\hat{u}_1, \quad \cdots, \quad \hat{u}_n\right)=\left(\lambda_1\hat{u}_1,\ldots,\lambda_n\hat{v}_n\right) \qquad (1)$$

where $\left(\hat{u}_1 \quad \cdots \quad \hat{u}_n\right)$ is the $n \times n$ matrix whose ith column is $\hat{u}_i$.

In our example, this says

$$\sqrt{A^*A}\begin{pmatrix} 1 & 0 \\ 0 & 1 \end{pmatrix} = \begin{pmatrix} \sqrt{2} & 0 \\ 0 & \sqrt{8} \end{pmatrix},$$

so that

$$\sqrt{A^*A} = \begin{pmatrix} \sqrt{2} & 0 \\ 0 & \sqrt{8} \end{pmatrix}\begin{pmatrix} 1 & 0 \\ 0 & 1 \end{pmatrix}^{-1} = \begin{pmatrix} \sqrt{2} & 0 \\ 0 & \sqrt{8} \end{pmatrix}.$$

Case 2. 0 is an eigenvalue for A^*A.

Let

$$A = \begin{pmatrix} 1 & 0 \\ 2 & 0 \end{pmatrix}.$$

Step 1. Compute A^*A and $\sqrt{A^*A}$.

$$A = \begin{pmatrix} 1 & 0 \\ 2 & 0 \end{pmatrix}, A^* = \begin{pmatrix} 1 & 2 \\ 0 & 0 \end{pmatrix}; \quad A^*A = \begin{pmatrix} 5 & 0 \\ 0 & 0 \end{pmatrix}, \sqrt{A^*A} = \begin{pmatrix} \sqrt{5} & 0 \\ 0 & 0 \end{pmatrix}.$$

Step 2. Find an orthonormal basis of eigenvectors and their eigenvalues for $\sqrt{A^*A}$.

In this example,

$$\hat{v}_1 = \begin{pmatrix} 1 \\ 0 \end{pmatrix}, \lambda_1 = \sqrt{5}; \; \hat{v}_2 = \begin{pmatrix} 0 \\ 1 \end{pmatrix}, \lambda_2 = 0.$$

Step 3. For the eigenvectors whose eigenvalues are non-zero, compute

$$\frac{1}{\lambda_i}A\hat{v}_i.$$

These will be columns for the matrix U.

In this example,

$$\frac{1}{\lambda_1}A\hat{v}_1 = \frac{1}{\sqrt{5}}\begin{pmatrix} 1 & 0 \\ 2 & 0 \end{pmatrix}\begin{pmatrix} 1 \\ 0 \end{pmatrix} = \frac{1}{\sqrt{5}}\begin{pmatrix} 1 \\ 2 \end{pmatrix};$$

Step 3'. This is where the algorithm differs from the previous case.

Since

$$A\hat{v}_2 = \begin{pmatrix} 0 \\ 0 \end{pmatrix},$$

the vectors $\{A\hat{v}_1, A\hat{v}_2\}$ do not form a basis. We must add a vector to

$$\frac{1}{\lambda_1}A\hat{v}_1 = \frac{1}{\sqrt{5}}\begin{pmatrix}1\\2\end{pmatrix} = \begin{pmatrix}\frac{1}{\sqrt{5}}\\\frac{2}{\sqrt{5}}\end{pmatrix}$$

to complete an orthonormal basis. One such vector is

$$\hat{u}_2 = \begin{pmatrix}\frac{-2}{\sqrt{5}}\\\frac{1}{\sqrt{5}}\end{pmatrix}.$$

Step 4. We now construct the matrix U.
   We have

$$U\hat{v}_i = \frac{1}{\lambda_i}A\hat{v}_i = \begin{pmatrix}\frac{1}{\sqrt{5}}\\\frac{2}{\sqrt{5}}\end{pmatrix},\quad U\hat{v}_2 = \hat{u}_2 = \begin{pmatrix}\frac{-2}{\sqrt{5}}\\\frac{1}{\sqrt{5}}\end{pmatrix}.$$

Thus

$$U = \begin{pmatrix}\frac{1}{\sqrt{5}} & \frac{-2}{\sqrt{5}}\\\frac{2}{\sqrt{5}} & \frac{1}{\sqrt{5}}\end{pmatrix}$$

and

$$U\sqrt{A^*A} = \begin{pmatrix}\frac{1}{\sqrt{5}} & \frac{-2}{\sqrt{5}}\\\frac{2}{\sqrt{5}} & \frac{1}{\sqrt{5}}\end{pmatrix}\begin{pmatrix}\sqrt{5} & 0\\0 & 0\end{pmatrix} = \begin{pmatrix}1 & 0\\2 & 0\end{pmatrix} = A$$

Exercises

1. Find the polar decomposition of

$$A = \begin{pmatrix}2 & 5\\-2 & 5\end{pmatrix}.$$

Ans $\begin{pmatrix}\sqrt{2}/2 & \sqrt{5}/2\\-\sqrt{2}/2 & \sqrt{5}/2\end{pmatrix}\begin{pmatrix}\sqrt{8} & 0\\0 & \sqrt{20}\end{pmatrix}$

2. Find the polar dcomposition of

$$A = \begin{pmatrix}1 & 0\\2 & 0\end{pmatrix}$$

3. Find the polar decomposition of

$$A = \begin{pmatrix} -3 & 4 \\ 6 & 4 \end{pmatrix}.$$

4. Find the polar decomposition of

$$A = \begin{pmatrix} 3 & 0 \\ 5 & 0 \end{pmatrix}$$

Section 8.2 Singular Value Decomposition of a Matrix

A decomposition of a matrix that is sometimes useful – especially in statistics – is the singular value decomposition. In the singular value decomposition, if A is an $m \times n$ matrix, we seek the express $A = U\Sigma V^T$

where U is an $m \times m$ unitary (orthogonal) matrix, V is an $n \times n$ unitary (orthogonal) matrix and £ is an $m \times n$ diagonal matrix; i.e., $\Sigma_{i,j} = 0$ unless $i = j$.

If A is an $m \times n$ matrix, then AA^T is an $m \times m$ symmetric matrix and $A^T A$ is an $n \times n$ symmetric matrix. By the spectral theorem, these matrices have an orthonormal basis of eigenvectors.

The matrices AA^T and $A^T A$ play a central role in the construction of the singular value decomposition. The next theorem says that these two matrices have the same eigenvalues.

Theorem 8.4:

If A is an $m \times n$ matrix and B is an $n \times m$ matrix, then AB and BA have the same eigenvalues.

Proof:

Suppose that $\lambda \neq 0$ is an eigenvalue of AB with eigenvector $\hat{x}$. Now

$$BA(B\hat{x}) = B(AB\hat{x}) = B(\lambda\hat{x}) = \lambda(B\hat{x}).$$

Thus $B\hat{x}$ is an eigenvector of BA with eigenvalue λ.

Note that the proof breaks down in the case $\lambda = 0$, because there is the possibility that $B\hat{x} = \hat{0}$ and $\hat{0}$ is not an eigenvector. However, the result does hold if $\lambda = 0$ is an eigenvalue, because $\lambda = 0$ is an eigenvalue for AB if and only if

$$\det(AB) = 0$$

and

$$\det(AB) = \det(BA).$$

Corollary:

For a matrix A, the matrices $A^T A$ and $A A^T$ have the same non-zero eigenvalues.

Theorem 8.5:

Any $m \times n$ matrix A can be expressed as

$$A = U \Sigma V^T$$

where U and V are unitary matrices and Σ is a diagonal matrix. In particular, U is the $m \times m$ matrix whose columns are the eigenvectors of $A A^T$, V is the $n \times n$ matrix whose columns are the eigenvectors of $A^T A$ and Σ is the $m \times n$ diagonal matrix whose diagonal elements are the square roots of the positive eigenvalues of $A A^T$ (which are the same as the square roots of the positive eigenvalues of $A^T A$).

Proof:

The matrix $A^T A$ has an orthonormal set of eigenvectors $\{\hat{v}_1, \dots, \hat{v}_n\}$. Suppose

$$A^T A \hat{v}_i = \lambda_i \hat{v}_i.$$

Note that $\lambda_i \geq 0$. Then

$$\hat{v}_i^T A^T A \hat{v}_j = \lambda_j \hat{v}_i^T \hat{v}_j = \lambda_j \delta_{ij}.$$

Let $\sigma_j = \sqrt{\lambda_j}$ for the $\lambda_j > 0$; say $j = 1, \dots, k$. Define

$$\hat{u}_i = \frac{1}{\sigma_i} A \hat{v}_i.$$

We show that $\{\hat{u}_1, \dots, \hat{u}_k\}$. is an orthonormal set. We have

$$\hat{u}_i^T \hat{u}_j = \left(\frac{1}{\sigma_i} A \hat{v}_i \right)^T \left(\frac{1}{\sigma_j} A \hat{v}_j \right) = \frac{1}{\sigma_i \sigma_j} \hat{v}_i^T A^T A \hat{v}_j = \frac{1}{\sigma_i \sigma_j} \lambda_j \hat{v}_i^T \hat{v}_j$$

$$= \begin{cases} 0 \text{ if } i \neq j \\ \\ \dfrac{1}{\sigma_j \sigma_j} \lambda_j = 1 \text{ if } i = j \end{cases}.$$

Extend $\{\hat{u}_1, \dots, \hat{u}_k\}$ to be an orthonormal set on $\mathbb{R}^m$. Define U to be the matrix whose ith column is $\hat{u}_i$; that is

$$U = (\hat{u}_1, \dots, \hat{u}_m)$$

and

$$V = (\hat{v}_1, \dots, \hat{v}_n).$$

Consider

$$\left(U^T A V\right)_{ij} = \left(\hat{u}_i\right)^T \left(A\hat{v}_j\right) = \left(\hat{u}_i\right)^T \left(\sigma_j \hat{u}_j\right) = \begin{cases} 0 \text{ if } j > k \\ \\ \sigma_j \left(\hat{u}_i\right)^T \hat{u}_j = \sigma_j \delta_{ij} 0 \text{ if } j \le k \end{cases}.$$

Thus

$$\Sigma = U^T A V$$

and so

$$U(U^T A V) V^T = U\Sigma V^T.$$

But U and V are unitary matrices, so

$$U\Sigma V^T = U(U^T A V) V^T = \left(UU^T\right) A \left(VV^T\right) = A.$$

Algorithm and Example:

Let

$$A = \begin{pmatrix} 2 & 0 & 1 \\ -1 & 1 & 2 \end{pmatrix}.$$

1. We form two matrices

$$U = AA^T \text{ and } V = A^T A$$

and find the eigenvalues and eigenvectors of each. Note that

a. Since U and V are symmetric matrices they each have an orthonormal basis of eigenvectors.

b. The eigenvalues of U and V are non-negative.

c. By Theorem 5, U and V have the same non-zero eigenvalues.

   In this example

$$U = AA^T = \begin{pmatrix} 5 & 0 \\ 0 & 6 \end{pmatrix} \text{ and } V = A^T A = \begin{pmatrix} 5 & -1 & 0 \\ -1 & 1 & 2 \\ 0 & 2 & 5 \end{pmatrix}.$$

2. Find the eigenvalues and eigenvectors of U and V.

For U the eigenvalues and corresponding eigenvectors are

$$\lambda_1 = 6, \hat{v}_1 = \begin{pmatrix} 0 \\ 1 \end{pmatrix}$$

$$\lambda_2 = 5, \hat{v}_2 = \begin{pmatrix} 1 \\ 0 \end{pmatrix}.$$

The matrix V is more complicated. The characteristic polynomial of V is

$$\lambda(\lambda - 5)(\lambda - 6)$$

so the eigenvalues are $\lambda_1 = 0$, $\lambda_2 = 5$, $\lambda_3 = 6$. The eigenvectors for each eigenvalue are

$$\lambda_1 = 0, \hat{v}_1 = \begin{pmatrix} -1 \\ -5 \\ 2 \end{pmatrix}$$

$$\lambda_2 = 5, \hat{v}_2 = \begin{pmatrix} 2 \\ 0 \\ 1 \end{pmatrix}$$

$$\lambda_3 = 6, \hat{v}_3 = \begin{pmatrix} -1 \\ 1 \\ 2 \end{pmatrix}.$$

3. Each group of eigenvectors forms an orthogonal basis for the appropriate vector space. Convert each set of vectors to an orthonormal basis. Form a matrix from these vectors with the first column being the normalized eigenvector with the largest eigenvalue, the second column the normalized eigenvector with the second largest eigenvalue, and so forth.

For the matrix U the eigenvalues and corresponding eigenvectors are

$$\lambda_1 = 6, \hat{v}_1 = \begin{pmatrix} 0 \\ 1 \end{pmatrix}$$

$$\lambda_2 = 5, \hat{v}_2 = \begin{pmatrix} 1 \\ 0 \end{pmatrix}.$$

Since the eigenvectors are already orthonormal, and the matrix we seek is

$$T = \begin{pmatrix} 0 & 1 \\ 1 & 0 \end{pmatrix}.$$

For the matrix V, the eigenvalues are $\lambda_1 = 0$, $\lambda_2 = 5$, $\lambda_3 = 6$. The eigenvectors for each eigenvalue are

$$\lambda_1 = 0, \hat{v}_1 = \begin{pmatrix} -1 \\ -5 \\ 2 \end{pmatrix}$$

$$\lambda_2 = 5, \hat{v}_2 = \begin{pmatrix} 2 \\ 0 \\ 1 \end{pmatrix}$$

$$\lambda_3 = 6, \hat{v}_3 = \begin{pmatrix} -1 \\ 1 \\ 2 \end{pmatrix}.$$

Since $\hat{v}_1$, $\hat{v}_2$, and $\hat{v}_3$ are orthogonal, we only need to normalize them. Beginning with the eigenvector with largest eigenvalue, we have

$$\hat{w}_3 = \frac{\hat{v}_3}{\langle \hat{v}_3 \rangle} = \begin{pmatrix} \dfrac{-1}{\sqrt{6}} \\ \dfrac{1}{\sqrt{6}} \\ \dfrac{2}{\sqrt{6}} \end{pmatrix}.$$

For the second vector we have

$$\hat{w}_2 = \frac{\hat{v}_2}{\langle \hat{v}_2 \rangle} = \begin{pmatrix} \dfrac{2}{\sqrt{5}} \\ 0 \\ \dfrac{1}{\sqrt{5}} \end{pmatrix}$$

and

$$\hat{w}_1 = \frac{\hat{v}_1}{\langle \hat{v}_1 \rangle} = \begin{pmatrix} \dfrac{-1}{\sqrt{30}} \\ \dfrac{-5}{\sqrt{30}} \\ \dfrac{2}{\sqrt{30}} \end{pmatrix}.$$

Let

$$W = \begin{pmatrix} \dfrac{-1}{\sqrt{6}} & \dfrac{2}{\sqrt{5}} & \dfrac{-1}{\sqrt{30}} \\ \dfrac{1}{\sqrt{6}} & 0 & \dfrac{-5}{\sqrt{30}} \\ \dfrac{2}{\sqrt{6}} & \dfrac{1}{\sqrt{5}} & \dfrac{2}{\sqrt{30}} \end{pmatrix}.$$

4. The matrix D is a matrix for which the diagonal entries are the only non-zero entries. The diagonal entries are the square root of the eigenvalues, beginning with the largest eigenvalue. The matrix is then completed with all 0s so as the dimension of the matrix enables the multiplication UDV^T to be performed.

In the example, D must be a 2×3 matrix, so

$$D = \begin{pmatrix} \sqrt{6} & 0 & 0 \\ 0 & \sqrt{5} & 0 \end{pmatrix}$$

and it can be verified that

$$\begin{pmatrix} 0 & 1 \\ 1 & 0 \end{pmatrix} \begin{pmatrix} \sqrt{6} & 0 & 0 \\ 0 & \sqrt{5} & 0 \end{pmatrix} \begin{pmatrix} \dfrac{-1}{\sqrt{6}} & \dfrac{2}{\sqrt{5}} & \dfrac{-1}{\sqrt{30}} \\ \dfrac{1}{\sqrt{6}} & 0 & \dfrac{-5}{\sqrt{30}} \\ \dfrac{2}{\sqrt{6}} & \dfrac{1}{\sqrt{5}} & \dfrac{2}{\sqrt{30}} \end{pmatrix}^T = \begin{pmatrix} 2 & 0 & 1 \\ -1 & 1 & 2 \end{pmatrix}.$$

Exercises

Find the singular value decomposition of the following matrices.

1. $A = \begin{pmatrix} 3 & 2 & 2 \\ 2 & 3 & -2 \end{pmatrix}$

Ans $\begin{pmatrix} 1/\sqrt{2} & 1/\sqrt{2} \\ 1/\sqrt{2} & -1/\sqrt{2} \end{pmatrix} \begin{pmatrix} 5 & 0 & 0 \\ 0 & 3 & 0 \end{pmatrix} \begin{pmatrix} 1/\sqrt{2} & 1/\sqrt{2} & 0 \\ 1/\sqrt{18} & -1/\sqrt{18} & 4/\sqrt{18} \\ 2/3 & -2/3 & -1/3 \end{pmatrix}$

2. $A = \begin{pmatrix} 2 & 2 \\ -1 & -1 \end{pmatrix}$

Ans $\begin{pmatrix} 2/\sqrt{5} & 1/\sqrt{5} \\ -1/\sqrt{5} & 2/\sqrt{5} \end{pmatrix} \begin{pmatrix} \sqrt{10} & 0 \\ 0 & 0 \end{pmatrix} \begin{pmatrix} 1/\sqrt{2} & 1/\sqrt{2} \\ 1/\sqrt{2} & -1/\sqrt{2} \end{pmatrix}$

3. $A = \begin{pmatrix} 1 & 1 \\ 0 & 1 \\ -1 & 1 \end{pmatrix}$

$$\text{Ans} \begin{pmatrix} 1/\sqrt{3} & 1/\sqrt{2} & 1/\sqrt{6} \\ 1/\sqrt{3} & 0 & -2/\sqrt{6} \\ 1/\sqrt{3} & -1/\sqrt{2} & 1/\sqrt{6} \end{pmatrix} \begin{pmatrix} \sqrt{3} & 0 \\ 0 & \sqrt{2} \\ 0 & 0 \end{pmatrix} \begin{pmatrix} 0 & 1 \\ 1 & 0 \end{pmatrix}$$

4. $A = \begin{pmatrix} 3 & 1 & 1 \\ -1 & 3 & 1 \end{pmatrix}$

$$\text{Ans} \begin{pmatrix} 1/\sqrt{2} & 1/\sqrt{2} \\ 1/\sqrt{2} & -1/\sqrt{2} \end{pmatrix} \begin{pmatrix} \sqrt{12} & 0 & 0 \\ 0 & \sqrt{10} & 0 \end{pmatrix} \begin{pmatrix} 1/\sqrt{6} & 2/\sqrt{6} & 1/\sqrt{6} \\ 2/\sqrt{5} & -1/\sqrt{5} & 0 \\ 1/\sqrt{30} & 2/\sqrt{30} & -5/\sqrt{30} \end{pmatrix}$$

5. *en.wikipedia.org/wiki/Singular_value_decomposition#Example*

$$A = \begin{pmatrix} 1 & 0 & 0 & 0 & 2 \\ 0 & 0 & 3 & 0 & 0 \\ 0 & 0 & 0 & 0 & 0 \\ 0 & 2 & 0 & 0 & 0 \end{pmatrix}$$

$$\text{Ans} \begin{pmatrix} 0 & 0 & 1 & 0 \\ 0 & 1 & 0 & 0 \\ 0 & 0 & 0 & -1 \\ 1 & 0 & 0 & 0 \end{pmatrix} \begin{pmatrix} 2 & 0 & 0 & 0 & 0 \\ 0 & 3 & 0 & 0 & 0 \\ 0 & 0 & \sqrt{5} & 0 & 0 \\ 0 & 0 & 0 & 0 & 0 \end{pmatrix} \begin{pmatrix} 0 & 1 & 0 & 0 & 0 \\ 0 & 0 & 1 & 0 & 0 \\ \sqrt{.2} & 0 & 0 & 0 & \sqrt{.8} \\ 0 & 0 & 0 & 1 & 0 \\ \sqrt{.8} & 0 & 0 & 0 & \sqrt{.2} \end{pmatrix}$$

Chapter 9

Determinants

Earlier in the text we presented the computations for calculating the determinant. In this chapter we develop the theory of determinants based on three axioms. Some of the computations are tedious, but they are presented for several reasons, including the derivation of the formula for the determinant given in more advanced settings.

Introduction

Determinants have a variety of uses in linear algebra, including solving systems of linear equations, determining whether a square matrix is invertible, and determining whether a collection of vectors form a basis. Computing a determinant can be quite tedious without a computer and understanding how the computation formulas arose can be nonintuitive. In this chapter we

1. Give the geometric intuition of the determinant.
2. From the intuition that has been established, we propose three axioms that the determinant function should satisfy.
3. Show that these axioms are sufficient to uniquely define a function

$$\det : \mathbb{R}^{n \times n} \to \mathbb{R}$$

which we call the determinant.
4. Show how this function yields the definition of the determinant that is given in more advanced settings.
5. Establish several theorems regarding determinants.

One approach to determinants relies on the geometric idea of measuring the change in the volume of a parallelepiped under a linear transformation. To be more specific, suppose that we are in $\mathbb{R}^n$ and have n vectors $\hat{v}_1, \ldots, \hat{v}_n$. These vectors form a (possibly degenerate) parallelepiped.

If

$$A : \mathbb{R}^n \to \mathbb{R}^n$$

is a linear transformation, then the vectors $A\hat{v}_1,\ldots,A\hat{v}_n$ also form a parallelepiped in $\mathbb{R}^n$. The determinant compares the volume and orientation (a term to be described shortly) of the parallelepipeds formed by $\hat{v}_1,\ldots,\hat{v}_n$ and $A\hat{v}_1,\ldots,A\hat{v}_n$. While this approach is intuitively comfortable, it does not give a formula that computes the determinant.

A second approach considers a function

$$A : \mathbb{R}^{n \times n} \to \mathbb{R}$$

that satisfies three postulates which are statements of the geometrical interpretation of the determinant. This method will enable us to prove the familiar properties of determinants. It also gives rise to a formula that describes how a determinant is computed and has some uses in advanced theoretical study.

Determinants from a Geometrical Point of View

The idea on which we base our analysis in the geometric approach to determinants is similar to what one does in calculus in converting from the Cartesian coordinate system to a different coordinate system to do integration. For example, in two dimensions one may want to convert the infinitesimal area element $dA = dx\,dy$ to an infinitesimal area element in a different coordinate system. In polar coordinates, this is $dA = r\,dr\,d\theta$ and r plays a role similar to the determinant.

One interpretation of an $n \times n$ matrix is the representation of a linear transformation on an n-dimensional vector space to itself with respect to some basis. We investigate determinants from the point of view of how a linear transformation changes the volume of the "unit cube". We first consider the two-dimensional case.

Suppose that

$$A : \mathbb{R}^2 \to \mathbb{R}^2$$

with

$$A\hat{e}_1 = \hat{u} = \begin{pmatrix} u_1 \\ u_2 \end{pmatrix} \quad \text{and} \quad A\hat{e}_2 = \hat{v} = \begin{pmatrix} v_1 \\ v_2 \end{pmatrix}.$$

The image is said to be positively oriented if one rotates $A\hat{e}_1$ to $A\hat{e}_2$ in the same direction as rotating $\hat{e}_1$ to get to $\hat{e}_2$ and negatively oriented if the rotations are in opposite directions. Figure 9.1(a) shows a positive orientation and Figure 9.1(b) a negative orientation.

The area of the rectangle formed by $\hat{e}_1$ and $\hat{e}_2$ is 1, and we want to find the area of the parallelogram formed by $A(\hat{e}_1)$ and $A(\hat{e}_2)$; that is, the area of the parallelogram formed by $\hat{u}$ and $\hat{v}$. We give two arguments to show this is

$$\|\hat{u}\|\,\|\hat{v}\| \sin\theta = \begin{cases} u_1 v_2 - u_2 v_1, & 0 < \theta < \dfrac{\pi}{2} \\[2ex] u_2 v_1 - u_1 v_2, & -\dfrac{\pi}{2} < \theta < 0 \end{cases}.$$

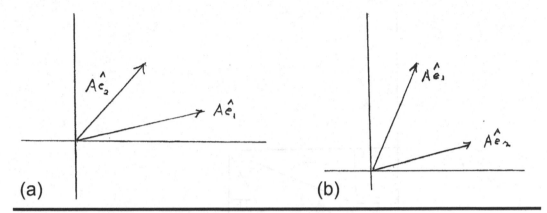

Figure 9.1 (a) Positive orientation. (b) Negative orientation.

The first argument is geometrical and the second is analytical.

In Figure 9.2, note that the area of the rectangle containing the parallelogram is

$$\left(u_1 + u_2\right)\left(v_1 + v_2\right) = u_1 v_1 + u_1 v_2 + u_2 v_1 + u_2 v_2,$$

the area of Triangles I and II are each

$$\frac{1}{2} v_1 v_2,$$

the area of Triangles III and IV are each

$$\frac{1}{2} u_1 u_2,$$

and the area of Rectangles V and VI are each $u_1 v_2$ if oriented as shown. If the orientation is reversed, the area is $u_2 v_1$.

Thus the area in the large square that is not in the parallelogram is

$$\begin{cases} v_1 v_2 + u_1 u_2 + 2 u_1 v_2 & \text{with the shown orientation} \\ \\ v_1 v_2 + u_1 u_2 + 2 u_2 v_1 & \text{with the reverse orientation} \end{cases}$$

and so the area of the parallelogram is

$$\begin{cases} u_2 v_1 - u_1 v_2 & \text{with the shown orientation} \\ \\ u_1 v_2 - u_2 v_1 & \text{with the reverse orientation} \end{cases}.$$

This gives the formula for computing a 2×2 determinant.

For the second argument we use that

$$\hat{u} \cdot \hat{v} = \|\hat{u}\| \, \|\hat{v}\| \, \cos\theta.$$

The area of the parallelogram is $\|\hat{u}\| \, \|\hat{v}\| \, \sin\theta$

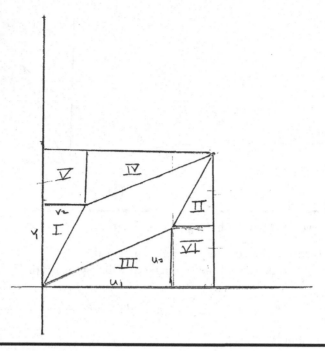

Figure 9.2 Finding the area of a parallelogram.

so

$$\cos\theta = \frac{u \cdot v}{\|u\| \|v\|} = \frac{u_1 v_1 + u_2 v_2}{\|u\| \|v\|} \quad \text{and} \quad \cos^2\theta = \frac{\left(u_1 v_1 + u_2 v_2\right)^2}{\|u\|^2 \|v\|^2}.$$

Thus

$$\left(\|\hat{u}\| \|\hat{v}\| \ \sin\theta\right)^2 = \|\hat{u}\|^2 \|\hat{v}\|^2 \sin^2\theta = \|\hat{u}^2\|\|\hat{v}^2\|\left(1 - \cos^2\theta\right)$$

$$= \|\hat{u}\|^2 \|\hat{v}\|^2 \left(1 - \frac{\left(u_1 v_1 + u_2 v_2\right)^2}{\|\hat{u}\|^2 \|\hat{v}\|^2}\right)$$

$$= \|\hat{u}\|^2 \|\hat{v}\|^2 \left[\frac{\|\hat{u}\|^2 \|\hat{v}\|^2 - \left(u_1 v_1 + u_2 v_2\right)^2}{\|\hat{u}\|^2 \|\hat{v}\|^2}\right]$$

$$= \|\hat{u}\|^2 \|\hat{v}\|^2 - \left(u_1 v_1 + u_2 v_2\right)^2$$

$$= \left(u_1^2 + u_2^2\right)\left(v_1^2 + v_2^2\right) - \left(u_1 v_1 + u_2 v_2\right)^2$$

$$= \left(u_1^2 v_1^2 + u_1^2 v_2^2 + u_2^2 v_1^2 + u_2^2 v_2^2\right)$$

$$\quad - \left(u_1^2 v_1^2 + 2u_1 v_1 u_2 v_2 + u_2^2 v_2^2\right)$$

$$= u_1^2 v_2^2 - 2u_1 v_1 u_2 v_2 + u_2^2 v_1^2 = \left(u_1 v_2 - u_2 v_1\right)^2.$$

Thus

$$\|\hat{u}\|\,\|\hat{v}\|\,\sin\theta = \begin{cases} u_1 v_2 - u_2 v_1,\; 0 < \theta < \dfrac{\pi}{2} \\[2em] u_2 v_1 - u_1 v_2,\; -\dfrac{\pi}{2} < \theta < 0 \end{cases}.$$

Three Postulates of the Determinant

We want the determinant to associate a number with an $n \times n$ matrix. Let *det* denote the determinant function, so that if the vector space is $\mathbb{R}^n$ then

$$det : \mathbb{R}^{n \times n} \to \mathbb{R}.$$

If $\hat{a}_i$ is the ith column of the matrix A, then we can describe A by

$$A = \left(\hat{a}_1, \ldots, \hat{a}_n \right).$$

We postulate three characteristics that the function *det* should possess based on the geometric interpretation that we have adopted.

1. If I is the identity matrix, then $\det(I) = 1$.
2. If A is a square matrix that has two identical columns, then $\det(A) = 0$.
3. If A is a square matrix $\left(\hat{a}_1, \ldots, \hat{a}_j + \lambda \hat{b}_j, \ldots, \hat{a}_n \right)$, then

$$\det\left(\hat{a}_1, \ldots, \hat{a}_j + \lambda \hat{b}_j, \ldots, \hat{a}_n \right) = \det\left(\hat{a}_1, \ldots, \hat{a}_j, \ldots, \hat{a}_n \right) + \lambda \det\left(\hat{a}_1, \ldots, \hat{b}_j, \ldots, \hat{a}_n \right).$$

The third postulate says that the determinant function is a linear function of each column.

We give the geometric intuition behind each of the postulates.

Postulate 1. If I is the identity matrix, then $\det(I) = 1$.

This is because the identity matrix maps the unit cube with basis $\{\hat{e}_1, \ldots, \hat{e}_n\}$ to itself.

Postulate 2. If A is a square matrix that has two identical columns, then $\det(A) = 0$.

The intuition of this postulate is that if the vectors $\{A\hat{e}_1, \ldots, A\hat{e}_n\}$ form a linearly dependent set, then the parallelepiped formed by these vectors has dimension at most $n - 1$. The volume of an $(n - 1)$ dimensional object in n-space is 0.

Postulate 3. If A is a square matrix $\left(\hat{a}_1, \ldots, \hat{a}_j + \lambda \hat{b}_j, \ldots, \hat{a}_n \right)$, then

$$\det\left(\hat{a}_1, \ldots, \hat{a}_j + \lambda \hat{b}_j, \ldots, \hat{a}_n \right) = \det\left(\hat{a}_1, \ldots, \hat{a}_j, \ldots, \hat{a}_n \right) + \lambda \det\left(\hat{a}_1, \ldots, \hat{b}_j, \ldots, \hat{a}_n \right).$$

To begin the discussion of the third postulate, we want to understand what happens to $\det(A)$ if we change one column of A in various ways.

Suppose we want to compare the determinants of the matrices

$$A = \begin{pmatrix} 2 & 0 & 0 \\ 0 & 3 & 0 \\ 0 & 0 & 4 \end{pmatrix} \text{ and } B = \begin{pmatrix} 2 & 0 & 0 \\ 0 & 3 & 0 \\ 0 & 0 & 4+5 \end{pmatrix}.$$

Then $\det(A) = 2 \times 3 \times 4$ and $\det(B) = 2 \times 3 \times (4 + 5)$. What we have done in modifying the third column is to change the parallelepiped in only the third dimension, so that we have kept the first two dimensions the same and the resulting volume can be seen as adding the volume $2 \times 3 \times 4$ to the volume $2 \times 3 \times 5$. Said another way,

$$\det \begin{pmatrix} 2 & 0 & 0 \\ 0 & 3 & 0 \\ 0 & 0 & 4+5 \end{pmatrix} = \det \begin{pmatrix} 2 & 0 & 0 \\ 0 & 3 & 0 \\ 0 & 0 & 4 \end{pmatrix} + \det \begin{pmatrix} 2 & 0 & 0 \\ 0 & 3 & 0 \\ 0 & 0 & 5 \end{pmatrix}.$$

This is an example of the abstract idea that

$$\det \left(\hat{a}_1, \ldots, \hat{a}_j + \hat{b}_j, \ldots, \hat{a}_n \right) = \det \left(\hat{a}_1, \ldots, \hat{a}_j, \ldots, \hat{a}_n \right) + \det \left(\hat{a}_1, \ldots, \hat{b}_j, \ldots, \hat{a}_n \right).$$

This does *not* say that if A and B are $n \times n$ matrices then

$$\det \left(A + B \right) = \det \left(A \right) + \det \left(B \right)$$

which is, in general, false.

Similarly, if we consider

$$A = \begin{pmatrix} 2 & 0 & 0 \\ 0 & 3 & 0 \\ 0 & 0 & 4 \times 5 \end{pmatrix}$$

we can view $\det(A)$ as the volume of a parallelepiped measuring 2 units long by 3 units wide by (4×5) units high. This has the same volume as 5 parallelepipeds each measuring 2 units long by 3 units wide by 4 units high. Thus,

$$\det \begin{pmatrix} 2 & 0 & 0 \\ 0 & 3 & 0 \\ 0 & 0 & 4 \times 5 \end{pmatrix} = 5 \det \begin{pmatrix} 2 & 0 & 0 \\ 0 & 3 & 0 \\ 0 & 0 & 4 \end{pmatrix}.$$

This is an example of the abstract idea that

$$\det \left(\hat{a}_1, \ldots, \lambda \hat{a}_j, \ldots, \hat{a}_n \right) = \lambda \det \left(\hat{a}_1, \ldots, \hat{a}_j, \ldots, \hat{a}_n \right).$$

If we combine the two ideas above, we get the third postulate.

The Determinant Based on the Three Postulates

We will show that the three postulates are sufficient to define a unique function

$$\det : \mathbb{R}^{n \times n} \to \mathbb{R}.$$

In later theorems we show how the three postulates imply the additional common rules for determinates. Specifically, we will show

1. Interchanging two columns of a matrix changes the sign of the determinant.
2. $\det(A) = 0$ if and only if the columns of a matrix A are a linearly dependent set of vectors.
3. If A and B are $n \times n$ matrices, then $\det(AB) = \det(A)\det(B)$.
4. A^{-1} exists if and only if the columns of a matrix A are a linearly independent set of vectors, and in that case

$$\det\left(A^{-1}\right) = \frac{1}{\det\left(A\right)}.$$

5. If A is an $n \times n$ matrix, then $\det(A) = \det(A^T)$.

Note: From Property 5. it follows that row operations on a matrix affect the determinant of a matrix in the same way that the column operations do.

Theorem 9.1:

Suppose that

$$A = \left(\hat{a}_1, \ldots, \hat{a}_i, .., \hat{a}_j, .., \hat{a}_n\right)$$

is a square matrix. Then

$$\det\left(\hat{a}_1, \ldots, \hat{a}_i, \ldots, \hat{a}_j, \ldots, \hat{a}_n\right) = -\det\left(\hat{a}_1, \ldots, \hat{a}_j, \ldots, \hat{a}_i, \ldots, \hat{a}_n\right).$$

That is, if two columns of a square matrix are interchanged, the resulting matrix is the negative of the original matrix.

Proof:

We give the proof for the 2×2 case and leave the $n \times n$ case as an exercise. We have

$$0 = \det\left(\hat{a}_1 + \hat{a}_2, \hat{a}_1 + \hat{a}_2\right) = \det\left(\hat{a}_1, \hat{a}_1 + \hat{a}_2\right) + \det\left(\hat{a}_2, \hat{a}_1 + \hat{a}_2\right)$$
$$= \det\left(\hat{a}_1, \hat{a}_1\right) + \det\left(\hat{a}_1, \hat{a}_2\right) + \det\left(\hat{a}_2, \hat{a}_1\right) + \det\left(\hat{a}_2, \hat{a}_2\right)$$
$$= \det\left(\hat{a}_1, \hat{a}_2\right) + \det\left(\hat{a}_2, \hat{a}_1\right)$$

since

$$\det\left(\hat{a}_i, \hat{a}_i\right) = 0.$$

Thus,

$$\det\left(\hat{a}_1,\hat{a}_2\right) = -\det\left(\hat{a}_2,\hat{a}_1\right).$$

Theorem 9.2:

Let

$$A : \mathbb{R}^n \to \mathbb{R}^n$$

be a linear transformation and let

$$\hat{v}_i = A\hat{e}_i, \quad i = 1,\ldots,n.$$

If $\left\{\hat{v}_1,\ldots,\hat{v}_n\right\}$ is a linearly dependent set, then

$$\det\left(\hat{v}_1,\ldots,\hat{v}_n\right) = 0.$$

Note: The reason that the theorem is stated in this fashion is

$$\det\left(\hat{v}_1,\ldots,\hat{v}_n\right) = \det\left(A\right).$$

Proof:

Since $\left\{\hat{v}_1,\ldots,\hat{v}_n\right\}$ is a linearly dependent set, one of the vectors can be written as a linear combination of the others. For notational convenience, suppose

$$\hat{v}_1 = \alpha_2\hat{v}_2 + \cdots + \alpha_n\hat{v}_n.$$

Then

$$\det\left(\hat{v}_1,\ldots\hat{v}_n\right) = \det\left(\alpha_2\hat{v}_2 + \cdots + \alpha_n\hat{v}_n,\hat{v}_2,\ldots,\hat{v}_n\right)$$

and by the third postulate

$$\det\left(\alpha_2\hat{v}_2 + \cdots + \alpha_n\hat{v}_n,\hat{v}_2 \ldots,\hat{v}_n\right)$$
$$= \alpha_2\det\left(\hat{v}_2,\hat{v}_2,\ldots\hat{v}_n\right) + \alpha_3\det\left(\hat{v}_3,\hat{v}_2,\hat{v}_3,\ldots\hat{v}_n\right) + \cdots$$
$$+ \alpha_n\det\left(\hat{v}_2,\ldots,\hat{v}_n,\hat{v}_n\right).$$

By the second postulate, each summand on the right-hand side is zero so the result follows.

Formula for Computing the Determinant

In this section we derive a formula, called the Leibniz formula, to compute the determinant from the postulates. The formula will be impractical to use in actual computations, but it does provide the basis of some interesting results and can be used as the definition of the determinant. In fact, many advanced texts use the formula we give here as the definition.

For A an $n \times n$ matrix over a field $\mathcal{F}$ (which is either $\mathbb{R}$ or $\mathbb{C}$) we show how the three postulates yield the formula for the determinant of A.

Case of a 2×2 Matrix

Let

$$A = \begin{pmatrix} a & c \\ b & d \end{pmatrix}$$

and

$$\hat{e}_1 = \begin{pmatrix} 1 \\ 0 \end{pmatrix}, \hat{e}_2 = \begin{pmatrix} 0 \\ 1 \end{pmatrix}, A_1 = \begin{pmatrix} a \\ b \end{pmatrix}, A_2 = \begin{pmatrix} c \\ d \end{pmatrix}.$$

Then, by the second postulate

$$
\begin{aligned}
\det(A) &= \det(A_1, A_2) = \det(a\hat{e}_1 + b\hat{e}_2, c\hat{e}_1 + d\hat{e}_2) \\
&= \det(a\hat{e}_1 + b\hat{e}_2, c\hat{e}_1) + \det(a\hat{e}_1 + b\hat{e}_2, d\hat{e}_2) \\
&= \det(a\hat{e}_1, c\hat{e}_1) + \det(b\hat{e}_2, c\hat{e}_1) + \det(a\hat{e}_1, d\hat{e}_2) + \det(b\hat{e}_2, d\hat{e}_2) \\
&= a \det(\hat{e}_1, c\hat{e}_1) + b \det(\hat{e}_2, c\hat{e}_1) + a \det(\hat{e}_1, d\hat{e}_2) + b \det(\hat{e}_2, d\hat{e}_2) \\
&= ac \det(\hat{e}_1, \hat{e}_1) + bc \det(\hat{e}_2, \hat{e}_1) + ad \det(\hat{e}_1, \hat{e}_2) + bd \det(\hat{e}_2, \hat{e}_2)
\end{aligned}
$$

By the first postulate, $\det(\hat{e}_1, \hat{e}_1) = 0$ and $\det(\hat{e}_2, \hat{e}_2) = 0$ so we have

$$\det(A) = bc \det(e_2, e_1) + ad \det(e_1, e_2).$$

Since

$$\det(\hat{e}_2, \hat{e}_1) = -\det(\hat{e}_1, \hat{e}_2)$$

we have

$$
\begin{aligned}
\det(A) &= bc \det(\hat{e}_2, \hat{e}_1) + ad \det(\hat{e}_1, \hat{e}_2) = -bc \det(\hat{e}_1, \hat{e}_2) + ad \det(\hat{e}_1, \hat{e}_2) \\
&= -bc \det(I) + ad \det(I) = -bc + ad
\end{aligned}
$$

which is the formula that one learns in high school.

The Case of a 3×3 Matrix

The case of a 3×3 matrix follows the same ideas as the 2×2 matrix but provides additional insight into the case of the $n \times n$ matrix.

Let

$$A = \begin{pmatrix} a & d & g \\ b & e & h \\ c & f & i \end{pmatrix}$$

and

$$\hat{e}_1 = \begin{pmatrix} 1 \\ 0 \\ 0 \end{pmatrix}, \hat{e}_2 = \begin{pmatrix} 0 \\ 1 \\ 0 \end{pmatrix}, \hat{e}_3 = \begin{pmatrix} 0 \\ 0 \\ 1 \end{pmatrix}, A_1 = \begin{pmatrix} a \\ b \\ c \end{pmatrix}, A_2 = \begin{pmatrix} d \\ e \\ f \end{pmatrix}, A_3 = \begin{pmatrix} g \\ h \\ i \end{pmatrix}.$$

$$\det\left(A_1, A_2, A_3\right) = \det\left(a e_1 + b e_2 + c e_3, d e_1 + e e_2 + f e_3, g e_1 + h e_2 + i \hat{e}_3\right).$$

Expanding as in the 2×2 case, we get

$$\det(A) = \det\left(a\hat{e}_1 + b\hat{e}_2 + c\hat{e}_3, d\hat{e}_1 + e\hat{e}_2 + f\hat{e}_3, g\hat{e}_1\right)$$
$$+ \det\left(a\hat{e}_1 + b\hat{e}_2 + c\hat{e}_3, d\hat{e}_1 + e\hat{e}_2 + f\hat{e}_3, h\hat{e}_2\right)$$
$$+ \det\left(a\hat{e}_1 + b\hat{e}_2 + c\hat{e}_3, d\hat{e}_1 + e\hat{e}_2 + f\hat{e}_3, i\hat{e}_3\right).$$

Now consider just the first summand in the expression above. We have

$$\det\left(a\hat{e}_1 + b\hat{e}_2 + c\hat{e}_3, d\hat{e}_1 + e\hat{e}_2 + f\hat{e}_3, g\hat{e}_1\right)$$
$$= \det\left(a\hat{e}_1 + b\hat{e}_2 + c\hat{e}_3, d\hat{e}_1, g\hat{e}_1\right) + \det\left(a\hat{e}_1 + b\hat{e}_2 + c\hat{e}_3, e\hat{e}_2, g\hat{e}_1\right)$$
$$+ \det\left(a\hat{e}_1 + b\hat{e}_2 + c\hat{e}_3, f\hat{e}_3, g\hat{e}_1\right)$$
$$= \det\left(a\hat{e}_1, d\hat{e}_1, g\hat{e}_1\right) + \det\left(b\hat{e}_2, d\hat{e}_1, g\hat{e}_1\right) + \det\left(c\hat{e}_3, d\hat{e}_1, g\hat{e}_1\right)$$
$$+ \det\left(a\hat{e}_1, e\hat{e}_2, g\hat{e}_1\right) + \det\left(b\hat{e}_2, e\hat{e}_2, g\hat{e}_1\right) + \det\left(c\hat{e}_3, e\hat{e}_2, g\hat{e}_1\right)$$
$$+ \det\left(a\hat{e}_1, f\hat{e}_3, g\hat{e}_1\right) + \det\left(b\hat{e}_2, f\hat{e}_3, g\hat{e}_1\right) + \det\left(c\hat{e}_3, f\hat{e}_3, g\hat{e}_1\right)$$
$$= adg \det\left(\hat{e}_1, \hat{e}_1, \hat{e}_1\right) + bdg \det\left(\hat{e}_2, \hat{e}_1, \hat{e}_1\right) + cdg \det\left(\hat{e}_3, \hat{e}_1, \hat{e}_1\right)$$
$$+ aeg \det\left(\hat{e}_1, \hat{e}_2, \hat{e}_1\right) + beg \det\left(\hat{e}_2, \hat{e}_2, \hat{e}_1\right) + ceg \det\left(\hat{e}_3, \hat{e}_2, \hat{e}_1\right)$$
$$+ afg \det\left(\hat{e}_1, \hat{e}_3, \hat{e}_1\right) + bfg \det\left(\hat{e}_2, \hat{e}_3, \hat{e}_1\right) + cfg \det\left(\hat{e}_3, \hat{e}_3, \hat{e}_1\right).$$

In the final expression, the only non-zero terms are

$$ceg \det\left(\hat{e}_3, \hat{e}_2, \hat{e}_1\right) \quad \text{and} \quad bfg \det\left(\hat{e}_2, \hat{e}_3, \hat{e}_1\right).$$

Similarly, in the expansion for

$$\det\left(a\hat{e}_1 + b\hat{e}_2 + c\hat{e}_3, d\hat{e}_1 + e\hat{e}_2 + f\hat{e}_3, h\hat{e}_2\right)$$

the only non-zero terms are

$$afh \det\left(\hat{e}_1, \hat{e}_3, \hat{e}_2\right) \quad \text{and} \quad cdh \det\left(\hat{e}_3, \hat{e}_1, \hat{e}_2\right)$$

and in the expansion for

$$\det\left(ae_1 + be_2 + ce_3, de_1 + ee_2 + cf_3, ie_3\right)$$

the only non-zero terms are

$$aei\,\det\left(e_1,e_2,e_3\right) \quad \text{and} \quad bdi\,\det\left(e_2,e_1,e_3\right).$$

We step back to contemplate what we have ascertained.

In the case of a 3×3 matrix, when the determinant is expanded the only terms that could possibly be non-zero are those associated with $\det\left(\hat{e}_l,\hat{e}_m,\hat{e}_n\right)$ where l, m and n are all different. This characteristic is true for $n\times n$ matrices where in the expansion of the determinant we have terms of the form $\det\left(\hat{e}_{i_1},\hat{e}_{i_2},\dots,\hat{e}_{i_n}\right)$ but the only terms that are possibly non-zero are those where $i_1,\dots,i_n$ are all different.

We return to the case of the 3×3 matrix to determine whether the value of $\det\left(\hat{e}_l,\hat{e}_m,\hat{e}_n\right)$ is 1 or -1. Now

$$\det\left(\hat{e}_1,\hat{e}_2,\hat{e}_3\right)=1$$

because $\left(\hat{e}_1,\hat{e}_2,\hat{e}_3\right)$ is the identity matrix.

To change $\left(\hat{e}_1,\hat{e}_2,\hat{e}_3\right)$ to $\left(\hat{e}_2,\hat{e}_1,\hat{e}_3\right)$ we switch the first and second columns; that is, we switch exactly two columns so $\det\left(e_2,e_1,e_3\right)=-1$.

To change $\left(\hat{e}_1,\hat{e}_2,\hat{e}_3\right)$ to $\left(\hat{e}_3,\hat{e}_1,\hat{e}_2\right)$ we first switch the first and third columns to get $\left(\hat{e}_3,\hat{e}_2,\hat{e}_1\right)$ and then switch the second and third columns of $\left(\hat{e}_3,\hat{e}_2,\hat{e}_1\right)$ to get $\left(\hat{e}_3,\hat{e}_1,\hat{e}_2\right)$. (The order of the switches does not matter. What is crucial is that two switches were involved.) Thus,

$$\det\left(\hat{e}_3,\hat{e}_1,\hat{e}_2\right)=\left(-1\right)^2=1.$$

Using this idea, we have

$$\det\left(\hat{e}_1,\hat{e}_2,\hat{e}_3\right)=\left(-1\right)^0=1$$

$$\det\left(\hat{e}_2,\hat{e}_1,\hat{e}_3\right)=\left(-1\right)^1=-1$$

$$\det\left(\hat{e}_1,\hat{e}_3,\hat{e}_2\right)=\left(-1\right)^1=-1$$

$$\det\left(\hat{e}_2,\hat{e}_3,\hat{e}_1\right)=\left(-1\right)^2=1$$

$$\det\left(\hat{e}_3,\hat{e}_1,\hat{e}_2\right)=\left(-1\right)^2=1$$

$$\det\left(\hat{e}_3,\hat{e}_2,\hat{e}_1\right)=\left(-1\right)^1=-1.$$

Thus, in computing the determinant of a 3×3 matrix, the non-zero terms were

$$ceg \det\left(\hat{e}_3, \hat{e}_2, \hat{e}_1\right) = -ceg$$

$$bfg \det\left(\hat{e}_2, \hat{e}_3, \hat{e}_1\right) = bfg$$

$$afh \det\left(\hat{e}_1, \hat{e}_3, \hat{e}_2\right) = -afh$$

$$cdh \det\left(\hat{e}_3, \hat{e}_1, \hat{e}_2\right) = cdh$$

$$aei \det\left(\hat{e}_1, \hat{e}_2, \hat{e}_3\right) = aei$$

$$bdi \det\left(\hat{e}_2, \hat{e}_1, \hat{e}_3\right) = -bdi.$$

and so,

$$\det\begin{pmatrix} a & d & g \\ b & e & h \\ c & f & i \end{pmatrix} = \left(aei - afh\right) + \left(bfg - bdi\right) + \left(cdh - ceg\right)$$

$$= a\left(ei - fh\right) - b\left(di - fg\right) + c\left(dh - eg\right).$$

We arranged the answer in this form to demonstrate that this gives the same result as expanding in cofactors. Applying that method along the first column, we have

$$\begin{vmatrix} a & d & g \\ b & e & h \\ c & f & i \end{vmatrix} = a\begin{vmatrix} e & h \\ f & i \end{vmatrix} - b\begin{vmatrix} d & g \\ f & i \end{vmatrix} + c\begin{vmatrix} d & g \\ e & h \end{vmatrix}$$

$$= a\left(ei - fh\right) - b\left(di - fg\right) + c\left(dh - eg\right).$$

The Case of an n×n Matrix

Consider the case

$$A = \begin{pmatrix} a_{11} & a_{12} & \cdots & a_{1n} \\ a_{21} & a_{22} & \cdots & a_{2n} \\ \vdots & \vdots & & \vdots \\ a_{n1} & a_{n2} & \cdots & a_{nn} \end{pmatrix}.$$

Let

$$\hat{e}_1 = \begin{pmatrix} 1 \\ 0 \\ \vdots \\ 0 \end{pmatrix}, \hat{e}_2 = \begin{pmatrix} 0 \\ 1 \\ \vdots \\ 0 \end{pmatrix}, \cdots, \hat{e}_n = \begin{pmatrix} 0 \\ \vdots \\ 0 \\ 1 \end{pmatrix}, A_1 = \begin{pmatrix} a_{11} \\ a_{21} \\ \vdots \\ a_{n1} \end{pmatrix}, \ldots, A_n = \begin{pmatrix} a_{1n} \\ a_{2n} \\ \vdots \\ a_{nn} \end{pmatrix}.$$

When det(A) is expanded as in the 3×3 case, the only terms that are possibly non-zero are those involving

$$\det\left(\hat{e}_{i_1}, \hat{e}_{i_2}, \ldots, \hat{e}_{i_n}\right)$$

where the $i_1, i_2, \ldots, i_n$ are all different numbers in the set $\{1, 2, \ldots, n\}$.

To arrive at the formula usually given for the determinant in the case of an $n \times n$ matrix, it will be convenient to establish some additional notation.

Permutations

A permutation of a finite set $A = \{1, 2, \ldots, n\}$ is a one-to-one and onto function from the set A to itself. Permutations are usually represented by Greek letters such as σ or τ. If $A = \{1, 2, \ldots, n\}$, then there are $n!$ permutations of A. One way to see this, is that we can construct a permutation by first selecting where we want to map the number 1, then the number 2, etc. If σ is the permutation, then we have n choices for $\sigma(1)$, $(n - 1)$ choices for $\sigma(2)$, $(n - 2)$ choices for $\sigma(3)$, etc., so that we have $n(n-1)(n-2)\cdots 1 = n!$ ways to define σ.

For our purposes, the major point is that for any permutation σ of $\{1, 2, \ldots, n\}$, we have

$$\sigma(i) \neq \sigma(j) \text{ if } i \neq j$$

so

$$\det\left(\hat{e}_{\sigma(1)}, \hat{e}_{\sigma(2)}, \ldots, \hat{e}_{\sigma(n)}\right) = \pm 1$$

and the only way

$$\det\left(\hat{e}_{i_1}, \hat{e}_{i_2}, \ldots, \hat{e}_{i_n}\right) \neq 0$$

is if $\{i_1, i_2, \ldots, i_n\}$ is a permutation of $\{1, 2, \ldots, n\}$.

The next question is, for a given permutation of σ of $\{1, 2, \ldots, n\}$ what is the value of

$$\det\left(\hat{e}_{\sigma(1)}, \hat{e}_{\sigma(2)}, \ldots, \hat{e}_{\sigma(n)}\right).$$

If we return to the case of a 3×3 matrix, but representing the matrix as

$$A = \begin{pmatrix} a_{11} & a_{12} & a_{13} \\ a_{21} & a_{22} & a_{23} \\ a_{31} & a_{32} & a_{33} \end{pmatrix}$$

then the non-zero terms in the determinant are

$$a_{31}a_{22}a_{13}\det\left(\hat{e}_3, \hat{e}_2, \hat{e}_1\right)$$

$$a_{21}a_{32}a_{13}\det\left(\hat{e}_2, \hat{e}_3, \hat{e}_1\right)$$

$$a_{11}a_{32}a_{23}\det\left(\hat{e}_1,\hat{e}_3,\hat{e}_2\right)$$

$$a_{31}a_{12}a_{23}\det\left(\hat{e}_3,\hat{e}_1,\hat{e}_2\right)$$

$$a_{11}a_{22}a_{33}\det\left(\hat{e}_1,\hat{e}_2,\hat{e}_3\right)$$

$$a_{21}a_{12}a_{33}\det\left(\hat{e}_2,\hat{e}_1,\hat{e}_3\right).$$

We want to associate each non-zero term with a permutation. There are six permutations of {1, 2, 3}:

$$\sigma_1:\begin{cases}1\to3\\2\to2\\3\to1\end{cases} \sigma_2:\begin{cases}1\to2\\2\to3\\3\to1\end{cases} \sigma_3:\begin{cases}1\to1\\2\to3\\3\to2\end{cases} \sigma_4:\begin{cases}1\to3\\2\to1\\3\to2\end{cases} \sigma_5:\begin{cases}1\to1\\2\to2\\3\to3\end{cases} \sigma_6:\begin{cases}1\to2\\2\to1\\3\to3\end{cases}$$

by which the table means $\sigma_1(1)=3$, $\sigma_1(2)=2$, $\sigma_1(3)=1$.

Then

$$a_{31}a_{22}a_{13}\det\left(\hat{e}_3,\hat{e}_2,\hat{e}_1\right)=a_{\sigma_1(1)1}a_{\sigma_1(2)2}a_{\sigma_1(3)3}\det\left(\hat{e}_3,\hat{e}_2,\hat{e}_1\right).$$

We next associate a permutation σ with either 1 or –1, which we call the sign of σ. In this particular case, we want to consider

$$\det\left(\hat{e}_3,\hat{e}_2,\hat{e}_1\right)=\det\left(\hat{e}_{\sigma_1(1)},\hat{e}_{\sigma_1(2)},\hat{e}_{\sigma_1(3)}\right).$$

Previously, we determined

$$\det\left(\hat{e}_3,\hat{e}_2,\hat{e}_1\right)=\left(-1\right)^n$$

where n was the number of times we had to switch two columns in the matrix $\left(\hat{e}_1,\hat{e}_2,\hat{e}_3\right)$ to obtain the matrix $\left(\hat{e}_3,\hat{e}_2,\hat{e}_1\right)$.

Here there was only one switch $1\leftrightarrow3$ so $n=1$. We thus assign the sign of value σ to be

$$sgn\left(\sigma_1\right)=\left(-1\right)^1=-1$$

and so

$$a_{31}a_{22}a_{13}\det\left(\hat{e}_3,\hat{e}_2,\hat{e}_1\right)=a_{\sigma_1(1)1}a_{\sigma_1(2)2}a_{\sigma_1(3)3}\det\left(\hat{e}_3,\hat{e}_2,\hat{e}_1\right)$$

$$=a_{\sigma_1(1)1}a_{\sigma_1(2)2}a_{\sigma_1(3)3}sgn\left(\sigma_1\right)=-a_{\sigma_1(1)1}a_{\sigma_1(2)2}a_{\sigma_1(3)3}.$$

We do this for each σ_i, $i=1,\ldots,6$ to obtain the other non-zero terms. We leave it as an exercise to check for $\sigma_2=\left(2,3,1\right)$,

$$a_{21}a_{32}a_{13}\det\left(\hat{e}_2,\hat{e}_3,\hat{e}_1\right)=a_{\sigma_2(1)1}a_{\sigma_2(2)2}a_{\sigma_2(3)3}sgn\left(\sigma_2\right)$$

for $\sigma_3 = (1,3,2)$,

$$a_{11}a_{32}a_{23}\det\left(\hat{e}_1,\hat{e}_3,\hat{e}_2\right) = a_{\sigma_3(1)1}a_{\sigma_3(2)2}a_{\sigma_3(3)3}sgn\left(\sigma_3\right)$$

for $\sigma_4 = (3,1,2)$,

$$a_{31}a_{12}a_{23}\det\left(\hat{e}_3,\hat{e}_1,\hat{e}_2\right) = a_{\sigma_4(1)1}a_{\sigma_4(2)2}a_{\sigma_4(3)3}sgn\left(\sigma_4\right)$$

for $\sigma_5 = (1,2,3)$,

$$a_{11}a_{22}a_{33}\det\left(\hat{e}_1,\hat{e}_2,\hat{e}_3\right) = a_{\sigma_5(1)1}a_{\sigma_5(2)2}a_{\sigma_5(3)3}sgn\left(\sigma_5\right)$$

and for $\sigma_6 = (2,1,3)$,

$$a_{21}a_{12}a_{33}\det\left(\hat{e}_1,\hat{e}_2,\hat{e}_3\right) = a_{\sigma_6(1)1}a_{\sigma_6(2)2}a_{\sigma_6(3)3}sgn\left(\sigma_6\right).$$

Thus, another way to express $\det(A)$ in the case of a 3×3 matrix is

$$\det(A) = \sum_{i=1}^{6} a_{\sigma_i(1)1}a_{\sigma_i(2)2}a_{\sigma_i(3)3}sgn\left(\sigma_i\right).$$

For the case of an $n\times n$ the formula is

$$\det(A) = \sum_{\sigma} a_{\sigma(1)1}a_{\sigma(2)2}\cdots a_{\sigma(n)n}sgn\left(\sigma\right)$$

where the sum is over all permutations of the set $\{1,2,\ldots,n\}$. We reiterate that this formula is derived from the postulates.

Further Properties of the Determinant

Some of the next theorems will be proven using elementary matrices. We now review that subject.
    In Chapter 1, we proved the following result:

Theorem 9.3:

Every invertible matrix can be expressed as a product of elementary matrices.

Theorem 9.4:

If A is a square matrix and E is an elementary matrix, then

$$\det(EA) = \det(E)\det(A).$$

Proof:

The proof amounts to checking the postulates for the three types of elementary matrices and is left as an exercise.

Theorem 9.5:

If A and B are $n \times n$ matrices, then $\det(AB) = \det(A)\det(B)$.

Proof:

We give two arguments. The first is heuristic and is based on the geometric interpretation of the determinant as a function that measures the change of the size of a region.

To that end, we regard AB as the composition of two linear transformations

$$B : \mathbb{R}^n \to \mathbb{R}^n \text{ followed by } A : \mathbb{R}^n \to \mathbb{R}^n.$$

Suppose B acts on a region Γ of volume V. Then the volume of $B(\Gamma)$ is $(\det B)V$. Now A maps every unit of volume 1 to 1 region of volume $(\det A)$ and so maps a region of volume $\det(B)$ to a region of volume $(\det A)(\det B)$. Thus the composition of A and B maps a region of volume 1 to a region of volume $(\det A)(\det B)$ so that $\det(AB) = (\det A)(\det B)$.

The second proof uses the idea of elementary matrices. Suppose that A is a nonsingular matrix, so that the columns of A are linearly independent. Then A can be expressed as the product of elementary matrices. Suppose that

$$A = E_1 E_2 \cdots E_k$$

and B is an $n \times n$ matrix. Then

$$\det(AB) = \det\left[\left(E_1 E_2 \cdots E_k\right)B\right] = \det\left[E_1\left(E_2 \cdots E_k B\right)\right].$$

Since E_1 is an elementary matrix, we have

$$\det\left[E_1\left(E_2 \cdots E_k B\right)\right] = \det\left(E_1\right)det\left(E_2 \cdots E_k B\right).$$

Continuing with this idea, we have

$$\det\left(E_2 \cdots E_k B\right) = \det\left(E_2\right)det\left(E_3 \cdots E_k B\right)$$

so that

$$\det\left[E_1\left(E_2 \cdots E_k B\right)\right] = \det\left(E_1\right)\det\left(E_2\right)\det\left(E_3 \cdots E_k B\right) = \cdots = \det\left(E_1\right)\det\left(E_2\right)\cdots\det\left(E_k\right)\det\left(B\right).$$

Likewise,

$$\det\left(A\right) = \det\left(E_1 \cdots E_k\right) = \det\left(E_1\right)\det\left(E_2 \cdots E_k\right) = \det\left(E_1\right)\det\left(E_2\right)\cdots\det\left(E_k\right)$$

so we have

$$\det(AB) = \left[\det(E_1)\det(E_2)\cdots\det(E_k)\right]\det(B) = \det(A)\det(B).$$

Corollary:

For A a square matrix, A^{-1}, exists if and only if the columns of A form a linearly independent set, and in that case

$$\det(A^{-1}) = \left[\det(A)\right]^{-1} = \frac{1}{\det(A)}.$$

Proof:

From our earlier work, we only need to show that

$$\det(A^{-1}) = \left[\det(A)\right]^{-1} = \frac{1}{\det(A)}.$$

If A^{-1} exists, then

$$1 = \det(I) = \det\left[A(A^{-1})\right] = \det(A)\det(A^{-1})$$

so

$$\det(A^{-1}) = \frac{1}{\det(A)}.$$

Next we relate the determinant of a matrix A with the determinant of its transpose, which we denote A^T. We will again use elementary matrices, and appeal to our earlier examples.

1. Interchanging two rows of a matrix.

   Consider the 4×4 elementary matrix that interchanges the second and fourth rows.

$$I_{24} = \begin{pmatrix} 1 & 0 & 0 & 0 \\ 0 & 0 & 0 & 1 \\ 0 & 0 & 1 & 0 \\ 0 & 1 & 0 & 0 \end{pmatrix}.$$

   Then

$$I_{24}{}^T = \begin{pmatrix} 1 & 0 & 0 & 0 \\ 0 & 0 & 0 & 1 \\ 0 & 0 & 1 & 0 \\ 0 & 1 & 0 & 0 \end{pmatrix}$$

and it is the case in general (because such matrices are symmetric) that $I_{ij} = I_{ij}^T$.

2. Elementary matrices that multiply a row by a number are symmetric because they are diagonal so each such matrix is its own transpose.

3. Adding a multiple of one row to another row.

Consider the example where we multiplied the second row of a 3×3 Identity matrix by 5 and the result to the third row. The matrix we used was

$$I_{23}(5) = \begin{pmatrix} 1 & 0 & 0 \\ 0 & 1 & 0 \\ 0 & 5 & 1 \end{pmatrix}.$$

Now

$$I_{a3}(5)^T = \begin{pmatrix} 1 & 0 & 0 \\ 0 & 1 & 5 \\ 0 & 0 & 1 \end{pmatrix}$$

which is the matrix that multiplies the third row of the matrix by 5 and adds the result to the second row. The determinant of such matrices is 1.

We conclude that the transpose of an elementary matrix is an elementary matrix with the same determinant.

Theorem 9.6:

If A is an $n \times n$ matrix, then $\det(A) = \det(A^T)$.

Proof:

We write A as the product of elementary matrices

$$A = E_1 E_2 \cdots E_k.$$

Then

$$A^T = \left(E_1 E_2 \cdots E_k \right)^T = E_k^T E_{k-1}^T \cdots E_1^T$$

so

$$\det(A) = \det\left(E_1 E_2 \cdots E_k \right) = \det(E_1)\,det(E_2)\cdots\det(E_k)$$
$$= \det(E_k)\det(E_{k-1})\cdots\det(E_1) = \det\left(E_k^T\right)\det\left(E_{k-1}^T\right)\cdots\det\left(E_1^T\right)$$
$$= \det\left(E_k^T E_{k-1}^T \cdots E_1^T\right) = \det\left(A^T\right).$$

Computing a Determinant

In Chapter 1 we gave a brief description of computing a matrix by cofactors. We now somewhat expand that explanation. We explain this method primarily by way of examples.

The simplest case is a 2×2 matrix. In that case, we have established that

$$\det \begin{pmatrix} a & b \\ c & d \end{pmatrix} \equiv \begin{vmatrix} a & b \\ c & d \end{vmatrix} \equiv ad - bc.$$

Suppose we have a 3×3 matrix. We assign a + sign to the upper left-hand entry, and alternate signs of the entries as we proceed either to the right or down. We then have

$$\begin{pmatrix} + & - & + \\ - & + & - \\ + & - & + \end{pmatrix}.$$

In expanding by cofactors, one chooses any row or any column, and breaks the determinant into pieces as in the next example.

Example

Consider

$$A = \begin{pmatrix} -2 & 3 & 1 \\ 0 & -1 & 2 \\ 4 & 5 & 3 \end{pmatrix}.$$

Suppose we elect to expand along the second column. We compute the determinate by expanding the determinate into three smaller pieces. This is done by using the entries in the second column and making the determinant of the 4 elements that are not in that particular row or column. So

$$\det(A) = \begin{vmatrix} -2 & 3 & 1 \\ 0 & -1 & 2 \\ 4 & 5 & 3 \end{vmatrix} = -3 \begin{vmatrix} 0 & 2 \\ 4 & 3 \end{vmatrix} - 1 \begin{vmatrix} -2 & 1 \\ 4 & 3 \end{vmatrix} - 5 \begin{vmatrix} -2 & 1 \\ 0 & 2 \end{vmatrix}.$$

What we have done is taken the entries in the second row and either kept or changed their sign according to table of + and – signs given earlier.

We complete the computation for the 2×2 determinants. So

$$-3 \begin{vmatrix} 0 & 2 \\ 4 & 3 \end{vmatrix} - 1 \begin{vmatrix} -2 & 1 \\ 4 & 3 \end{vmatrix} - 5 \begin{vmatrix} -2 & 1 \\ 0 & 2 \end{vmatrix} = -3(0 - 8) - 1(-6 - 4) - 5(-4 - 0)$$

$$= 24 + 10 + 20 = 54.$$

Clearly, this becomes exponentially more tedious as the size of the matrix increases.

If one must compute the determinant of a matrix larger than 3×3 by hand, the usual approach is to apply row operations to convert the matrix to triangular form. This can greatly reduce the number of arithmetic operations and is the method computers use in evaluating determinants.

To apply this technique, one must keep in mind how row operations affect the determinant.

Operation	Effect on the Determinant
Interchange rows	Multiply by –1
Add a multiple of one row to another	No change
Multiply a row by a constant k	Multiplies the determinant by k

An example of the last rule is

$$\begin{vmatrix} ka_{11} & \cdots & ka_{1n} \\ a_{21} & \cdots & a_{2n} \\ \vdots & & \vdots \\ a_{n1} & & a_{nn} \end{vmatrix} = k \begin{vmatrix} a_{11} & \cdots & a_{1n} \\ a_{21} & \cdots & a_{2n} \\ \vdots & & \vdots \\ a_{n1} & & a_{nn} \end{vmatrix}.$$

Example

We evaluate

$$\begin{vmatrix} 2 & 4 & -6 & 0 \\ 1 & 5 & 3 & 6 \\ 3 & 1 & 4 & 5 \\ 2 & 6 & -7 & -3 \end{vmatrix} = 2 \begin{vmatrix} 1 & 2 & -3 & 0 \\ 1 & 5 & 3 & 6 \\ 3 & 1 & 4 & 5 \\ 2 & 6 & -7 & -3 \end{vmatrix} = 2 \begin{vmatrix} 1 & 2 & -3 & 0 \\ 0 & 3 & 6 & 6 \\ 0 & -5 & 13 & 5 \\ 0 & 2 & -1 & -3 \end{vmatrix}$$

$$= 2(3) \begin{vmatrix} 1 & 2 & -3 & 0 \\ 0 & 1 & 2 & 2 \\ 0 & -5 & 13 & 5 \\ 0 & 2 & -1 & -3 \end{vmatrix} = 2(3) \begin{vmatrix} 1 & 2 & -3 & 0 \\ 0 & 1 & 2 & 2 \\ 0 & 0 & 23 & 15 \\ 0 & 0 & -5 & -7 \end{vmatrix}$$

$$= 2(3)(23) \begin{vmatrix} 1 & 2 & -3 & 0 \\ 0 & 1 & 2 & 2 \\ 0 & 0 & 1 & \frac{15}{23} \\ 0 & 0 & -5 & -7 \end{vmatrix} = 2(3)(23) \begin{vmatrix} 1 & 2 & -6 & 0 \\ 0 & 1 & 2 & 2 \\ 0 & 0 & 1 & \frac{15}{23} \\ 0 & 0 & 0 & -\frac{86}{23} \end{vmatrix}$$

$$= 2(3)(23) \left(-\frac{86}{23} \right) = -516.$$

Exercise

Prove or disprove: If A and B are 2×2 matrices then

$$\det \begin{pmatrix} A & B \\ B & A \end{pmatrix} = \det(A + B) \det(A - B).$$

Chapter 10

Jordan Canonical Form

Historical Note

Camille Jordan (born 5 January 1838, died 20 January 1922)

Camille Jordan was born in Lyon, France, and educated at the École Polytechnique. He was an engineer by profession. Later in life he taught at the École Polytechnique and the Collège de France.

Jordan was a mathematician who worked in a wide variety of different areas contributing to essentially every mathematical topic that was studied at that time.

He is remembered now by name in a number of results:

- The Jordan curve theorem, a topological result required in complex analysis.
- The Jordan normal form and the Jordan matrix in linear algebra.
- In mathematical analysis, Jordan measure (or *Jordan content*) is an area measure that predates measure theory.
- In group theory, the Jordan–Hölder theorem on composition series is a basic result.
- Jordan's theorem on finite linear groups.

His work did much to bring Galois theory into the mainstream. His *Traité des substitutions*, on permutation groups, was published in 1870. This treatise won the 1870 *Prix Poncelet*. In *Traité des Substitutions et des Equations Algébriques* (1870), he gave the definition of what is today called the Jordan normal form, although in 1868 Weierstrass had defined an equivalent normal form.

Jordan was particularly interested in the theory of finite groups. Before Jordan began his research in this area there was little to no theory of finite groups. It was Jordan who was the first to develop a systematic approach to the topic. It was not until Liouville republished Galois's original work in 1846 that its significance was noticed at all. It was Jordan who was the first to formulate the direction the subject would take.

Jordan is best remembered today among analysts and topologists for his proof that a simply closed curve divides a plane into exactly two regions, now called the *Jordan curve theorem*. He also originated the concept of functions of bounded variation and is known especially for his definition of the length of a curve. These concepts appear in his *Cours d'analyse de l'École Polytechnique* first published in three volumes between 1882 and 1887.

Introduction

Let V be an n-dimensional vector space, and $T : V \rightarrow V$ be a linear transformation. A linear transformation on an n-dimensional vector space can be expressed as a multiplication of vectors by an $n \times n$ matrix, but the matrix depends on the choice of basis.

From several points of view – including computations – the ideal representation of a linear transformation is a diagonal matrix, but this is not always possible. It is possible if and only if there is a basis of eigenvectors.

If there is not a basis of eigenvectors, it may be possible to choose a basis so that the matrix representation of T is a matrix whose diagonal entries are the eigenvalues and the only other entries that are non-zero are 1s which occur, if at all, only on the upper diagonal. This representation is called the Jordan canonical form. Figure 10.1 gives an example of the Jordan canonical form. This representation is possible if and only if the characteristic polynomial factors into linear factors. If the scalar field is the complex numbers, then this is always the case.

The chapter is organized so that we first show how to compute the Jordan canonical form and how to determine the associated basis of the vector space. Later in the chapter we give the mathematical theory that justifies the algorithms that we develop.

Section 10.1 Finding the Jordan Canonical Form and the Associated Basis

The central ideas in the construction of the Jordan canonical form are generalized eigenvectors and chains of generalized eigenvectors.

Definition:

For $A : V \rightarrow V$ a linear transformation, a non-zero vector $\hat{v} \in V$ is a generalized eigenvector of A of rank r with eigenvalue λ if

$$\left(A - \lambda I \right)^r \hat{v} = \hat{0}, \text{ but } \left(A - \lambda I \right)^{r-1} \hat{v} \neq \hat{0}.$$

$$\begin{pmatrix} 2 & 1 & 0 & 0 & 0 & 0 \\ 0 & 2 & 1 & 0 & 0 & 0 \\ 0 & 0 & 2 & 1 & 0 & 0 \\ 0 & 0 & 0 & 2 & 0 & 0 \\ 0 & 0 & 0 & 0 & 2 & 1 \\ 0 & 0 & 0 & 0 & 0 & 2 \end{pmatrix}$$

Figure 10.1   Example of a matrix in the Jordan canonical form.

One could refer to an eigenvector of A as being a generalized eigenvector of A with rank 1.

Example

Consider the matrix

$$A = \begin{pmatrix} 3 & 1 \\ 0 & 3 \end{pmatrix}.$$

We have the characteristic polynomial

$$p_A(\lambda) = (\lambda - 3)^2,$$

so the only eigenvalue of A is $\lambda = 3$.

It is easy to check that the eigenspace has dimension 1 and

$$\hat{v}_1 = \begin{pmatrix} 1 \\ 0 \end{pmatrix}$$

is an eigenvector. The vector

$$\hat{v}_2 = \begin{pmatrix} 0 \\ 1 \end{pmatrix}$$

has the property that

$$(A - 3I)\hat{v}_2 = \hat{v}_1 \text{ and } (A - 3I)^2 \hat{v}_2 = \begin{pmatrix} 0 \\ 0 \end{pmatrix}$$

so that $\hat{v}_2$ is a generalized eigenvector of rank 2. The set $\{\hat{v}_1, \hat{v}_2\}$ is a basis for $\mathbb{R}^2$.

Definition:

For A a linear transformation with eigenvalue λ, the subspace

$$V_\lambda = \left\{ \hat{x} \,|\, (A - \lambda I)^k \, \hat{x} = \hat{0} \text{ for some positve integer } k \right\}$$

is called the generalized eigenspace of λ.

We leave it as an exercise to show that V_λ is a subspace.

An important fact that will be proven later is that the vector space of a linear transformation can be written as the direct sum of its generalized eigenspaces.

That is,

$$V = V_{\lambda_1} \oplus \cdots \oplus V_{\lambda_n}$$

where $\lambda_1, \ldots, \lambda_n$ are the distinct eigenvalues of the linear transformation.

Suppose that $\hat{v}$ is a generalized eigenvector in V_λ and that p is the smallest positive integer for which $(A-\lambda I)^p \hat{v} = \hat{0}$. Then $(A-\lambda I)^{p-1}\hat{v}, (A-\lambda I)^{p-2}\hat{v}, \cdots (A-\lambda I)\hat{v}, \hat{v}$ are all generalized eigenvectors for λ and, of these vectors, only $(A-\lambda I)^{p-1}\hat{v}$ is an eigenvector for λ.

Definition:

If p is the smallest positive integer for which $(A-\lambda I)^p \hat{v} = \hat{0}$, then the ordered collection of vectors

$$\left\{ (A-\lambda I)^{p-1}\hat{v}, (A-\lambda I)^{p-2}\hat{v}, \ldots, (A-\lambda I)\hat{v}, \hat{v} \right\}$$

is called a chain (or a string or a cycle) of generalized eigenvectors. The vector $(A-\lambda I)^{p-1}\hat{v}$ is the initial vector in the chain and $\hat{v}$ is the final vector in the chain.

For a given eigenvalue λ, there may be several linearly independent eigenvectors; say $\lambda_1, \ldots, \lambda_k$. With each of these eigenvectors there will be a chain of generalized eigenvectors. We will show that

(i) each chain will be disjoint from the others.
(ii) The union of the disjoint chains will be a basis for V_λ.
(iii) The union of all the disjoint chains for all the different eigenvalues is a basis for V.
(iv) This basis, when properly ordered, yields the Jordan canonical form for the linear transformation.

Example

This example shows why the chains of generalized eigenvectors yield the Jordan canonical form.

Suppose that A is a 5×5 matrix with a single eigenvalue of 3. Suppose that there are exactly two linearly independent eigenvectors, $\hat{v}_1$ and $\hat{w}_1$, and we have the following chains of generalized eigenvectors that together form a basis.

$$\hat{v}_3 \xrightarrow{A-3I} \hat{v}_2 \xrightarrow{A-3I} \hat{v}_1 \xrightarrow{A-3I} \hat{0} \qquad \hat{w}_2 \xrightarrow{A-3I} \hat{w}_1 \xrightarrow{A-3I} \hat{0}.$$

We want to find the representation of A with respect to this basis.

Recall that in the representation of a matrix A with respect to a basis $\{\hat{u}_1, \hat{u}_2, \hat{u}_3, \hat{u}_4, \hat{u}_5\}$, the ith column of A is determined by the coefficients of the image of $\hat{u}_i$ with respect to this basis. For example, if

$$A\hat{u}_4 = 2\hat{u}_1 + 7\hat{u}_2 - 4\hat{u}_3 + 1\hat{u}_4 + 0\hat{u}_5$$

then the fourth column of A in this basis is

$$\begin{pmatrix} 2 \\ 7 \\ -4 \\ 1 \\ 0 \end{pmatrix}.$$

Suppose we assume there are two linearly independent eigenvectors, and one chain of length two and one of length 3. (Later we describe how to determine this.) Now for the vectors $\{\hat{v}_1, \hat{v}_2, \hat{v}_3, \hat{w}_1, \hat{w}_2\}$ we have

$$(A-3I)\hat{v}_1 = \hat{0}, \text{ or } A\hat{v}_1 = 3\hat{v}_1 + 0\hat{v}_2 + 0\hat{v}_3 + 0\hat{w}_1 + 0\hat{w}_2$$

so the first column of A is

$$\begin{pmatrix} 3 \\ 0 \\ 0 \\ 0 \\ 0 \end{pmatrix}.$$

Similarly,

$$(A-3I)\hat{v}_2 = \hat{v}_1, \text{ or } A\hat{v}_2 = 1\hat{v}_1 + 3\hat{v}_2 + 0\hat{v}_3 + 0\hat{w}_1 + 0\hat{w}_2$$

so the second column of A is

$$\begin{pmatrix} 1 \\ 3 \\ 0 \\ 0 \\ 0 \end{pmatrix}$$

and other columns are computed in a similar manner, as follows.

$$(A-3I)\hat{v}_3 = \hat{v}_2, \text{ or } A\hat{v}_3 = 0\hat{v}_1 + 1\hat{v}_2 + 3\hat{v}_3 + 0\hat{w}_1 + 0\hat{w}_2$$

so the third column of A is

$$\begin{pmatrix} 0 \\ 1 \\ 3 \\ 0 \\ 0 \end{pmatrix}.$$

Also,

$$(A-3I)\hat{w}_1 = \hat{0}, \text{ or } A\hat{w}_1 = 3\hat{w}_1 = 0\hat{v}_1 + 0\hat{v}_2 + 0\hat{v}_3 + 3\hat{w}_1 + 0\hat{w}_2$$

so the fourth column of A is

$$\begin{pmatrix} 0 \\ 0 \\ 0 \\ 3 \\ 0 \end{pmatrix}$$

and

$$\left(A-3I\right)\hat{w}_2 = \hat{w}_1, \text{ or } A\hat{w}_2 = 0\hat{v}_1 + 0\hat{v}_2 + 0\hat{v}_3 + 1\hat{w}_1 + 3\hat{w}_2$$

so the fifth column of A is

$$\begin{pmatrix} 0 \\ 0 \\ 0 \\ 1 \\ 3 \end{pmatrix}.$$

Thus, the matrix of A with respect to the basis $\{\hat{v}_1, \hat{v}_2, \hat{v}_1, \hat{w}_1, \hat{w}_2\}$ is

$$A = \begin{pmatrix} 3 & 1 & 0 & 0 & 0 \\ 0 & 3 & 1 & 0 & 0 \\ 0 & 0 & 3 & 0 & 0 \\ 0 & 0 & 0 & 3 & 1 \\ 0 & 0 & 0 & 0 & 3 \end{pmatrix},$$

which is in Jordan canonical form.

The next example demonstrates the process of finding the Jordan canonical form and the associated basis when there is a single eigenvalue.

Example

Consider the matrix

$$A = \begin{pmatrix} 2 & 0 & 0 & 0 & -1 & 1 \\ -2 & 0 & 2 & 2 & -2 & 3 \\ -1 & -1 & 3 & 2 & -2 & 3 \\ 0 & 0 & 0 & 2 & 0 & 0 \\ 1 & 1 & -1 & -1 & 3 & -1 \\ 0 & 0 & 0 & 0 & 0 & 2 \end{pmatrix}.$$

The characteristic polynomial for A is

$$p_A\left(\lambda\right) = \left(\lambda - 2\right)^6.$$

Thus, $\lambda = 2$ is the only eigenvalue.

We find the independent eigenvectors of $A - 2I$.

Now

$$A - 2I = \begin{pmatrix} 0 & 0 & 0 & 0 & -1 & 1 \\ -2 & -2 & 2 & 2 & -2 & 3 \\ -1 & -1 & 1 & 2 & -2 & 3 \\ 0 & 0 & 0 & 0 & 0 & 0 \\ 1 & 1 & -1 & -1 & 1 & -1 \\ 0 & 0 & 0 & 0 & 0 & 0 \end{pmatrix}.$$

We want

$$(A-2I)\begin{pmatrix} x_1 \\ x_2 \\ x_3 \\ x_4 \\ x_5 \\ x_6 \end{pmatrix} = \begin{pmatrix} 0 \\ 0 \\ 0 \\ 0 \\ 0 \\ 0 \end{pmatrix},$$

When $A - 2I$ is row reduced the result is

$$\begin{pmatrix} 1 & 1 & -1 & 0 & 0 & 0 \\ 0 & 0 & 0 & 1 & 0 & 0 \\ 0 & 0 & 0 & 0 & 1 & 0 \\ 0 & 0 & 0 & 0 & 0 & 1 \\ 0 & 0 & 0 & 0 & 0 & 0 \\ 0 & 0 & 0 & 0 & 0 & 0 \end{pmatrix}$$

which yields the equations

$$x_1 + x_2 - x_3 = 0, \quad x_4 = x_5 = x_6 = 0.$$

This means we have 2 free variables, and thus 2 independent eigenvectors. If we set $x_1 = r$ and $x_2 = s$, then an eigenvector is of the form

$$r\begin{pmatrix} 1 \\ 0 \\ 1 \\ 0 \\ 0 \\ 0 \end{pmatrix} \text{ or } s\begin{pmatrix} 0 \\ 1 \\ 1 \\ 0 \\ 0 \\ 0 \end{pmatrix}$$

and we set

$$\hat{v}_1 = \begin{pmatrix} 1 \\ 0 \\ 1 \\ 0 \\ 0 \\ 0 \end{pmatrix}, \ \hat{w}_1 = \begin{pmatrix} 0 \\ 1 \\ 1 \\ 0 \\ 0 \\ 0 \end{pmatrix}.$$

Each of these vectors will have its own chain.

Later in this chapter we show how to determine the structure of the chains. Since we have a 6×6 matrix, the sum of the lengths of the two chains must be 6. The possible lengths of the chains are: 5 and 1; 4 and 2; 3 and 3. Later, we will see how to determine that 4 and 2 is the correct pairing in this example.

At this point we do not know which eigenvector generates a chain of length 4 and which a chain of length 2. Said another way, at this point we do not know which of the sequences of diagrams

$$\hat{v}_4 \xrightarrow{A-2I} \hat{v}_3 \xrightarrow{A-2I} \hat{v}_2 \xrightarrow{A-2I} \hat{v}_1 \xrightarrow{A-2I} \hat{0} \qquad \hat{w}_2 \xrightarrow{A-2I} \hat{w}_1 \xrightarrow{A-2I} \hat{0}$$

or

$$\hat{w}_4 \xrightarrow{A-2I} \hat{w}_3 \xrightarrow{A-2I} \hat{w}_2 \xrightarrow{A-2I} \hat{w}_1 \xrightarrow{A-2I} \hat{0} \qquad \hat{v}_2 \xrightarrow{A-2I} \hat{v}_1 \xrightarrow{A-2I} \hat{0}$$

is correct.

We will guess that

$$\hat{v}_4 \xrightarrow{A-2I} \hat{v}_3 \xrightarrow{A-2I} \hat{v}_2 \xrightarrow{A-2I} \hat{v}_1 \xrightarrow{A-2I} \hat{0} \qquad \hat{w}_2 \xrightarrow{A-2I} \hat{w}_1 \xrightarrow{A-2I} \hat{0}$$

is correct and see whether the calculations validate our guess.

Consider the chain

$$\hat{v}_4 \xrightarrow{A-2I} \hat{v}_3 \xrightarrow{A-2I} \hat{v}_2 \xrightarrow{A-2I} \hat{v}_1 \xrightarrow{A-2I} \hat{0}.$$

If we can find a vector $\hat{v}_4$ for which $(A-2I)^3 \hat{v}_4 = \hat{v}_1$, then we know the whole chain because

$$(A-2I)\hat{v}_4 = \hat{v}_3 \text{ and } (A-2I)^2 \hat{v}_4 = \hat{v}_2.$$

Thus, we seek a vector

$$\begin{pmatrix} a \\ b \\ c \\ d \\ e \\ f \end{pmatrix} \text{ for which } (A-2I)^3 \begin{pmatrix} a \\ b \\ c \\ d \\ e \\ f \end{pmatrix} = \hat{v}_1 = \begin{pmatrix} 1 \\ 0 \\ 1 \\ 0 \\ 0 \\ 0 \end{pmatrix}.$$

Expanding

$$(A-2I)^3 \begin{pmatrix} a \\ b \\ c \\ d \\ e \\ f \end{pmatrix}$$

gives

$$(A-2I)^3 \begin{pmatrix} a \\ b \\ c \\ d \\ e \\ f \end{pmatrix} = \begin{pmatrix} 2d \\ 0 \\ 2d \\ 0 \\ 0 \end{pmatrix} = \hat{v}_1 = \begin{pmatrix} 1 \\ 0 \\ 1 \\ 0 \\ 0 \end{pmatrix}$$

so $d = 1/2$ and

$$\hat{v}_4 = \begin{pmatrix} 1/2 \\ 0 \\ 1/2 \\ 0 \\ 0 \\ 0 \end{pmatrix}; \ \hat{v}_3 = (A-2I)\,\hat{v}_4 = \begin{pmatrix} 0 \\ 1 \\ 3/2 \\ 0 \\ -1/2 \\ 0 \end{pmatrix}; \ \hat{v}_2 = (A-2I)^2 = \begin{pmatrix} 1/2 \\ 2 \\ 3/2 \\ 0 \\ -1 \\ 0 \end{pmatrix}.$$

If we had guessed that $\hat{w}_1$ gave rise to the length 4 chain then we would have tried to solve

$$(A-2I)^3 \begin{pmatrix} a \\ b \\ c \\ d \\ e \\ f \end{pmatrix} = \begin{pmatrix} 2d \\ 0 \\ 2d \\ 0 \\ 0 \end{pmatrix} = \hat{w}_1 = \begin{pmatrix} -1 \\ 1 \\ 0 \\ 0 \\ 0 \\ 0 \end{pmatrix}$$

for which there is no solution. This computation shows that there was really no need to guess which eigenvector yields the chain of length 4.

As an exercise, one can check that

$$\hat{w}_2 = \begin{pmatrix} 0 \\ -1 \\ -1 \\ 0 \\ 1 \\ 1 \end{pmatrix}.$$

Section 10.2 Constructing the Basis That Gives the Jordan Canonical Form for the Matrix

The procedure is similar to diagonalizing a matrix. We use the generalized eigenvectors as the basis, ordered by their chains.

In the previous example, we have the chain of generalized eigenvectors associated with the eigenvector $\hat{v}_1$ as

$$\hat{v}_1 = \begin{pmatrix} 1 \\ 0 \\ 1 \\ 0 \\ 0 \\ 0 \end{pmatrix}, \; \hat{v}_2 = \begin{pmatrix} 0 \\ 2 \\ 1 \\ 0 \\ -1 \\ 0 \end{pmatrix}, \; \hat{v}_3 = \begin{pmatrix} 0 \\ 0 \\ 1 \\ 0 \\ 0 \\ 0 \end{pmatrix}, \; \hat{v}_4 = \begin{pmatrix} 0 \\ 0 \\ -1 \\ 1 \\ 0 \\ 0 \end{pmatrix},$$

and the chain of generalized eigenvectors associated with the eigenvector $\hat{w}_1$ as

$$\hat{w}_1 = \begin{pmatrix} 0 \\ 1 \\ 1 \\ 0 \\ 0 \\ 0 \end{pmatrix}, \; \hat{w}_2 = \begin{pmatrix} 0 \\ -1 \\ -1 \\ 0 \\ 1 \\ 1 \end{pmatrix}.$$

The vectors in the two strings comprise a basis that converts the original matrix to the Jordan canonical form. If we concatenate the two strings and order the columns as $\left(\hat{v}_1, \hat{v}_2, \hat{v}_3, \hat{v}_4, \hat{w}_1, \hat{w}_2 \right)$ we obtain the matrix

$$S = \begin{pmatrix} 1 & 0 & 0 & 0 & 0 & 0 \\ 0 & 2 & 0 & 0 & 1 & -1 \\ 1 & 1 & 1 & -1 & 1 & -1 \\ 0 & 0 & 0 & 1 & 0 & 0 \\ 0 & -1 & 0 & 0 & 0 & 1 \\ 0 & 0 & 0 & 0 & 0 & 1 \end{pmatrix}$$

and

$$S^{-1}AS = \begin{pmatrix} 2 & 1 & 0 & 0 & 0 & 0 \\ 0 & 2 & 1 & 0 & 0 & 0 \\ 0 & 0 & 2 & 1 & 0 & 0 \\ 0 & 0 & 0 & 2 & 0 & 0 \\ 0 & 0 & 0 & 0 & 2 & 1 \\ 0 & 0 & 0 & 0 & 0 & 2 \end{pmatrix}$$

is in Jordan canonical form.

At this point we have shown how to find the Jordan canonical form and the associated basis when there is a single eigenvalue.

Example

In this example we compute the change of basis matrix in the case where there are two distinct eigenvalues.

Let

$$A = \begin{pmatrix} 6 & -1 & 1 & 0 & 1 \\ 2 & 5 & 3 & 0 & 1 \\ 0 & 1 & 5 & 0 & -1 \\ -2 & 0 & -2 & 6 & 0 \\ 0 & 0 & 0 & -2 & 4 \end{pmatrix}.$$

The characteristic polynomial of A is

$$P(\lambda) = (\lambda - 4)^2 (\lambda - 6)^3.$$

When $A - 6I$ is row reduced the result is

$$\begin{pmatrix} 1 & 0 & 1 & 0 & 0 \\ 0 & 1 & -1 & 0 & -1 \\ 0 & 0 & 0 & 1 & 1 \\ 0 & 0 & 0 & 0 & 0 \\ 0 & 0 & 0 & 0 & 0 \end{pmatrix}.$$

This yields the equations

$$x_1 + x_3 = 0; \quad x_2 - x_3 - x_5 = 0; \quad x_4 + x_5 = 0.$$

The free variables are x_3 and x_5, and we have

$$x_1 = -x_3; \quad x_2 = x_3 + x_5; \quad x_4 = -x_5. \quad x_3 = r, \quad x_5 = s.$$

Thus

$$\begin{pmatrix} x_1 \\ x_2 \\ x_3 \\ x_4 \\ x_5 \end{pmatrix} = \begin{pmatrix} -r \\ r+s \\ r \\ -s \\ s \end{pmatrix} = r \begin{pmatrix} -1 \\ 1 \\ 1 \\ 0 \\ 0 \end{pmatrix} + s \begin{pmatrix} 0 \\ 1 \\ 0 \\ -1 \\ 1 \end{pmatrix}$$

and two linearly independent eigenvector for $\lambda = 6$ are

$$\hat{w}_1 = \begin{pmatrix} -1 \\ 1 \\ 1 \\ 0 \\ 0 \end{pmatrix} \text{ and } \hat{v}_1 = \begin{pmatrix} 0 \\ 1 \\ 0 \\ -1 \\ 1 \end{pmatrix}.$$

Since the algebraic dimension of $\lambda = 6$ is 3, there will be a generalized eigenvector and exactly one of the equations

$$\left(A-6I\right)\hat{x}=\hat{v}_1 \text{ or } \left(A-6I\right)\hat{x}=\hat{w}_1$$

will have a solution.
   Now

$$A-6I=\begin{pmatrix} 0 & -1 & 1 & 0 & 1 \\ 2 & -1 & 3 & 0 & 1 \\ 0 & 1 & -1 & 0 & -1 \\ -2 & 0 & -2 & 0 & 0 \\ 0 & 0 & 0 & -2 & -2 \end{pmatrix}$$

and

$$\left(A-6I\right)\begin{pmatrix} 1/2 \\ 0 \\ 0 \\ -1/2 \\ 0 \end{pmatrix}=\begin{pmatrix} 0 & -1 & 1 & 0 & 1 \\ 2 & -1 & 3 & 0 & 1 \\ 0 & 1 & -1 & 0 & -1 \\ -2 & 0 & -2 & 0 & 0 \\ 0 & 0 & 0 & -2 & -2 \end{pmatrix}\begin{pmatrix} 1/2 \\ 0 \\ 0 \\ -1/2 \\ 0 \end{pmatrix}=\begin{pmatrix} 0 \\ 1 \\ 0 \\ -1 \\ 1 \end{pmatrix}=\hat{v}_1$$

and

$$\left(A-6I\right)\begin{pmatrix} a \\ b \\ c \\ d \\ e \end{pmatrix}=\hat{w}_1=\begin{pmatrix} -1 \\ 1 \\ 1 \\ 0 \\ 0 \end{pmatrix}$$

has no solution.
   Thus, we take

$$\hat{v}_2=\begin{pmatrix} 1/2 \\ 0 \\ 0 \\ -1/2 \\ 0 \end{pmatrix}$$

and we have the chain $\left(A-6I\right)\hat{v}_2=\hat{v}_1$.
   We now analyze the case $\lambda=4$.
   When $A-4I$ is row reduced the result is

$$\begin{pmatrix} 1 & 0 & 1 & 0 & 0 \\ 0 & 1 & 1 & 0 & 0 \\ 0 & 0 & 0 & 1 & 0 \\ 0 & 0 & 0 & 0 & 1 \\ 0 & 0 & 0 & 0 & 0 \end{pmatrix}$$

and

$$\hat{u}_1 = \begin{pmatrix} -1 \\ -1 \\ 1 \\ 0 \\ 0 \end{pmatrix}$$

is a basis for the $\lambda = 4$ eigenspace.

One can check that

$$(A - 4I) \begin{pmatrix} 0 \\ 0 \\ 0 \\ 0 \\ -1 \end{pmatrix} = \begin{pmatrix} -1 \\ -1 \\ 1 \\ 0 \\ 0 \end{pmatrix}$$

So

$$\hat{u}_2 = \begin{pmatrix} 0 \\ 0 \\ 0 \\ 0 \\ -1 \end{pmatrix}, \quad \hat{u}_1 = \begin{pmatrix} -1 \\ -1 \\ 1 \\ 0 \\ 0 \end{pmatrix}$$

is a chain of generalized eigenvectors for $\lambda = 4$.

A change of basis matrix that yields the Jordan canonical form is $(\hat{v}_1, \hat{v}_2, \hat{w}_1, \hat{u}_1, \hat{u}_2)$ so

$$S = \begin{pmatrix} 0 & \dfrac{1}{2} & -1 & -1 & 0 \\ 1 & 0 & 1 & -1 & 0 \\ 0 & 0 & 1 & 1 & 0 \\ -1 & -\dfrac{1}{2} & 0 & 0 & 0 \\ 1 & 0 & 0 & 0 & -1 \end{pmatrix}$$

and

$$S^{-1}AS = \begin{pmatrix} 6 & 1 & 0 & 0 & 0 \\ 0 & 6 & 0 & 0 & 0 \\ 0 & 0 & 6 & 0 & 0 \\ 0 & 0 & 0 & 4 & 1 \\ 0 & 0 & 0 & 0 & 4 \end{pmatrix}.$$

Section 10.3 The Theory That Justifies the Algorithms We Have Developed

Let $T : V \to V$ be a linear transformation and let A be the matrix associated with T with respect to some basis. We will also refer to A as a linear transformation. A non-zero vector $\hat{v}$ is an eigenvector of A with eigenvalue λ if $A\hat{v} = \lambda\hat{v}$. This is the same as saying $(A - \lambda I)\hat{v} = \hat{0}$ or that $\hat{v}$ is in the null space of $(A - \lambda I)$. The last statement is denoted by

$$\hat{v} \in \mathcal{N}(A - \lambda I).$$

The null space of a linear transformation is also called the kernel of the transformation.

We recall some facts from earlier chapters.

1. The characteristic polynomial of A, denoted $p_A(\lambda)$ or $p(A)$, is the determinant of $(A - \lambda I)$.
2. The eigenvalues of A are the values of λ for which the characteristic polynomial is zero.
3. If

$$p(A) = \prod_{j=1}^{k} (x - \lambda_j)^{n_j}$$

where the λ_j are distinct, then n_j is the algebraic dimension of λ_j and $n_1 + \cdots + n_k = n$, where n is the dimension of the vector space.

4. The geometric dimension of λ_j is the dimension of $\mathcal{N}(A - \lambda_j)$ which is the number of linearly independent eigenvectors with eigenvalue λ_j.
5. The geometric dimension of λ_j is less than or equal to the algebraic dimension of λ_j.
6. There will be a basis of V consisting of the eigenvectors of A if and only if the geometric dimension of λ_j is equal to the algebraic dimension of λ_j, for each eigenvalue λ_j.
7. Eigenvectors for different eigenvalues are linearly independent.

A proof of Statement 5 is as follows.

Theorem 10.1:

The geometric multiplicity of an eigenvalue is less than or equal to the algebraic multiplicity.

Proof:

Let $T : V \rightarrow V$ be a linear transformation and let λ be an eigenvalue of T with geometric multiplicity m. Let

$$\hat{v}_1, \ldots, \hat{v}_m$$

be linearly independent eigenvectors associated with λ; that is, $T\hat{v}_i = \lambda \hat{v}_i$.

Extend $\{\hat{v}_1, \ldots, \hat{v}_m\}$ to a basis of V by adding the vectors $\hat{u}_{m+1}, \ldots, \hat{u}_n$.

To get the matrix associated with T with respect to this basis, we note that for $i = 1, \ldots, m$

$$T\hat{v}_i = \lambda \hat{v}_i = 0\hat{v}_1 + \cdots + 0\hat{v}_{i-1} + \lambda \hat{v}_i + 0\hat{v}_{i+1} + \cdots + 0\hat{u}_n$$

so the ith column of this matrix for $i = 1, \ldots, m$ is

$$\begin{pmatrix} 0 \\ \vdots \\ 0 \\ \lambda \\ 0 \\ \vdots \\ 0 \end{pmatrix}$$

where λ occurs in the ith position.

Thus the matrix of T with respect to this basis is

$$\begin{pmatrix} \lambda I_m & B \\ 0 & C \end{pmatrix}$$

and

$$p_T(x) = (x - \lambda)^m p_C(x)$$

so the algebraic multiplicity of λ is at least m.

Theorem 10.2: Schur's Theorem

By a change of basis, a square matrix may be put in an upper triangular form where the diagonal entries are the eigenvalues, repeated according to multiplicity.

A proof of Schur's Theorem may be found in *An Introduction to Linear Algebra*, by Gilbert Strang.

From Schur's Theorem, there exist a unitary matrix U and an upper triangular matrix M such that $A = UMU^T$ and

$$xI - A = xI - UMU^T = U(xI)U^T - UMU^T = U(xI - M)U^T$$

so

$$\det(xI - A) = \det\left[U(xI - M)U^T\right]$$
$$= \det(U)\det(xI - M)\det(U^T) = \det(xI - M).$$

Thus, the characteristic polynomial of $(xI - A)$ equals the characteristic polynomial of $(xI - M)$.

If the eigenvalues of A are $\lambda_1, \ldots, \lambda_n$ (including multiplicities) then the characteristic polynomial of $(xI - A)$ is

$$(x - \lambda_1)\cdots(x - \lambda_n).$$

Since M is a triangular matrix, $(xI - M)$ is a triangular matrix and the characteristic polynomial of $(xI - M)$ is

$$(x - m_1)\cdots(x - m_n)$$

where the m_i's are the diagonal entries of M. Thus, we may choose M to be

$$\begin{pmatrix} \lambda_1 & * & * & * \\ 0 & \lambda_2 & * & * \\ \vdots & \vdots & \ddots & * \\ 0 & 0 & \cdots & \lambda_n \end{pmatrix}.$$

Theorem 10.3:

If A is an $n \times n$ matrix and λ is an eigenvalue of A with algebraic multiplicity k, then for $1 \leq j \leq k$ the dimension of $N\left[\left(A - \lambda I\right)^{j}\right]$ is at least j but never more than k, and for $j \geq k$ the dimension of $N\left[\left(A - \lambda I\right)^{j}\right]$ is k.

The ideas in the following example can be formalized to give proof of this result.

Example

Suppose that

$$
A = \begin{pmatrix}
\lambda & a_{12} & a_{13} & a_{14} & a_{15} & a_{16} \\
0 & \lambda & a_{23} & a_{24} & a_{25} & a_{26} \\
0 & 0 & \lambda & a_{34} & a_{35} & a_{36} \\
0 & 0 & 0 & \lambda & a_{45} & a_{46} \\
0 & 0 & 0 & 0 & \alpha & a_{56} \\
0 & 0 & 0 & 0 & 0 & \beta
\end{pmatrix}
$$

where $\alpha, \beta \neq \lambda$.
   Then

$$
A - \lambda I = \begin{pmatrix}
0 & a_{12} & a_{13} & a_{14} & a_{15} & a_{16} \\
0 & 0 & a_{23} & a_{24} & a_{25} & a_{26} \\
0 & 0 & 0 & a_{34} & a_{35} & a_{36} \\
0 & 0 & 0 & 0 & a_{45} & a_{46} \\
0 & 0 & 0 & 0 & \gamma & a_{56} \\
0 & 0 & 0 & 0 & 0 & \delta
\end{pmatrix} = \begin{pmatrix} D & B \\ O & \mathcal{E} \end{pmatrix}
$$

where $\gamma = \alpha - \lambda \neq 0$, $\delta = \beta - \lambda \neq 0$ and

$$
D = \begin{pmatrix}
0 & a_{12} & a_{13} & a_{14} \\
0 & 0 & a_{23} & a_{24} \\
0 & 0 & 0 & a_{34} \\
0 & 0 & 0 & 0
\end{pmatrix}, \quad B = \begin{pmatrix}
a_{15} & a_{16} \\
a_{25} & a_{26} \\
a_{35} & a_{36} \\
a_{45} & a_{46}
\end{pmatrix}
$$

$$
O = \begin{pmatrix}
0 & 0 & 0 & 0 \\
0 & 0 & 0 & 0
\end{pmatrix}, \quad \mathcal{E} = \begin{pmatrix} \gamma & a_{56} \\ 0 & \delta \end{pmatrix}.
$$

Thus,

$$
(A - \lambda I)^{k} = \begin{pmatrix} D & B \\ O & \mathcal{E} \end{pmatrix}^{k} = \begin{pmatrix} D^{k} & * \\ O & \mathcal{E}^{k} \end{pmatrix}
$$

where * is a 4×2 matrix whose entries are inconsequential.
   Now

$$
\mathcal{E}^{k} = \begin{pmatrix} \gamma^{k} & \# \\ 0 & \delta^{k} \end{pmatrix}
$$

and $\gamma^k \neq 0$, $\delta^k \neq 0$.

Also,

$$D^4 = \begin{pmatrix} 0 & a_{12} & a_{13} & a_{14} \\ 0 & 0 & a_{23} & a_{24} \\ 0 & 0 & 0 & a_{34} \\ 0 & 0 & 0 & 0 \end{pmatrix}^4 = \begin{pmatrix} 0 & 0 & 0 & 0 \\ 0 & 0 & 0 & 0 \\ 0 & 0 & 0 & 0 \\ 0 & 0 & 0 & 0 \end{pmatrix}.$$

These two facts together show that the dimension of $\mathcal{N}\left((A - \lambda I)^k\right)$ is 4 if $k \geq 4$.

For $k = 2, 3$ we show that the dimension of $\mathcal{N}\left((A - \lambda I)^k\right)$ depends on the non-zero entries of D. If

$$D = \begin{pmatrix} 0 & a & b & c \\ 0 & 0 & d & e \\ 0 & 0 & 0 & f \\ 0 & 0 & 0 & 0 \end{pmatrix}$$

then

$$D^2 = \begin{pmatrix} 0 & 0 & ad & ae+bc \\ 0 & 0 & 0 & df \\ 0 & 0 & 0 & 0 \\ 0 & 0 & 0 & 0 \end{pmatrix}$$

so that the dimension of

$$\mathcal{N}\left((A - \lambda I)^2\right)$$

is at least 2, but could be 3 or 4 depending on the entries of D.

Likewise,

$$D^3 = \begin{pmatrix} 0 & 0 & 0 & adf \\ 0 & 0 & 0 & 0 \\ 0 & 0 & 0 & 0 \\ 0 & 0 & 0 & 0 \end{pmatrix}$$

so that the dimension of

$$\mathcal{N}\left((A - \lambda I)^3\right)$$

is at least 3, but could be 4.

Corollary:

If $A : V \to V$ is a linear transformation and λ_i is an eigenvalue of A, then the dimension of V_{λ_i} = the algebraic dimension of λ_i.

Definition:

Suppose that $A : V \to V$ is a linear transformation and W is a subspace of V. We say that A is invariant on W if $A(W) \subset W$.

Theorem 10.4:

If $A : V \to V$ is a linear transformation and λ is an eigenvalue of A, then for any positive integer k

(a.) A is invariant on the null space $\mathcal{N}\left[\left(A - \lambda I\right)^k\right]$.

(b.) A is invariant on range space $\mathcal{R}\left[\left(A - \lambda I\right)^k\right]$

Proof:

(a.) Suppose

$$\hat{v} \in \mathcal{N}\left[\left(A - \lambda I\right)^k\right].$$

We must show

$$A\hat{v} \in \mathcal{N}\left[\left(A - \lambda I\right)^k\right].$$

Now

$$\left(A - \lambda I\right)^k \hat{v} = \hat{0}, \text{ so } A\left[\left(A - \lambda I\right)^k \hat{v}\right] = \hat{0},$$

but then

$$\hat{0} = A\left[\left(A - \lambda I\right)^k \hat{v}\right] = \left[\left(A - \lambda I\right)^k\right]\left(A\hat{v}\right)$$

so

$$A\hat{v} \in \mathcal{N}\left[\left(A - \lambda I\right)^k\right].$$

(b.) Suppose

$$\hat{v} \in \mathcal{R}\left[\left(A - \lambda I\right)^k\right].$$

We must show

$$A\hat{v} \in \mathcal{R}\left[(A-\lambda I)^k\right].$$

Since $\hat{v} \in \mathcal{R}\left[(A-\lambda I)^k\right]$, and $\mathcal{R}\left[(A-\lambda I)^k\right]$ is a subspace, then $\lambda\hat{v} \in \mathcal{R}\left[(A-\lambda I)^k\right]$.

We claim $(A-\lambda I)\hat{v} \in \mathcal{R}\left[(A-\lambda I)^k\right]$. Since $\hat{v} \in \mathcal{R}\left[(A-\lambda I)^k\right]$, there is a $\hat{w}$ with $\hat{v} = (A-\lambda I)^k\hat{w}$.

So

$$(A-\lambda I)v = (A-\lambda I)\left[(A-\lambda I)^k\hat{w}\right]$$
$$= (A-\lambda I)^k\left[(A-\lambda I)\hat{w}\right] \in \mathcal{R}\left[(A-\lambda I)^k\right].$$

Since $\mathcal{R}\left[(A-\lambda I)^k\right]$ is a subspace with $\lambda\hat{v} \in \mathcal{R}\left[(A-\lambda I)^k\right]$ and $(A-\lambda I)\hat{v} \in \mathcal{R}\left[(A-\lambda I)^k\right]$, then $(A-\lambda I)\hat{v} + \lambda\hat{v} = A\hat{v} \in \mathcal{R}\left[(A-\lambda I)^k\right]$ so A is invariant on $\mathcal{R}\left[(A-\lambda I)^k\right]$.

Corollary:

If $A:V \to V$ is a linear transformation and λ is an eigenvalue of A, then A is invariant on V_λ, where V_λ is the generalized eigenspace of V_λ.

Proof:

This follows from part (a) of the theorem.

Theorem 10.5:

Suppose $A:V \to V$ is a linear transformation with λ_1 and λ_2 eigenvalues of A, with $\lambda_1 \neq \lambda_2$. Let $V(\lambda_1) =$ eigenspace of λ_1 and $V_{\lambda_2} =$ generalized eigenspace of λ_2.
Then

$$V(\lambda_1) \cap V_{\lambda_2} = \{\hat{0}\}.$$

Proof:

Suppose that $\hat{v} \neq \hat{0}$ and $\hat{v} \in V(\lambda_1) \cap V_{\lambda_2}$. Then $A\hat{v} = \lambda_1\hat{v}$ and there is a $k > 1$ with $\left((A-\lambda_2 I)\right)^{k-1}\hat{v} \neq \hat{0}$ and $(A-\lambda_2 I)^k\hat{v} = \hat{0}$. Now

$$\left(A-\lambda_2 I\right)\hat{v}=A\hat{v}-\lambda_2\hat{v}=\left(\lambda_1-\lambda_2\right)\hat{v}$$

$$\left(A-\lambda_2 I\right)^2\hat{v}=\left(A-\lambda_2 I\right)\left[\left(A-\lambda_2 I\right)\hat{v}\right]=\left(A-\lambda_2 I\right)\left(\lambda_1-\lambda_2\right)\hat{v}$$

$$=\left(\lambda_1-\lambda_2\right)\left(A-\lambda_2 I\right)\hat{v}=\left(\lambda_1-\lambda_2\right)^2\hat{v}$$

$$\vdots$$

$$\left(A-\lambda_2 I\right)^k\hat{v}=\left(\lambda_1-\lambda_2\right)^k\hat{v}$$

but

$$\left(A-\lambda_2 I\right)^k\hat{v}=\hat{0}\text{ and }\left(\lambda_1-\lambda_2\right)^k\hat{v}\neq\hat{0}$$

so we have a contradiction.

Theorem 10.6:

Suppose $A:V\to V$ is a linear transformation with λ_1 and λ_2 eigenvalues of A, with $\lambda_1\neq\lambda_2$. Let $V_{\lambda_i}=$ generalized eigenspace of λ_i.
   Then

$$V_{\lambda_1}\cap V_{\lambda_2}=\left\{\hat{0}\right\}.$$

Proof:

Generalized eigenspaces are invariant under A so if $\hat{v}$ is in $\mathcal{N}\left[\left(A-\lambda_1 I\right)^{d_1}\right]$ then so are $A\hat{v}$ and $c\hat{v}$ for any constant c. Thus

$$\left(A-\lambda_2 I\right)\hat{v}\in\mathcal{N}\left[\left(A-\lambda_1 I\right)^{d_1}\right].$$

Suppose $\hat{v}\in\mathcal{N}\left[\left(A-\lambda_2 I\right)^{d_2}\right]$. Then

$$\hat{0}=\left(A-\lambda_2 I\right)^{d_2}\hat{v}=\left(A-\lambda_2 I\right)\left(A-\lambda_2 I\right)^{d_2-1}\hat{v}$$

$$=A\left(A-\lambda_2 I\right)^{d_2-1}\hat{v}-\lambda_2\left(A-\lambda_2 I\right)^{d_2-1}\hat{v}$$

so

$$A\left(A-\lambda_2 I\right)^{d_2-1}\hat{v}=\lambda_2\left(A-\lambda_2 I\right)^{d_2-1}\hat{v}.$$

Now

$$\left(A-\lambda_1 I\right)^{d_1}\left(A-\lambda_2 I\right)^{d_2-1}\hat{v}=\left(A-\lambda_1 I\right)^{d_1-1}\left(A-\lambda_1 I\right)\left(A-\lambda_2 I\right)^{d_2-1}\hat{v}$$

$$=\left(A-\lambda_1 I\right)^{d_1-1}\left[\lambda_2\left(A-\lambda_2 I\right)^{d_2-1}-\lambda_1\left(A-\lambda_2 I\right)^{d_2-1}\right]\hat{v}$$

$$=\left(\lambda_2-\lambda_1\right)\left(A-\lambda_1 I\right)^{d_1-1}\left(A-\lambda_2 I\right)^{d_2-1}\hat{v}.$$

Repeating the process several times, we get

$$\left(A-\lambda_1 I\right)^{d_1}\left(A-\lambda_2 I\right)^{d_2-1}\hat{v}=\left(\lambda_2-\lambda_1\right)^{d_1}\left(A-\lambda_2 I\right)^{d_2-1}\hat{v}.$$

Now suppose

$$\hat{v}\in\mathcal{N}\left[\left(A-\lambda_1 I\right)^{d_1}\right]\cap\mathcal{N}\left[\left(A-\lambda_2 I\right)^{d_2}\right].$$

Then

$$\hat{0}=\left(A-\lambda_1 I\right)^{d_1}\left(A-\lambda_2 I\right)^{d_2-1}\hat{v}=\left(\lambda_2-\lambda_1\right)^{d_1}\left(A-\lambda_2 I\right)^{d_2-1}\hat{v}$$

Since $\lambda_1\neq\lambda_2$, we have $\left(A-\lambda_2 I\right)^{d_2-1}\hat{v}=\hat{0}$.

Thus

$$\hat{v}\in\mathcal{N}\left[\left(A-\lambda_2 I\right)^{d_2-1}\right].$$

The argument can be repeated to get

$$\hat{v}\in\mathcal{N}\left(A-\lambda_2 I\right).$$

Thus

$$\hat{v}\in\mathcal{N}\left[\left(A-\lambda_1 I\right)^{d_1}\right]\cap\mathcal{N}\left(A-\lambda_2 I\right)$$

and so, by the previous theorem, $\hat{v}=\hat{0}$.

Since the dimensions of the generalized eigenspaces sum to the dimension of the vector space, from the previous theorem we get the following result.

Theorem 10.7:

Suppose $A:V\to V$ is a linear transformation with distinct eigenvalues $\lambda_1,\ldots,\lambda_k$. Then

$$V=V_{\lambda_1}\oplus\cdots\oplus V_{\lambda_k}.$$

Theorem 10.7 supplies a major fact that we need to justify the construction of the Jordan canonical form. We now turn to the problem of determining the block structure of the Jordan canonical form.

Section 10.4 Determining the Block Structure of the Jordan Canonical Form

Theorem 10.8:

Let $A:V\to V$ be a linear transformation with eigenvalue λ. Let

$$d_k = \begin{cases} 0 & \text{if } k = 0 \\ \text{dimension of } N\left(A - \lambda I\right)^k & \text{if } k > 0 \end{cases}.$$

Then

(a.) the number of Jordan blocks for λ of size k or larger is $d_k - d_{k-1}$.
(b.) The number of blocks of exactly size k is the number of blocks of size k or larger minus the number of blocks of size $(k-1)$ or larger.

This may be easier to remember in the following description:

$d_1 - d_0 = d_1 =$ number of eigenvectors $=$ number of Jordan blocks
$d_2 - d_1 =$ number of Jordan blocks of size 2 or larger
$d_3 - d_2 =$ number of Jordan blocks of size 3 or larger

$$\vdots$$

An Algorithm for Determining the Jordan Blocks for a Particular Eigenvalue λ

1. Determine the eigenvalue λ.
   The crucial factor in determining the number and sizes of chains of generalized eigenvectors is the dimension of $N\left(A - \lambda I\right)^k$.
2. Determine the dimension of $N\left(A - \lambda I\right)$.
   To do this, row reduce $A - \lambda I$. The rank of $(A - \lambda I)$ is the number of leading 1s in the row reduced form of $(A - \lambda I)$. The dimension of $N\left(A - \lambda I\right)$ (or the nullity of $A - \lambda I$) is the number of columns of $(A - \lambda I)$ – rank of $(A - \lambda I)$
   So
   dimension of $N\left(A - \lambda I\right) =$ Number of columns of $\left(A - \lambda I\right)$ minus number of leading 1s in the row reduced form of $A - \lambda I$.
   This number is d_1 which is the number of chains there are.
3. Determine the dimension of $N\left[\left(A - \lambda I\right)^2\right]$. To do this apply the procedures of step 2 to $(A - \lambda I)^2$. Then the dimension of $N\left[\left(A - \lambda I\right)^2\right] =$ Number of columns of $\left(A - \lambda I\right)^2$ – number of leading 1s in the row reduced form of $(A - \lambda I)^2$.
   This number is d_2.
   The number of chains of length 2 or more is $d_2 - d_1$.
4. Continue the process to find the dimensions of $N\left(A - \lambda I\right)^k$ for all k.
   The number of chains of length k or more is $d_k - d_{k-1}$.
   It may be helpful to draw a diagram of what has been determined. Suppose

$$d_1 = 3, d_2 = 5, d_3 = 6, d_4 = 7$$

Since $d_1 = 3$, there are 3 chains, which we could represent as
*
*
*

Since $d_2 = 5$, there are $5 - 3 = 2$ chains of size greater than or equal to 2 and 1 chain of length 1.

We could represent our knowledge to this point as

Chain of length 2 or more   * *
Chain of length 2 or more   * *
Chain of length 1   *

Since $d_3 = 6$, there are $6 - 5 = 1$ chain of size greater than or equal to 3 and 1 chain of length 2.

We could represent our knowledge to this point as

    * * *
     * *
      *

Since $d_4 = 6$, there are $7 - 6 = 1$ chain of size greater than or equal to 4

We could represent our knowledge to this point as

  * * * *
   * *
    *

If this is the termination of the process, we see there are 3 chains; one of length 1, one of length 2 and one of length 4.

Example

Suppose that $\dim \mathcal{N}(A - \lambda I) = 3$, $\dim \mathcal{N}(A - \lambda I)^2 = 5$, $\dim \mathcal{N}(A - \lambda I)^3 = 6$.

That is,

$$d_1 = 3, d_2 = 5, d_3 = 6.$$

Since $d_1 = 3$ there are three Jordan blocks.

Because $d_2 - d_1 = 2$ there are two Jordan blocks of size 2 or larger.
Because $d_3 - d_2 = 1$ there is one Jordan block of size 3 or larger.
Thus we have the following diagram

$$\hat{w}_1 \xrightarrow{A-\lambda I} \hat{u}_1 \xrightarrow{A-\lambda I} \hat{v}_1 \xrightarrow{A-\lambda I} \hat{0}$$
$$\hat{u}_2 \xrightarrow{A-\lambda I} \hat{v}_2 \xrightarrow{A-\lambda I} \hat{0}$$
$$\hat{v}_3 \xrightarrow{A-\lambda I} \hat{0}$$

Thus, there is one block of size 3, one block of size 2 and one block of size 1.

Exercises

For the matrices below

(i) find the eigenvalues.
(ii) For each eigenvalue, find the chain(s) of generalized eigenvectors.
(iii) Give the block structure of the Jordan canonical form.
(iv) Find the matrix that converts the given matrix to the Jordan canonical form.
(v) Find the Jordan canonical form.

1. $\begin{pmatrix} 2 & -3 \\ 3 & -4 \end{pmatrix}$

2. $\begin{pmatrix} -1 & -1 & 0 \\ 0 & -1 & -2 \\ 0 & 0 & -1 \end{pmatrix}$

3. $\begin{pmatrix} -2 & 2 & 1 \\ -7 & 4 & 2 \\ 5 & 0 & 0 \end{pmatrix}$

4. $\begin{pmatrix} 0 & 1 & 1 \\ 2 & 1 & -1 \\ -6 & -5 & -3 \end{pmatrix}$

5. $\begin{pmatrix} 2 & 0 & 0 & 0 \\ 0 & 2 & 1 & 0 \\ 0 & 0 & 2 & 0 \\ 1 & 0 & 0 & 2 \end{pmatrix}$

6. $\begin{pmatrix} 0 & 1 & 0 & 0 \\ 11 & 6 & -4 & -4 \\ 22 & 15 & -8 & -9 \\ -3 & -2 & 1 & 2 \end{pmatrix}$

7. $\begin{pmatrix} 4 & -1 & -2 & -2 \\ 7 & -4 & -12 & 10 \\ -6 & 6 & 13 & -8 \\ -3 & 3 & 5 & -1 \end{pmatrix}$

8. $\begin{pmatrix} 3 & 6 & 0 & 0 & 4 \\ 0 & 2 & 0 & 0 & 0 \\ 0 & 0 & 1 & 0 & 0 \\ 0 & 0 & 0 & 1 & 0 \\ 0 & 0 & 0 & 0 & 1 \end{pmatrix}$

Chapter 11

Applications of the Jordan Canonical Form

Section 11.1 Finding the Square Root of a Matrix

For B an $n \times n$ matrix with positive eigenvalues, the Jordan canonical form can be used to find a matrix A for which $A^2 = B$. The procedure is as follows:

(a.) Find the Jordan canonical blocks for the matrix B. For a single Jordan block, this will be of the form $\lambda I + N$, where $\lambda > 0$ and N is nilpotent; that is, there is a positive integer k for which N^k is the zero matrix.

(b.) Find the Maclaurin series for $\sqrt{\lambda + x}$

(c.) In the Maclaurin series for $\sqrt{\lambda + x}$, replace λ by λI and replace x by N.
   1. Since N is nilpotent, there is a positive integer k for which N^k is the zero matrix, so the series has only finitely many terms.

(d.) For P the matrix whose columns are the generalized eigenvectors, the matrix can be converted back to the standard basis via

$$P\left(\sqrt{\lambda I + N}\right)P^{-1}.$$

Example

We find the square root of

$$A = \begin{pmatrix} -1 & -2 & -1 \\ 2 & 4 & -1 \\ 6 & 3 & 6 \end{pmatrix}.$$

The only eigenvalue is $\lambda = 3$, and a basis for the eigenspace is

$$\hat{v}_1 = \begin{pmatrix} -1 \\ 2 \\ 0 \end{pmatrix}.$$

To find the generalized eigenvectors, we solve $(A - 3I)\hat{v}_2 = \hat{v}_1$, and find that

$$\hat{v}_2 = \begin{pmatrix} 1 \\ -1 \\ -1 \end{pmatrix}.$$

We also solve $(A - 3I)\hat{v}_3 = \hat{v}_2$, and find that

$$\hat{v}_3 = \begin{pmatrix} -2/3 \\ 2/3 \\ 1/3 \end{pmatrix}.$$

Thus,

$$P = \begin{pmatrix} -1 & 1 & -2/3 \\ 2 & -1 & 2/3 \\ 0 & -1 & 1/3 \end{pmatrix}$$

and

$$P^{-1}AP = \begin{pmatrix} -1 & 1 & -\dfrac{2}{3} \\ 2 & -1 & \dfrac{2}{3} \\ 0 & -1 & \dfrac{1}{3} \end{pmatrix}^{-1} \begin{pmatrix} -1 & -2 & -1 \\ 2 & 4 & -1 \\ 6 & 3 & 6 \end{pmatrix} \begin{pmatrix} -1 & 1 & -\dfrac{2}{3} \\ 2 & -1 & \dfrac{2}{3} \\ 0 & -1 & \dfrac{1}{3} \end{pmatrix}$$

$$= \begin{pmatrix} 3 & 1 & 0 \\ 0 & 3 & 1 \\ 0 & 0 & 3 \end{pmatrix} = 3I + N$$

where

$$N = \begin{pmatrix} 0 & 1 & 0 \\ 0 & 0 & 1 \\ 0 & 0 & 0 \end{pmatrix}.$$

We note that

$$N^2 = \begin{pmatrix} 0 & 0 & 1 \\ 0 & 0 & 0 \\ 0 & 0 & 0 \end{pmatrix} \text{ and } N^3 = \begin{pmatrix} 0 & 0 & 0 \\ 0 & 0 & 0 \\ 0 & 0 & 0 \end{pmatrix}.$$

The first three terms of the Maclaurin series of $\sqrt{3+x}$ are

$$\sqrt{3} + \frac{\sqrt{3}}{6}x - \frac{\sqrt{3}}{72}x^2$$

so we form the sum

$$\sqrt{3}I + \frac{\sqrt{3}}{6}N - \frac{\sqrt{3}}{72}N^2 = \sqrt{3} \begin{pmatrix} 1 & 1/6 & -1/72 \\ 0 & 1 & 1/6 \\ 0 & 0 & 1 \end{pmatrix}.$$

One can check that it is indeed the case that

$$\left[\sqrt{3}\begin{pmatrix}1 & 1/6 & -1/72 \\ 0 & 1 & 1/6 \\ 0 & 0 & 1\end{pmatrix}\right]^2 = \begin{pmatrix}3 & 1 & 0 \\ 0 & 3 & 1 \\ 0 & 0 & 3\end{pmatrix}.$$

Thus, we have found $\sqrt{A}$ with respect to the basis $\{\hat{v}_1,\hat{v}_2,\hat{v}_3\}$. To find $\sqrt{A}$ with respect to the standard basis, we compute

$$P\left[\sqrt{3}\begin{pmatrix}1 & 1/6 & -1/72 \\ 0 & 1 & 1/6 \\ 0 & 0 & 1\end{pmatrix}\right]P^{-1} = \begin{pmatrix}\sqrt{3}/4 & -3\sqrt{3}/8 & -5\sqrt{3}/24 \\ \sqrt{3}/2 & 5\sqrt{3}/4 & -\sqrt{3}/12 \\ \sqrt{3} & \sqrt{3}/2 & 3\sqrt{3}/2\end{pmatrix}.$$

One can check it is indeed the case that

$$\begin{pmatrix}\sqrt{3}/4 & -3\sqrt{3}/8 & -5\sqrt{3}/24 \\ \sqrt{3}/2 & 5\sqrt{3}/4 & -\sqrt{3}/12 \\ \sqrt{3} & \sqrt{3}/2 & 3\sqrt{3}/2\end{pmatrix}^2 = \begin{pmatrix}-1 & -2 & -1 \\ 2 & 4 & -1 \\ 6 & 3 & 6\end{pmatrix}.$$

One can extend this idea to finding a function of A as long as the function has a Taylor series. Note that applying the Taylor series to the original matrix will normally not be beneficial if the matrix is not nilpotent.

Example

In this example, we find the square root of a matrix that has two eigenvalues.

Let

$$A = \begin{pmatrix}3 & 1 & -1 & 1 & -2 \\ 0 & 4 & 0 & 0 & -1 \\ 0 & 1 & 3 & -1 & 0 \\ 0 & 2 & 1 & 1 & 0 \\ 0 & 1 & 0 & 0 & 2\end{pmatrix}.$$

The Jordan canonical form of A is

$$\begin{pmatrix}3 & 1 & 0 & 0 & 0 \\ 0 & 3 & 0 & 0 & 0 \\ 0 & 0 & 3 & 0 & 0 \\ 0 & 0 & 0 & 2 & 1 \\ 0 & 0 & 0 & 0 & 2\end{pmatrix}.$$

We consider the Jordan blocks separately.

Let

$$B = \begin{pmatrix}3 & 1 & 0 \\ 0 & 3 & 0 \\ 0 & 0 & 3\end{pmatrix} = \begin{pmatrix}3 & 1 & 0 \\ 0 & 3 & 0 \\ 0 & 0 & 3\end{pmatrix} + \begin{pmatrix}0 & 1 & 0 \\ 0 & 0 & 0 \\ 0 & 0 & 0\end{pmatrix}$$

and

$$C = \begin{pmatrix} 2 & 1 \\ 0 & 2 \end{pmatrix} = \begin{pmatrix} 2 & 0 \\ 0 & 2 \end{pmatrix} + \begin{pmatrix} 0 & 1 \\ 0 & 0 \end{pmatrix}.$$

For the matrix B,

$$\lambda I = \sqrt{3} \begin{pmatrix} 1 & 0 & 0 \\ 0 & 1 & 0 \\ 0 & 0 & 1 \end{pmatrix} = \begin{pmatrix} \sqrt{3} & 0 & 0 \\ 0 & \sqrt{3} & 0 \\ 0 & 0 & \sqrt{3} \end{pmatrix} \text{ and } N = \begin{pmatrix} 0 & 1 & 0 \\ 0 & 0 & 0 \\ 0 & 0 & 0 \end{pmatrix}.$$

The Taylor expansion of $\sqrt{3+x}$ to three terms is

$$\sqrt{3} + \frac{\sqrt{3}x}{6} - \frac{\sqrt{3}x^2}{72}.$$

Thus we compute

$$\begin{pmatrix} \sqrt{3} & 0 & 0 \\ 0 & \sqrt{3} & 0 \\ 0 & 0 & \sqrt{3} \end{pmatrix} + \frac{1}{6} \begin{pmatrix} \sqrt{3} & 0 & 0 \\ 0 & \sqrt{3} & 0 \\ 0 & 0 & \sqrt{3} \end{pmatrix} \begin{pmatrix} 0 & 1 & 0 \\ 0 & 0 & 0 \\ 0 & 0 & 0 \end{pmatrix}$$

$$- \frac{1}{72} \begin{pmatrix} \sqrt{3} & 0 & 0 \\ 0 & \sqrt{3} & 0 \\ 0 & 0 & \sqrt{3} \end{pmatrix} \begin{pmatrix} 0 & 1 & 0 \\ 0 & 0 & 0 \\ 0 & 0 & 0 \end{pmatrix}^2 = \begin{pmatrix} \sqrt{3} & \frac{\sqrt{3}}{6} & 0 \\ 0 & \sqrt{3} & 0 \\ 0 & 0 & \sqrt{3} \end{pmatrix}$$

and note that

$$\begin{pmatrix} \sqrt{3} & \frac{\sqrt{3}}{6} & 0 \\ 0 & \sqrt{3} & 0 \\ 0 & 0 & \sqrt{3} \end{pmatrix}^2 = \begin{pmatrix} 3 & 1 & 0 \\ 0 & 3 & 0 \\ 0 & 0 & 3 \end{pmatrix}.$$

We follow the same procedure for $C = \begin{pmatrix} 2 & 1 \\ 0 & 2 \end{pmatrix}$.

The Taylor expansion of $\sqrt{2+x}$ to two terms is

$\sqrt{2} + \frac{\sqrt{2}x}{4}$ so with $\lambda I = \begin{pmatrix} \sqrt{2} & 0 \\ 0 & \sqrt{2} \end{pmatrix}$ and $N = \begin{pmatrix} 0 & 1 \\ 0 & 0 \end{pmatrix}$,

$$\lambda I + \frac{1}{4}\lambda IN = \begin{pmatrix} \sqrt{2} & 0 \\ 0 & \sqrt{2} \end{pmatrix} + \frac{1}{4} \begin{pmatrix} \sqrt{2} & 0 \\ 0 & \sqrt{2} \end{pmatrix} \begin{pmatrix} 0 & 1 \\ 0 & 0 \end{pmatrix} = \begin{pmatrix} \sqrt{2} & \frac{\sqrt{2}}{4} \\ 0 & \sqrt{2} \end{pmatrix}$$

and note that

$$\begin{pmatrix} \sqrt{2} & \dfrac{\sqrt{2}}{4} \\ 0 & \sqrt{2} \end{pmatrix}^2 = \begin{pmatrix} 2 & 1 \\ 0 & 2 \end{pmatrix}.$$

Now, if we let

$$P = \begin{pmatrix} 1 & 0 & 0 & 0 & 1 \\ 1 & 1 & 1 & 0 & 0 \\ 0 & 1 & 0 & 1 & 1 \\ 1 & 1 & 1 & 1 & 0 \\ 1 & 0 & 1 & 0 & 0 \end{pmatrix}, \text{ then } P^{-1} = \begin{pmatrix} 1 & 0 & -1 & 1 & -1 \\ 0 & 1 & 0 & 0 & -1 \\ -1 & 0 & 1 & -1 & 2 \\ 0 & -1 & 0 & 1 & 0 \\ 0 & 0 & 1 & -1 & 1 \end{pmatrix}.$$

So

$$P^{-1} = \begin{pmatrix} 1 & 0 & -1 & 1 & -1 \\ 0 & 1 & 0 & 0 & -1 \\ -1 & 0 & 1 & -1 & 2 \\ 0 & -1 & 0 & 1 & 0 \\ 0 & 0 & 1 & -1 & 1 \end{pmatrix} \begin{pmatrix} \sqrt{3} & \dfrac{\sqrt{3}}{6} & 0 & 0 & 0 \\ 0 & \sqrt{3} & 0 & 0 & 0 \\ 0 & 0 & \sqrt{3} & 0 & 0 \\ 0 & 0 & 0 & \sqrt{2} & \dfrac{\sqrt{2}}{4} \\ 0 & 0 & 0 & 0 & \sqrt{2} \end{pmatrix} \begin{pmatrix} 1 & 0 & -1 & 1 & -1 \\ 0 & 1 & 0 & 0 & -1 \\ -1 & 0 & 1 & -1 & 2 \\ 0 & -1 & 0 & 1 & 0 \\ 0 & 0 & 1 & -1 & 1 \end{pmatrix}$$

$$\begin{pmatrix} \sqrt{3} & \dfrac{\sqrt{3}}{6} & \sqrt{2}-\sqrt{3} & \sqrt{3}-\sqrt{2} & \sqrt{2}-\dfrac{7\sqrt{3}}{6} \\ 0 & \dfrac{7\sqrt{3}}{6} & 0 & 0 & -\dfrac{\sqrt{3}}{6} \\ 0 & \sqrt{3}-\sqrt{2} & \dfrac{5\sqrt{2}}{4} & -\dfrac{\sqrt{2}}{4} & \dfrac{5\sqrt{2}}{4}-\sqrt{3} \\ 0 & \dfrac{7\sqrt{3}}{6}-\sqrt{2} & \dfrac{\sqrt{2}}{4} & \dfrac{3\sqrt{2}}{4} & \dfrac{\sqrt{2}}{4}-\dfrac{\sqrt{3}}{6} \\ 0 & \dfrac{\sqrt{3}}{6} & 0 & 0 & \dfrac{5\sqrt{3}}{6} \end{pmatrix}$$

and

$$\begin{pmatrix} \sqrt{3} & \dfrac{\sqrt{3}}{6} & \sqrt{2}-\sqrt{3} & \sqrt{3}-\sqrt{2} & \sqrt{2}-\dfrac{7\sqrt{3}}{6} \\ 0 & \dfrac{7\sqrt{3}}{6} & 0 & 0 & -\dfrac{\sqrt{3}}{6} \\ 0 & \sqrt{3}-\sqrt{2} & \dfrac{5\sqrt{2}}{4} & -\dfrac{\sqrt{2}}{4} & \dfrac{5\sqrt{2}}{4}-\sqrt{3} \\ 0 & \dfrac{7\sqrt{3}}{6}-\sqrt{2} & \dfrac{\sqrt{2}}{4} & \dfrac{3\sqrt{2}}{4} & \dfrac{\sqrt{2}}{4}-\dfrac{\sqrt{3}}{6} \\ 0 & \dfrac{\sqrt{3}}{6} & 0 & 0 & \dfrac{5\sqrt{3}}{6} \end{pmatrix}^2 = \begin{pmatrix} 3 & 1 & -1 & 1 & -2 \\ 0 & 4 & 0 & 0 & -1 \\ 0 & 1 & 3 & -1 & 0 \\ 0 & 2 & 1 & 1 & 0 \\ 0 & 1 & 0 & 0 & 2 \end{pmatrix}$$

$$= A.$$

Exercises

1. Find the square root of the matrices below.

(a.)
$$\begin{pmatrix} 3 & 0 & 1 \\ 1 & 2 & 2 \\ 1 & -1 & 4 \end{pmatrix}$$

(b.)
$$\begin{pmatrix} 5 & 0 & -1 \\ 0 & 4 & 0 \\ 1 & 0 & 3 \end{pmatrix}$$

(c.)
$$\begin{pmatrix} 2 & 0 & 1 & 0 \\ -1 & 3 & -1 & 1 \\ 0 & 0 & 2 & 0 \\ 1 & -1 & 2 & 1 \end{pmatrix}$$

(d.)
$$\begin{pmatrix} 3 & 0 & 1 & 0 \\ 1 & 3 & -1 & 0 \\ 0 & 0 & 3 & 0 \\ 0 & 0 & 0 & 3 \end{pmatrix}$$

(e.)
$$\begin{pmatrix} 7/2 & 1/2 & 1/2 & 0 & -1/2 \\ 0 & 3 & 0 & 1 & 0 \\ 0 & 0 & 4 & 0 & 0 \\ 0 & 0 & 1 & 4 & 0 \\ -1/2 & -1/2 & 1/2 & 1 & 7/2 \end{pmatrix}.$$

2. Extending the ideas for the square root of a matrix, find the cube root of the matrices above.

Section 11.2 The Cayley–Hamilton Theorem, Minimal Polynomials, Characteristic Polynomials, and the Jordan Canonical Form.

The goal in this section is to prove the Cayley–Hamilton theorem and to determine the minimal polynomial of a square matrix. The Cayley–Hamilton theorem says that if A is a square matrix, then A satisfies its characteristic equation. However, there may be a polynomial of lesser degree that also satisfies the characteristic equation. (Any such polynomial will divide the characteristic polynomial.)

Historical Note

Arthur Cayley (born16 August 1821, died 26 January 1895)

When Cayley was 17, he was sent to Trinity College in Cambridge where he studied mathematics and languages, excelling in both. After his graduation he then enrolled at Lincoln's Inn to study law and started his legal practice in 1849.

While training to be a lawyer, Cayley went to Dublin to hear William Rowan Hamilton's lecture on quaternions. Cayley became a good friend of Hamilton's although the two disagreed as to the importance of the quaternions in the study of geometry. Another of Cayley's friends was James Joseph Sylvester, who was also in the legal profession.

Cayley took up a career in academics by accepting the position of Sadleirian Professor of Pure Mathematics at Cambridge University in 1863.

In linear algebra, the Cayley–Hamilton theorem states that every square matrix over a commutative ring (such as the real or complex field) satisfies its own characteristic equation:

$$p(\lambda) = \det(\lambda I_n - A)$$

The theorem was first proved in 1853 by Hamilton in the special case of certain 4×4 real or 2×2 complex matrices. Cayley in 1858 stated it for 3×3 and smaller matrices, but only published a proof for the 2×2 case. The general case was first proved by Frobenius in 1878.

He was the first to define the concept of a group in the modern way – as a set with a binary operation satisfying certain laws. Formerly, when mathematicians spoke of "groups", they had meant permutation groups. Cayley tables and Cayley graphs as well as Cayley's theorem are named in honor of Cayley.

Definition:

The minimal polynomial of a square matrix is the monic polynomial of least degree that satisfies the characteristic equation.

Theorem 11.1: (Cayley–Hamilton Theorem)

(a.) Let A be an $n \times n$ matrix. Then A satisfies its characteristic equation

$$\det(\lambda I - A) = 0.$$

(b.) The minimal polynomial of A is of the form

$$(\lambda_1 - A)^{n_1} (\lambda_2 - A)^{n_2} \cdots (\lambda_k - A)^{n_k}$$

where λ_i are the distinct eigenvalues of A and n_i is the size of the largest Jordan block of the eigenvalue λ_i.

Proof:

Consider a matrix that is in Jordan canonical form with a single eigenvalue, λ.

We claim that if the largest sub-block is of size $n \times n$, then $(\lambda I - A)^n = 0$

but $(\lambda I - A)^{n-1} \neq 0$.

Consider

$$A = \begin{pmatrix} 2 & 1 & 0 & 0 & 0 & 0 \\ 0 & 2 & 1 & 0 & 0 & 0 \\ 0 & 0 & 2 & 0 & 0 & 0 \\ 0 & 0 & 0 & 2 & 1 & 0 \\ 0 & 0 & 0 & 0 & 2 & 0 \\ 0 & 0 & 0 & 0 & 0 & 2 \end{pmatrix}.$$

Then

$$2I - A = \begin{pmatrix} 0 & -1 & 0 & 0 & 0 & 0 \\ 0 & 0 & -1 & 0 & 0 & 0 \\ 0 & 0 & 0 & 0 & 0 & 0 \\ 0 & 0 & 0 & 0 & -1 & 0 \\ 0 & 0 & 0 & 0 & 0 & 0 \\ 0 & 0 & 0 & 0 & 0 & 0 \end{pmatrix}$$

We can describe A as

$$A = \begin{pmatrix} B_1 & 0 & 0 \\ 0 & B_2 & 0 \\ 0 & 0 & B_3 \end{pmatrix} \text{ where } B_1 = \begin{pmatrix} 2 & 1 & 0 \\ 0 & 2 & 1 \\ 0 & 0 & 2 \end{pmatrix}, \; B_2 = \begin{pmatrix} 2 & 1 \\ 0 & 2 \end{pmatrix}, \; B_3 = (2)$$

and

$$2I - A = \begin{pmatrix} C_1 & 0 & 0 \\ 0 & C_2 & 0 \\ 0 & 0 & C_3 \end{pmatrix} \text{ where } C_1 = \begin{pmatrix} 0 & -1 & 0 \\ 0 & 0 & -1 \\ 0 & 0 & 0 \end{pmatrix}, \; C_2 = \begin{pmatrix} 0 & -1 \\ 0 & 0 \end{pmatrix}, \; C_3 = (0).$$

Note that for any positive integer n

$$(2I - A)^n = \begin{pmatrix} C_1 & 0 & 0 \\ 0 & C_2 & 0 \\ 0 & 0 & C_3 \end{pmatrix}^n = \begin{pmatrix} C_1^n & 0 & 0 \\ 0 & C_2^n & 0 \\ 0 & 0 & C_3^n \end{pmatrix}.$$

Also

$$C_1^2 = \begin{pmatrix} 0 & 0 & 1 \\ 0 & 0 & 0 \\ 0 & 0 & 0 \end{pmatrix}, \; C_1^3 = \begin{pmatrix} 0 & 0 & 0 \\ 0 & 0 & 0 \\ 0 & 0 & 0 \end{pmatrix}, \; C_2^2 = \begin{pmatrix} 0 & 0 \\ 0 & 0 \end{pmatrix}$$

$$(2I - A)^2 = \begin{pmatrix} C_1^2 & 0 & 0 \\ 0 & C_2^2 & 0 \\ 0 & 0 & C_3^2 \end{pmatrix} = \begin{pmatrix} 0 & 0 & 1 & 0 & 0 & 0 \\ 0 & 0 & 0 & 0 & 0 & 0 \\ 0 & 0 & 0 & 0 & 0 & 0 \\ 0 & 0 & 0 & 0 & 0 & 0 \\ 0 & 0 & 0 & 0 & 0 & 0 \\ 0 & 0 & 0 & 0 & 0 & 0 \end{pmatrix}$$

$$(2I - A)^3 = \begin{pmatrix} C_1^3 & 0 & 0 \\ 0 & C_2^3 & 0 \\ 0 & 0 & C_3^3 \end{pmatrix} = \begin{pmatrix} 0 & 0 & 0 & 0 & 0 & 0 \\ 0 & 0 & 0 & 0 & 0 & 0 \\ 0 & 0 & 0 & 0 & 0 & 0 \\ 0 & 0 & 0 & 0 & 0 & 0 \\ 0 & 0 & 0 & 0 & 0 & 0 \\ 0 & 0 & 0 & 0 & 0 & 0 \end{pmatrix}$$

This example illustrates that when there is a single eigenvalue λ, then $\lambda I - A$ is of the form

$$\begin{pmatrix} B_1 & 0 & \cdots & 0 \\ 0 & B_2 & \cdots & 0 \\ \vdots & \vdots & \cdots & 0 \\ 0 & 0 & \cdots & B_n \end{pmatrix}$$

where each B_i is a square matrix of the form

$$\begin{pmatrix} 0 & -1 & 0 & \cdots & 0 \\ 0 & 0 & -1 & \cdots & 0 \\ 0 & 0 & \cdots & -1 & 0 \\ 0 & 0 & \cdots & 0 & -1 \\ 0 & 0 & \cdots & 0 & 0 \end{pmatrix}$$

If B_i is a $k \times k$ matrix, then $B_i^k = 0$, but $B_i^{k-1} \neq 0$. Thus, if the largest block is of size $k \times k$, then $(\lambda I - A)^k = 0$, but $(\lambda I - A)^{k-1} \neq 0$.

Note that the characteristic polynomial of the example is

$$c(x) = (2 - x)^6$$

where 6 is the algebraic dimension of 2, and 6 is the sum of the dimensions of the Jordan blocks. The largest Jordan block is 3×3 and the minimal polynomial is

$$m(x) = (2 - x)^3$$

We now consider the case where there is more than one eigenvalue. The example

$$A = \begin{pmatrix} 2 & 1 & 0 & 0 & 0 \\ 0 & 2 & 0 & 0 & 0 \\ 0 & 0 & 3 & 1 & 0 \\ 0 & 0 & 0 & 3 & 0 \\ 0 & 0 & 0 & 0 & 3 \end{pmatrix}$$

will guide our thinking.

Now

$$2I - A = \begin{pmatrix} 0 & -1 & 0 & 0 & 0 \\ 0 & 0 & 0 & 0 & 0 \\ 0 & 0 & -1 & -1 & 0 \\ 0 & 0 & 0 & -1 & 0 \\ 0 & 0 & 0 & 0 & -1 \end{pmatrix}.$$

We will think of this as a matrix of the form

$$\begin{pmatrix} B & 0 \\ 0 & C \end{pmatrix}$$

where

$$B = \begin{pmatrix} 0 & -1 \\ 0 & 0 \end{pmatrix}$$

and C is a 3×3 matrix whose entries will not concern us.

Also

$$3I - A = \begin{pmatrix} 1 & -1 & 0 & 0 & 0 \\ 0 & 1 & 0 & 0 & 0 \\ 0 & 0 & 0 & -1 & 0 \\ 0 & 0 & 0 & 0 & 0 \\ 0 & 0 & 0 & 0 & 0 \end{pmatrix}.$$

We will think of this as a matrix of the form

$$\begin{pmatrix} D & 0 \\ 0 & E \end{pmatrix}$$

where

$$E = \begin{pmatrix} 0 & -1 & 0 \\ 0 & 0 & 0 \\ 0 & 0 & 0 \end{pmatrix}$$

and D is a 2×2 matrix whose entries will not concern us.

We note that the matrix E Can also be expressed

$$E = \begin{pmatrix} F & 0 \\ 0 & 0 \end{pmatrix} \text{ where } F = \begin{pmatrix} -1 & 0 \\ 0 & 0 \end{pmatrix}.$$

The maximum size of block associated with $\lambda = 2$ is 2×2 as is the case for

$$\lambda = 3.$$

Consider the form of

$$(2I - A)^2 (3I - A)^2.$$

We can write this as

$$\begin{pmatrix} B & 0 \\ 0 & C \end{pmatrix}^2 \begin{pmatrix} D & 0 \\ 0 & E \end{pmatrix}^2 = \begin{pmatrix} B^2 & 0 \\ 0 & C^2 \end{pmatrix} \begin{pmatrix} D^2 & 0 \\ 0 & E^2 \end{pmatrix}.$$

Since B is a 2×2 matrix and D is a 2×2 matrix, the matrix
$B^2 D^2$ is defined and is a 2×2 matrix.
Similarly, $C^2 E^2$ is defined and is a 3×3 matrix.
Since $B^2 = 0$ and $E^2 = 0$, we have

$$\begin{pmatrix} B & 0 \\ 0 & C \end{pmatrix}^2 \begin{pmatrix} D & 0 \\ 0 & E \end{pmatrix}^2 = 0.$$

We have given the structure of the proof of the following theorem.

Theorem 11.2:

Suppose that A is a square matrix in Jordan canonical form with distinct eigenvalues $\lambda_1, \ldots, \lambda_k$.
Suppose also that the largest Jordan block for λ_i is of size n_i. Then

$$(\lambda_1 I - A)^{n_1} \cdots (\lambda_k I - A)^{n_k} = 0.$$

Furthermore, the exponents are the smallest values for which this is true.

Corollary:
    If A is a square matrix and if A is similar to B, then for any polynomial $f(x)$,
    we have if $f(A) = 0$, then $f(B) = 0$. Thus the theorem holds for any square matrix.

Proof:

Suppose that $f(x) = a_n x^n + a_{n-1} x^{n-1} + \cdots + a_1 x + a_0$.

If $f(A) = a_n(A)^n + a_{n-1}(A)^{n-1} + \cdots + a_1 A + a_0 I = 0$ and $B = S^{-1}AS$ then

$$f(B) = a_n(S^{-1}AS)^n + a_{n-1}(S^{-1}AS)^{n-1} + \cdots + a_1 S^{-1}AS + a_0 I =$$

$$a_n S^{-1}(A)^n S + a_{n-1}S^{-1}(A)^{n-1} S + \cdots + a_1 S^{-1}AS + a_0 I =$$

$$S^{-1}\left(a_n(A)^n + a_{n-1}(A)^{n-1} + \cdots + a_1 A + a_0 I\right)S = S^{-1}(0)S = 0.$$

If the field is the real numbers, the characteristic polynomial may not factor into linear factors, but the Cayley–Hamilton theorem is still valid. Here is why.

Solve the problem in the complex numbers. The factors that are irreducible in the real numbers will factor in the complex numbers as

$$(x - \lambda)(x - \bar{\lambda})$$

where $\bar{\lambda}$ is the complex conjugate of λ. We then have the factors

$$(\lambda I - A)(\bar{\lambda} I - A) = |\lambda|^2 I - (\lambda + \bar{\lambda})A + A^2$$

and $\lambda + \bar{\lambda}$ is real.

Exercises

1. Show that similar matrices have the same characteristic polynomial and minimal polynomial.
2. Find the characteristic polynomial and minimal polynomial for the matrices below.

(a.) $\begin{pmatrix} 3 & 1 & 0 \\ 0 & 3 & 0 \\ 0 & 0 & 2 \end{pmatrix}$

(b.) $\begin{pmatrix} 4 & 0 & 0 \\ 0 & 4 & 0 \\ 0 & 0 & 1 \end{pmatrix}$

(c.) $\begin{pmatrix} 5 & 1 & 0 & 0 \\ 0 & 5 & 0 & 0 \\ 0 & 0 & 3 & 0 \\ 0 & 0 & 0 & -3 \end{pmatrix}$

(d.) $\begin{pmatrix} 7 & 0 & 0 & 0 \\ 0 & 7 & 0 & 0 \\ 0 & 0 & 7 & 0 \\ 0 & 0 & 0 & 7 \end{pmatrix}$

(e.) $\begin{pmatrix} 6 & 1 & 0 & 0 \\ 0 & 6 & 0 & 0 \\ 0 & 0 & 3 & 1 \\ 0 & 0 & 0 & 3 \end{pmatrix}$

(f.) $\begin{pmatrix} -3 & 1 & 0 & 0 & 0 \\ 0 & -3 & 0 & 0 & 0 \\ 0 & 0 & 6 & 0 & 0 \\ 0 & 0 & 0 & 1 & 0 \\ 0 & 0 & 0 & 0 & 0 \end{pmatrix}$

Chapter 12

The Perron–Frobenius Theorem

The Perron–Frobenius theorem has applications in many fields, including Markov chains, population dynamics, and commodity pricing. While the proof of the theorem is beyond the scope of the text, we will state the theorem, demonstrate the proof in the case of a 2×2 matrix, and give an application.

Theorem 12.1: Perron–Frobenius Theorem

Let A be an $n \times n$ matrix, all of whose entries are positive. Then

1. There is a positive eigenvalue λ_{PF} of A such that if λ is any other eigenvalue of A then $|\lambda| < \lambda_{PF}$.
2. The algebraic and geometric dimensions of the eigenspace of λ_{PF} are 1.
3. The matrix A has a left eigenvector, all of whose components are positive, and a right eigenvector, all of whose components are positive.
4. If $\hat{v}_{PF}$ is an eigenvector with eigenvalue λ_{PF} and $\hat{u}$ is an eigenvector of A with all positive entries, then $\hat{u}$ is a scalar multiple of $\hat{v}_{PF}$.

(See Lawler *An Introduction to Stochastic Processes* p. 40 for the steps of a proof.)

To demonstrate this in the 2×2 case, let

$$A = \begin{pmatrix} a & b \\ c & d \end{pmatrix}.$$

Then the characteristic polynomial of A is

$$\begin{vmatrix} a - \lambda & b \\ c & d - \lambda \end{vmatrix} = \lambda^2 - (a + d)\lambda + ad - bc$$

and so the eigenvalues of A are

$$\lambda = \frac{a+d \pm \sqrt{(a+d)^2 - 4(ad-bc)}}{2} = \frac{a+d \pm \sqrt{(a-d)^2 + 4bc}}{2}$$

or

$$\lambda_1 = \frac{a+d + \sqrt{(a-d)^2 + 4bc}}{2}$$

$$\lambda_2 = \frac{a+d - \sqrt{(a-d)^2 + 4bc}}{2}.$$

Since all entries of A are positive, $(a-d)^2 + 4bc > 0$, so both eigenvalues are real and $\lambda_1 > \lambda_2$. Thus, $\lambda_{PF} = \lambda_1$. Also note that $|\lambda_2| < \lambda_1$. This is because

$$\lambda_1 = \frac{|a+d| + \left|\sqrt{(a-d)^2 + 4bc}\right|}{2}$$

and $$|\lambda_2| = \frac{\left\|a+d\right| - \left|\sqrt{(a-d)^2 + 4bc}\right\|}{2}$$

We demonstrate that λ_1 has an eigenvector, both of whose entries are positive.

Let $\begin{pmatrix} v_1 \\ v_2 \end{pmatrix}$ be an eigenvector of A with eigenvalue λ_1. Then

$$\begin{pmatrix} a & b \\ c & d \end{pmatrix} \begin{pmatrix} v_1 \\ v_2 \end{pmatrix} = \lambda_1 \begin{pmatrix} v_1 \\ v_2 \end{pmatrix}$$

so

$$av_1 + bv_2 = \lambda_1 v_1$$

$$cv_1 + dv_2 = \lambda_1 v_2$$

which yields

$$v_1 = \frac{b}{\lambda_1 - a} v_2 \text{ and } v_1 = \frac{\lambda_1 - d}{c} v_2.$$

Now

$$\lambda_1 - a = \frac{a + d + \sqrt{(a-d)^2 + 4bc}}{2} - a = \frac{d - a + \sqrt{(a-d)^2 + 4bc}}{2} > 0$$

since $\sqrt{(a-d)^2 + 4bc} > \sqrt{(a-d)^2} = |a - d|$

and $b > 0$, then v_1, and v_2 have the same sign, so it is possible to choose and eigenvector with all positive entries.

An Application of the Perron–Frobenius Theorem

In a Markov chain, there is a "process" that is in exactly one of the "states" $s_1, \ldots, s_n$ at any given time. At each unit of time the process chooses to change states or remain in its present state, and the decision is governed by a set of transition probabilities P_{ij}, $i, j = 1, \ldots, n$ where
  P_{ij} = the probability the process in the state s_i at time t
is in the state s_j at time $t + 1$.

  Thus $P = (P_{ij})$ is an $n \times n$ matrix called the transition matrix, that has the properties

$$P_{ij} \geq 0 \text{ and } \sum_{j=1}^{n} P_{ij} = 1, \ i = 1, \ldots, n.$$

We use an $n - vector$ $\hat{x}(t) = (x_1, \ldots, x_n)$ where
  x_i = the probability the system is in state s_i at time t.
  The crucial relationship is

$$\hat{x}(t + 1) = \hat{x}(t) P.$$

If it is the case that there is a vector $\hat{x}(t)$ for which

$$\hat{x}(t) = \hat{x}(t) P$$

then $\hat{x}(t)$ is said to be an equilibrium state. Finding equilibrium states is a central problem of Markov processes.
  If P is a transition matrix, then

$$P \begin{pmatrix} 1 \\ \vdots \\ 1 \end{pmatrix} = \begin{pmatrix} 1 \\ \vdots \\ 1 \end{pmatrix} \text{ so } \begin{pmatrix} 1 \\ \vdots \\ 1 \end{pmatrix}$$

is an eigenvector of P with eigenvalue 1. The example below demonstrates that one way to find the equilibrium state of a Markov process is to find the eigenvector of P^T whose eigenvalue is 1.
  To take advantage of the Perron–Frobenius theorem, we will use the equivalent expression

$$P^T \left[\hat{x}(t) \right]^T = \left[\hat{x}(t + 1) \right]^T.$$

Suppose there are three restaurants that we designate as *A, B, C*. A marketing study has found that

20% of the customers that eat at *A* will eat at *A* their next time out

30% of the customers that eat at *A* will eat at *B* their next time out

50% of the customers that eat at *A* will eat at *C* their next time out

30% of the customers that eat at *B* will eat at *A* their next time out

30% of the customers that eat at *B* will eat at *B* their next time out

40% of the customers that eat at *B* will eat at *C* their next time out

40% of the customers that eat at *C* will eat at *A* their next time out

30% of the customers that eat at *C* will eat at *B* their next time out

30% of the customers that eat at *C* will eat at *C* their next time out.

The states of the process are the restaurants and the transition matrix for this process is

$$P = \begin{pmatrix} .2 & .3 & .5 \\ .3 & .3 & .4 \\ .4 & .3 & .3 \end{pmatrix},$$

The eigenvalues of *P*, which are also the eigenvalues of P^T, are $\lambda = 1$, 0, and –1/5.
The eigenvectors for P^T that correspond to the eigenvalues are

$$\lambda_1 = 1, \quad \hat{e}_1 = \begin{pmatrix} \frac{37}{47} \\ \frac{36}{47} \\ 1 \end{pmatrix}; \quad \lambda_2 = 0, \quad \hat{e}_2 = \begin{pmatrix} 1 \\ -2 \\ 1 \end{pmatrix};$$

$$\lambda_3 = -\frac{1}{5}, \quad \hat{e}_3 = \begin{pmatrix} -1 \\ 0 \\ 1 \end{pmatrix}.$$

We choose an arbitrary initial distribution, say

$$\hat{v} = \begin{pmatrix} .5 \\ .4 \\ .1 \end{pmatrix}.$$

Now

$$\hat{v} = \frac{47}{120}\hat{e}_1 - \frac{1}{20}\hat{e}_2 - \frac{29}{120}\hat{e}_3$$

so

$$\left(P^T\right)\hat{v} = \left(P^T\right)\left(\frac{47}{120}\hat{e}_1 - \frac{1}{20}\hat{e}_2 - \frac{29}{120}\hat{e}_3\right)$$

$$= \frac{47}{120}\left(P^T\right)\hat{e}_1 - \left(P^T\right)\frac{1}{20}\hat{e}_2 - \frac{29}{120}\left(P^T\right)\hat{e}_3$$

$$= \frac{47}{120}\lambda_1\hat{e}_1 - \lambda_2\frac{1}{20}\hat{e}_2 - \frac{29}{120}\lambda_3\hat{e}_3.$$

Similarly,

$$\left(P^T\right)^n\hat{v} = \frac{47}{120}(\lambda_1)^n\hat{e}_1 - (\lambda_2)^n\frac{1}{20}\hat{e}_2 - \frac{29}{120}(\lambda_3)^n\hat{e}_3.$$

Now

$$(\lambda_1)^n = 1, \lim_{n\to\infty}(\lambda_2)^n = 0, \lim_{n\to\infty}(\lambda_3)^n = 0$$

So

$$\lim_{n\to\infty}\left(P^T\right)^n\hat{v} = \frac{47}{120}\hat{e}_1 = \begin{pmatrix} \dfrac{37}{120} \\[2mm] \dfrac{36}{120} \\[2mm] \dfrac{47}{120} \end{pmatrix}.$$

The vector

$$\begin{pmatrix} \dfrac{37}{120} \\[2mm] \dfrac{36}{120} \\[2mm] \dfrac{47}{120} \end{pmatrix}$$

is a probability vector; that is, all entries are non-negative and sum to 1. Furthermore, one can check that

$$P^T\begin{pmatrix} \dfrac{37}{120} \\[2mm] \dfrac{36}{120} \\[2mm] \dfrac{47}{120} \end{pmatrix} = \begin{pmatrix} \dfrac{37}{120} \\[2mm] \dfrac{36}{120} \\[2mm] \dfrac{47}{120} \end{pmatrix}$$

so that

$$
\left[P^T \left(\begin{array}{c} \dfrac{37}{120} \\[2mm] \dfrac{36}{120} \\[2mm] \dfrac{47}{120} \end{array} \right) \right]^T = \left(\begin{array}{c} \dfrac{37}{120} \\[2mm] \dfrac{36}{120} \\[2mm] \dfrac{47}{120} \end{array} \right)^T
$$

and thus,

$$
\left(\frac{37}{120}, \frac{36}{120}, \frac{47}{120} \right) P = \left(\frac{37}{120}, \frac{36}{120}, \frac{47}{120} \right).
$$

We give another application of the Perron–Frobenius theorem.

The Perron–Frobenius Theorem in Population Dynamics

We divide the life cycle of an organism into stages. The organism is in exactly one stage at each point in time. We denote the stages of life by $s_1, \ldots, s_n$. We divide time into discrete units and make the simplifying assumption that any change that occurs in the interval $(t_0, t_0 + 1]$ occurs at $t_0 + 1$. The number of individuals in stage k at time t_0 is denoted $n_k(t_0)$.

Knowing the number of organisms in each of the stages at time t_0 we want to predict the number of organisms in each stage at time $t_0 + 1$. To do this, we must consider the ways an individual in stage s_i at time t_0 can produce an individual an stage s_j at time $t_0 + 1$. This occurs by either growth or reproduction.

We denote the probability that an organism moves from stage i to stage j in one unit of time by p_{ij} and the expected number of reproductions that an individual from stage i contributes to stage j in one unit of time by f_{ij}. Thus, each individual on average contributes $p_{ij} + f_{ij}$ to state j. Summing over all i we get

$$
n_j(t_0 + 1) = \sum_i n_i(t_0) \left[p_{ij} + f_{ij} \right].
$$

For the ginseng plant, the stages of life are seed, seedling, one-leaf plant, two-leaf plant, three-leaf plant, and four-leaf plant. We denote these stages of life 1 through 6, respectively. The diagram in Figure 12.1 describes the movement between stages in one unit of time.

We create a matrix A whose (i, j) entry is the probability that a plant in state i passes to state j in one unit of time plus the expected number of seedlings that a single plant produces. Figure 12.1 says that a seed becomes a seedling in one unit of time with probability .15 and the remainder of the seeds die. We create a matrix A whose (i, j) entry is the probability that a plant in state i passes to state j in one unit of time. Thus $A_{1,2} = .15$. We assume that all other seeds die, so $A_{1,j} = 0$ if j $\neq$ 2. Some stages of life can produce seeds. With those stages, the expected number of seeds that a single plant produces is incorporated into the matrix entry.

The model that we are considering is that mature plants can produce seeds. The numbers attached to the arrows from a mature plant to seeds is the expected number of seeds that a single plant reproduces in one unit of time. For example, the $A_{1,6}$ entry is 40, which says that a single four-leaved planted produces on average 40 seeds in one unit of time.

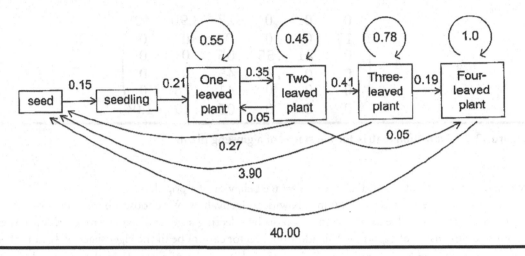

Figure 12.1   Life process of a ginseng plant.

The matrix entries, except in the first row, are transition probabilities. In the first row, the values represent the average number of seeds made by plants in other stages. It is worth noting that there is not a death stage. This is because the numbers are best estimates of available data. Additional years of gathering data would provide a more realistic estimate.

Figure 12.1 gives a diagram of the life process and Figure 12.2 gives the associated matrix.

Not all the entries of the matrix A are positive. To ensure that the Perron–Frobenius theorem can be used, we show that some power of the matrix A is positive. In fact,

$$A^5 = \begin{pmatrix} .04727 & .7475 & 6.2298 & 15.873 & 23.214 & 40.119 \\ .0004 & .0427 & .5339 & 1.8309 & 3.0004 & 6.0000 \\ .0061 & .0258 & .1197 & .3387 & .7246 & 2.3562 \\ .0085 & .0397 & .1281 & .0831 & .1603 & .8820 \\ .0080 & .0650 & .3191 & .3597 & .3063 & .1808 \\ .0020 & .0261 & .1926 & .4391 & .6164 & 1.0220 \end{pmatrix}.$$

Since A^5 is positive, then A is regular, and the Perron–Frobenius theorem applies.

We denote the largest eigenvalue guaranteed by the Perron–Frobenius theorem λ_{PF}. We will see that the normalized eigenvector for λ_{PF} gives the relative distribution of the long-term distribution and the value of λ_{PF} describes the growth or decay rate of the size of the population.

The eigenvalues of A are 1.184103325, $0.59159 - .2165i$, $0.05916 + .2165i$,

$$0.7597 - .4019i, \quad 0.7597 + .4019i \text{ so } \lambda_{PF} = 1.184103325.$$

An eigenvector for λ_{PF} with positive entries is

$$\begin{pmatrix} .806747 \\ .102972 \\ .035167 \\ .016767 \\ .017011 \\ .022110 \end{pmatrix}.$$

$$A = \begin{pmatrix} 0 & 0 & 0 & .27 & 3.90 & 40 \\ .15 & 0 & 0 & 0 & 0 & 0 \\ 0 & .21 & .55 & .05 & 0 & 0 \\ 0 & 0 & .35 & .45 & 0 & 0 \\ 0 & 0 & 0 & .41 & .78 & 0 \\ 0 & 0 & 0 & .05 & .19 & 1.0 \end{pmatrix}$$

Figure 12.2 Transition matrix for life stages of a ginseng plant.

We describe a method for finding the long-range behavior of a population.

In this example, we have a 6×6 matrix with 5 eigenvalues. We choose a basis using an eigenvector for each eigenvalue and expand to a basis by selecting a generalized eigenvector. Since the geometric dimension of λ_{PF} is one, this additional vector cannot be in the eigenspace of λ_{PF}. Let the basis so constructed be denoted $\{\hat{v}_{PF}, \hat{v}_2, \hat{v}_3, \hat{v}_4, \hat{v}_5, \hat{v}_6\}$. Let $\hat{v}$ be a vector. Then there are scalars c_1, ..., c_6 for which $\hat{v} = c_1 \hat{v}_{PF} + c_2 \hat{v}_2 + \cdots + c_6 \hat{v}_6$. Then

$$A\hat{v} = A\left(c_1 \hat{v}_{PF} + c_2 \hat{v}_2 + \cdots + c_6 \hat{v}_6\right) = c_1 \lambda_{PF} \hat{v}_{PF} + c_2 \lambda_2 \hat{v}_2 + \cdots + c_6 \lambda_6 \hat{v}_6$$

and

$$A^k \hat{v} = c_1 \lambda_{PF}{}^k \hat{v}_{PF} + c_2 \lambda_2{}^k \hat{v}_2 + \cdots + c_6 \lambda_6{}^k \hat{v}_6.$$

Then

$$\frac{1}{\lambda_{PF}{}^k} A^k \hat{v} = c_1 \hat{v}_{PF} + c_2 \left(\frac{\lambda_2}{\lambda_{PF}}\right)^k \hat{v}_2 + \cdots + c_6 \left(\frac{\lambda_6}{\lambda_{PF}}\right)^k \hat{v}_6$$

so

$$\lim_{k \to \infty} \frac{1}{\lambda_{PF}{}^k} A^k \hat{v} = c_1 \hat{v}_{PF}.$$

Suppose that our initial population is given by the vector

$$n(t_0) = \begin{pmatrix} 800 \\ 90 \\ 56 \\ 23 \\ 31 \\ 11 \end{pmatrix}.$$

Then the predicted population at a time $t_0 + 1$ *is*

$$A[n(t_0)] = \begin{pmatrix} 0 & 0 & 0 & .27 & 3.90 & 40 \\ .15 & 0 & 0 & 0 & 0 & 0 \\ 0 & .21 & .55 & 0.05 & 0 & 0 \\ 0 & 0 & .35 & .45 & 0 & 0 \\ 0 & 0 & 0 & .41 & .78 & 0 \\ 0 & 0 & 0 & .05 & .19 & 1.0 \end{pmatrix} \begin{pmatrix} 800 \\ 90 \\ 56 \\ 23 \\ 31 \\ 11 \end{pmatrix} = \begin{pmatrix} 567.11 \\ 120 \\ 50.95 \\ 29.95 \\ 33.61 \\ 18.04 \end{pmatrix}.$$

We want to investigate the long-term behavior of the process. We demonstrate that the relative frequency distribution of the states approaches a steady state and that the population grows by a factor λ_{PF} each generation. We have

$$\frac{A^{100}}{\lambda_{PF}{}^{100}} \begin{pmatrix} 800 \\ 90 \\ 56 \\ 23 \\ 31 \\ 11 \end{pmatrix} = \begin{pmatrix} 840.99659 \\ 106.53588 \\ 36.66038 \\ 17.47865 \\ 17.73369 \\ 23.04866 \end{pmatrix}.$$

Normalizing the vector gives

$$\begin{pmatrix} .806747 \\ .102972 \\ .035167 \\ .016767 \\ .017011 \\ .022110 \end{pmatrix}$$

$$\frac{A^{200}}{\lambda_{PF}{}^{200}} \begin{pmatrix} 800 \\ 90 \\ 56 \\ 23 \\ 31 \\ 11 \end{pmatrix} = \begin{pmatrix} 840.99659 \\ 106.53588 \\ 36.66038 \\ 17.47865 \\ 17.73369 \\ 23.04866 \end{pmatrix}.$$

Normalizing this vector again gives

$$\begin{pmatrix} .806747 \\ .102972 \\ .035167 \\ .016767 \\ .017011 \\ .022110 \end{pmatrix}$$

Also,

$$A\begin{pmatrix} .806747 \\ .102972 \\ .035167 \\ .016767 \\ .017011 \\ .022110 \end{pmatrix} = \begin{pmatrix} .95527 \\ .12101 \\ .04164 \\ .01985 \\ .02014 \\ .02618 \end{pmatrix} \quad \text{and} \quad 1.1841033\begin{pmatrix} .806747 \\ .102972 \\ .035167 \\ .016767 \\ .017011 \\ .022110 \end{pmatrix} = \begin{pmatrix} .95527 \\ .12101 \\ .04164 \\ .01985 \\ .02014 \\ .02618 \end{pmatrix}.$$

Several other applications of the Perron–Frobenius theorem can be found at https://epubs.siam.org/doi/pdf/10.1137/

Exercises

1. For the projection matrices given below
   (a.) Find the Perron–Frobenius eigenvalue.
   (b.) Find the normalized Perron–Frobenius eigenvector.
   (c.) Find the equilibrium state.
   (d.) Find the rate of growth/decay.

   (i) $\begin{pmatrix} .4 & .3 & 0 \\ 1.2 & .5 & .2 \\ 0 & 0 & .6 \end{pmatrix}$

   (ii) $\begin{pmatrix} .3 & 1.8 & 4 \\ .6 & 0 & 0 \\ 0 & .7 & 0 \end{pmatrix}$

   (iii) $\begin{pmatrix} 0 & .75 & 2.3 \\ .6 & .5 & 0 \\ 0 & .5 & .7 \end{pmatrix}$

2. For the matrices below, find
   (a.) The positive eigenvalue λ_{PF} of A such that if λ is any other eigenvalue of A then $|\lambda| < \lambda_{PF}$.
   (b.) A left eigenvector, all of whose components are positive, and a right eigenvector, all of whose components are positive.

   (i) $\begin{pmatrix} .1 & .2 & .5 & .2 \\ .3 & .2 & .2 & .3 \\ .4 & .3 & .2 & .1 \\ .6 & .1 & .1 & .2 \end{pmatrix}$

   (ii) $\begin{pmatrix} 2 & 1 & .5 & 1 \\ 4 & 2 & 1 & 6 \\ 3 & 4 & 1 & 1 \\ .6 & .1 & .1 & .2 \end{pmatrix}$

Chapter 13

Bilinear Forms

In this chapter we discuss bilinear forms and quadratic forms. In a sense, bilinear forms are an extension of the inner product, but with some significant differences. In dealing with bilinear forms, some difficulties arise if the scalar field has characteristic 2. Likewise, if the scalar field is the complex numbers, some complications exist. To avoid these issues, *we will consider the scalar field to be the real numbers in this chapter.*

Section 13.1 Bilinear Forms

Suppose V is a finite dimensional vector space over the field $\mathbb{R}$. A function

$$B : V \times V \to \mathbb{R}$$

is a bilinear form if

(i) $B\left(\alpha \hat{u} + \beta \hat{v}, \hat{w}\right) = \alpha B\left(\hat{u}, \hat{w}\right) + \beta B\left(\hat{v}, \hat{w}\right)$ for $\hat{u}, \hat{v}, \hat{w} \in V; \alpha, \beta \in \mathbb{R}$

(ii) $B\left(\hat{u}, \alpha \hat{v} + \beta \hat{w}\right) = \alpha B\left(\hat{u}, \hat{v}\right) + \beta B\left(\hat{u}, \hat{w}\right)$ for $\hat{u}, \hat{v}, \hat{w} \in V; \alpha, \beta \in \mathbb{R}$.

In other words, a bilinear form is a function of two arguments that is linear in each argument.

Example

The inner product on $\mathbb{R}^n$ is a bilinear form.

Theorem 13.1:

For B a bilinear form on the finite dimensional vector space V and an ordered basis $\mathcal{D}$ of V there is a unique square matrix $M(\mathcal{D})$ for which

$$B\left(\hat{x}, \hat{y}\right) = \hat{x}^T \left[M\left(\mathcal{D}\right) \hat{y} \right] \text{ for } \hat{x}, \hat{y} \in V,$$

where if $\mathcal{D} = \left\{ \hat{e}_1, \ldots, \hat{e}_n \right\}$, then $M\left(\mathcal{D}\right)_{ij} = B\left(\hat{e}_i, \hat{e}_j\right)$.

Proof:

Let $\mathcal{D} = \{\hat{e}_1,\ldots,\hat{e}_n\}$ and choose $\hat{x}, \hat{y} \in V$. Suppose

$$\hat{x} = \sum_{i=1}^{n} x_i \hat{e}_i, \qquad \hat{y} = \sum_{j=1}^{n} y_j \hat{e}_j.$$

Then

$$B(\hat{x},\hat{y}) = B\left(\sum_{i=1}^{n} x_i \hat{e}_i, \sum_{j=1}^{n} y_j \hat{e}_j\right)$$

$$= \sum_{i=1}^{n} x_i B\left(\hat{e}_i, \sum_{j=1}^{n} y_j \hat{e}_j\right) \text{(by linearity of B in the first component)}$$

$$= \sum_{i=1}^{n} x_i \sum_{j=1}^{n} y_j B\left(\hat{e}_i,\hat{e}_j\right) \text{(by linearity of B in the second component)}.$$

Define the $n \times n$ matrix $M(\mathcal{D})$ by

$$\left[M(\mathcal{D})\right]_{i,j} = B\left(\hat{e}_i,\hat{e}_j\right)$$

so that the entries of $M(\mathcal{D})$ are determined by both the bilinear form and the basis. Then

$$\hat{x}^T\left[M(\mathcal{D})\hat{y}\right] = (x_1,\ldots,x_n)\begin{pmatrix} B(\hat{e}_1,\hat{e}_1) & \cdots & B(\hat{e}_1,\hat{e}_n) \\ \vdots & \vdots & \vdots \\ B(\hat{e}_n,\hat{e}_1) & \cdots & B(\hat{e}_n,\hat{e}_n) \end{pmatrix}\begin{pmatrix} y_1 \\ \vdots \\ y_n \end{pmatrix}$$

$$= (x_1,\ldots,x_n)\begin{pmatrix} B(\hat{e}_1,\hat{e}_1) y_1 + \cdots + B(\hat{e}_1,\hat{e}_n) y_n \\ \vdots \\ B(\hat{e}_n,\hat{e}_1) y_1 + \cdots + B(\hat{e}_n,\hat{e}_n) y_n \end{pmatrix}$$

$$= x_1\left[B(\hat{e}_1,\hat{e}_1) y_1 + \cdots + B(\hat{e}_1,\hat{e}_n) y_n\right] + \cdots + x_n\left[B(\hat{e}_n,\hat{e}_1) y_1 + \cdots + B(\hat{e}_n,\hat{e}_n) y_n\right]$$

$$= \sum_{i=1}^{n} x_i\left[B(\hat{e}_i,\hat{e}_1) y_1 + \cdots + B(\hat{e}_i,\hat{e}_n) y_n\right] = \sum_{i=1}^{n} x_i\left[\sum_{j=1}^{n} y_j B(\hat{e}_i,\hat{e}_j)\right].$$

Note that

$$(\hat{x}_1 + \hat{x}_2)^T\left[M(\mathcal{D})\hat{y}\right] = (\hat{x}_1)^T\left[M(\mathcal{D})\hat{y}\right] + (\hat{x}_2)^T\left[M(\mathcal{D})\hat{y}\right]$$

and

$$(\alpha\hat{x})^T\left[M(\mathcal{D})\hat{y}\right] = \alpha\left(\hat{x}^T\left[M(\mathcal{D})\hat{y}\right]\right)$$

$$B(\alpha\hat{u} + \beta\hat{v},\hat{w}) = \alpha B(\hat{u},\hat{w}) + \beta B(\hat{v},\hat{w}) \text{ for } \hat{u},\hat{v},\hat{w} \in V; \alpha,\beta \in \mathbb{R}.$$

Likewise, there is linearity in the second argument.

The representation of $\hat{u}$ and $\hat{v}$ in the calculation of $B\left(\hat{u},\hat{w}\right)$ above depends on the basis chosen for V.

For a different basis, we must use a different matrix in the computation. Later in the chapter, we show that $B\left(\hat{u},\hat{w}\right)$ is independent of the basis.

Definition:

We say B is nondegenerate if whenever $B\left(\hat{x},\hat{y}\right)=0$ for every $\hat{x}$, then $\hat{y}=0$ and if $B\left(\hat{x},\hat{y}\right)=0$ for every $\hat{y}$, then $\hat{x}=\hat{0}$.

Theorem 13.2:

Let V be a finite dimensional vector space with bilinear form B. Then B is nondegenerate if and only if the matrix of B is invertible.

Proof: Let $[B]$ denote the matrix for B with respect to some basis. Now $[B]$ is not invertible if and only if $[B]\,\hat{x}=\hat{0}$ for some $\hat{x}\neq\hat{0}$. This is true if and only if $\hat{v}^T\left[B\right]\hat{x}=\hat{0}$ for all $\hat{v}\ \varepsilon\ V$. Thus $[B]$ is not invertible if and only if B is not nondegenerate.

Definition:

A bilinear form B is symmetric if $B\left(\hat{u},\hat{v}\right)=B\left(\hat{v},\hat{u}\right)$ for all $\hat{u},\hat{v}\in V$.
A bilinear form B is anti-symmetric if $B\left(\hat{u},\hat{v}\right)=-B\left(\hat{v},\hat{u}\right)$ for all $\hat{u},\hat{v}\in V$.
A bilinear form B is alternating if $B\left(\hat{u},\hat{u}\right)=0$ for all $\hat{u}\in V$.
A bilinear form B is symmetric (skew-symmetric, alternating) if and only if the matrix representation is symmetric (skew-symmetric, alternating) as we now show.

Theorem 13.3:

If the matrix of a bilinear form is symmetric in one basis, then it is symmetric in every basis.

Proof:

Suppose that B is a bilinear form and B has the symmetric matrix M as its representation in the basis $\mathcal{B}_1$. If $\mathcal{B}_2$ is another basis for V, then there is a nonsingular matrix C (the change of basis matrix) for which the matrix representation in the $\mathcal{B}_2$ basis is C^TMC.

Now $\left(C^TMC\right)^T=\left(C\right)^T\left(M^T\right)(C^T)^T=C^TMC$ since M is symmetric.
An almost identical argument shows that if B is a bilinear form and B has the anti-symmetric matrix M in one basis, then a representation in any basis in anti-symmetric.

Note: We are considering only the cases where the scalar field is the real numbers. In bilinear and quadratic forms (a term to be defined later that is related to bilinear forms) special restrictions arise if the scalar field has characteristic 2.

Theorem 13.4:

A bilinear form is skew-symmetric if and only if it is alternating.

Proof:

For B an alternating bilinear form on V we have

$$0 = B(\hat{u} + \hat{v}, \hat{u} + \hat{v}) = B(\hat{u},\hat{u}) + B(\hat{u},\hat{v}) + B(\hat{v},\hat{u}) + B(\hat{v},\hat{v})$$
$$= B(\hat{u},\hat{v}) + B(\hat{v},\hat{u})$$

$$\text{so } B(\hat{u},\hat{v}) = -B(\hat{v},\hat{u}).$$

For B a skew-symmetric bilinear form on V we have

$$B(\hat{u},\hat{u}) = -B(\hat{u},\hat{u}) \text{ so } B(\hat{u},\hat{u}) = 0.$$

Theorem 13.5:

Every bilinear form can be uniquely written as the sum of a symmetric bilinear form and an alternating form.

Proof:

Suppose that B *is a* bilinear form on V. For $\hat{u}, \hat{v} \in V$ define

$$B_1(\hat{u},\hat{v}) = \frac{B(\hat{u},\hat{v}) + B(\hat{v},\hat{u})}{2} \text{ and } B_2(\hat{u},\hat{v}) = \frac{B(\hat{u},\hat{v}) - B(\hat{v},\hat{u})}{2}$$

then

$$B_1(\hat{v},\hat{u}) = \frac{B(\hat{v},\hat{u}) + B(\hat{u},\hat{v})}{2} = B_1(\hat{u},\hat{v})$$

and

$$B_2(\hat{v},\hat{u}) = \frac{B(\hat{v},\hat{u}) - B(\hat{u},\hat{v})}{2} = -\frac{B(\hat{u},\hat{v}) - B(\hat{v},\hat{u})}{2} = -B_2(\hat{u},\hat{v}).$$

Note that this is consistent with the fact that every $n \times n$ matrix can be written as the sum of a symmetric and a skew-symmetric matrix.

Theorem 13.6:

A symmetric bilinear form is determined by the values on the diagonal.

Proof:

Let B be a bilinear form on the vector space, and suppose $\hat{u}, \hat{v} \in V$. Then

$$B(\hat{u}+\hat{v},\hat{u}+\hat{v}) - B(\hat{u},\hat{u}) - B(\hat{v},\hat{v})$$
$$= \left[B(\hat{u},\hat{u}) + B(\hat{u},\hat{v}) + B(\hat{v},\hat{u}) + B(\hat{v},\hat{v})\right] - B(\hat{u},\hat{u}) - B(\hat{v},\hat{v})$$
$$= B(\hat{u},\hat{v}) + B(\hat{v},\hat{u}) = 2B(\hat{u},\hat{v}).$$

Change of Basis

Suppose that V is a finite dimensional vector space over $\mathbb{R}$.

Given a bilinear form B and an ordered basis $\mathcal{B}$, there is a matrix $M_{\mathcal{B}}$ for which $B(\hat{x},\hat{y}) = \hat{x}^T M_{\mathcal{B}} \hat{y}$ for all $\hat{x}, \hat{y} \in V$. The matrix $M_{\mathcal{B}}$ is defined by

$(M_{\mathcal{B}})_{i,j} = B(e_i, e_j)$ if $\mathcal{B}$ is the ordered basis $\{\hat{e}_1, \ldots, \hat{e}_n\}$.

Suppose that $\mathcal{B}_1$ is another ordered basis of V. If $\hat{u} \in V$ then $\hat{u}$ has a representation in the basis $\mathcal{B}$, which we denote $[\hat{u}]_{\mathcal{B}}$ and a representation in the basis $\mathcal{B}_1$, which we denote $[\hat{u}]_{\mathcal{B}_1}$.

There is a unitary matrix $P\left(P^T = P^{-1}\right)$, the change of basis matrix, for which $[\hat{u}]_{\mathcal{B}} = P[\hat{u}]_{\mathcal{B}_1}$.
Now

$$B(\hat{x},\hat{y}) = \left[\hat{x}^T\right]_{\mathcal{B}} M_{\mathcal{B}} [\hat{y}]_{\mathcal{B}} = \left[(P\hat{x})^T\right]_{\mathcal{B}_1} M_{\mathcal{B}} P[\hat{y}]_{\mathcal{B}_1}$$
$$= \left[(\hat{x})^T\right]_{\mathcal{B}_1} P^T M_{\mathcal{B}} P[\hat{y}]_{\mathcal{B}_1}.$$

Note that since $M_{\mathcal{B}}$ is symmetric,

$$\left(P^T M_{\mathcal{B}} P\right)^T = (P^T)(M_{\mathcal{B}})^T (P^T)^T = P^T M_{\mathcal{B}} P$$

so $P^T M_{\mathcal{B}} P$ is symmetric and

$$M_{\mathcal{B}_1} = P^T M_{\mathcal{B}} P.$$

By the Spectral theorem, if $M_{\mathcal{B}}$ is a symmetric matrix, then there is a basis $\mathcal{B}_1$ for which the operator represented by $M_{\mathcal{B}}$ is diagonal.

The Spectral Theorem says, that If A is a symmetric matrix, then there is a unitary matrix P and a diagonal matrix D whose diagonal entries are the eigenvalues of A for which

$$A = PDP^T.$$

The columns of P are the normalized eigenvectors of A.

Exercises

Find the bilinear form corresponding to the matrices below, using the standard basis.

(a.) $\begin{pmatrix} 1 & 0 \\ 2 & 0 \end{pmatrix}$

(b.) $\begin{pmatrix} 0 & -1 \\ 1 & 0 \end{pmatrix}$

$$(c.) \begin{pmatrix} 3 & 2 \\ -1 & 4 \end{pmatrix}$$

$$(d.) \begin{pmatrix} 1 & 2 & -1 \\ 3 & 0 & 2 \\ 0 & 0 & 0 \end{pmatrix}$$

Section 13.2 Orthogonality on Bilinear Forms

Bilinear forms generalize the dot product, and we will define what it means for vectors to be orthogonal with respect to a bilinear form. There will be some important differences to keep in mind as we note below.

Definition:

If B is a bilinear form on the vector space V and $\hat{v}$ and $\hat{w}$ are vectors in V with $B(\hat{v}, \hat{w}) = 0$, then we say that $\hat{v}$ is orthogonal to $\hat{w}$ which is denoted $\hat{v} \perp \hat{w}$.

It might be better to say that $\hat{v}$ is orthogonal to $\hat{w}$ with respect to B but this is normally clear from the context and is not usually done.

There are some things to be cautious about with bilinear forms, because of the contrast with dot products. One is that $\hat{v} \perp \hat{w}$ does not necessarily mean that $\hat{w} \perp \hat{v}$ (although it does if the bilinear form is symmetric). Another is that it is possible for $\hat{v} \perp \hat{v}$ and $\hat{v} \neq \hat{0}$.

Theorem 13.7:

The relation $\hat{v} \perp \hat{w}$ implies $\hat{w} \perp \hat{v}$ if and only if the bilinear form is symmetric or alternating.

Proof:

Exercise:

In discussing orthogonality, it is most often the case that the bilinear form is symmetric. In this case, we define orthogonal subspaces.

Definition:

If B is a symmetric bilinear form on V and if W is a subspace of V, we define $W^\perp = \{\hat{v} \in V | B(\hat{v}, \hat{w})$
$= 0$ for all $\hat{w} \in W\}$.

The set $W^\perp$ is often called the orthogonal complement of W although it is not always the case that $W + W^\perp = V$. The next theorem gives the basic results on orthogonality.

Theorem 13.8:

Let V be a finite dimensional vector space and B be a symmetric bilinear form on V. Let W be a subspace of V. Then

(1.) $W^\perp$ is a subspace of V.

(2.) $V = \{\hat{0}\}^\perp$.

(3.) $V^\perp = \{\hat{0}\}$ if and only if B is nondegenerate.

(4.) If $\dim V = n$ and $\dim W = m$, then $\dim W^\perp \geq n - m$.

(5.) $B|_W$ is nondegenerate if and only if $W \cap W^\perp = \{\hat{0}\}$.

(6.) $V = W \oplus W^\perp$ if any only if $B|_W$ is nondegenerate.

Proof:

(1.) and (2.) The proof is left as an exercise.

(3.) Suppose $V^\perp = \{\hat{0}\}$.

Now $V^\perp = \{\hat{v} \in V | B(\hat{v}, \hat{y}) = 0 \text{ for all } \hat{y} \in V\} = \{\hat{0}\}$; *i.e.*,

$B(\hat{v}, \hat{y}) = 0$ for all $\hat{v}, \hat{y}$, so $\hat{v} = \hat{0}$. Since B is symmetric, $B(\hat{v}, \hat{y}) = B(\hat{y}, \hat{v})$ for all $\hat{v}, \hat{y}$. Thus, $B(\hat{v}, \hat{y}) = 0$ for all $\hat{v} \in V$, so $\hat{y} = \hat{0}$. Thus B is nondegenerate.

Note that what we have done is to show there is not a non-zero vector $\hat{v}$ for which $B(\hat{v}, \hat{v}) = 0$.

(4.) Let $\{\hat{w}_1, \ldots, \hat{w}_m\}$ be a basis for W. Define a function $T : V \to \mathbb{R}^m$ by

$$T(\hat{x}) = \left(B(\hat{x}, \hat{w}_1), \ldots, B(\hat{x}, \hat{w}_m) \right).$$

If $T(\hat{x}) = (0, \ldots, 0)$, then $B(\hat{x}, \hat{w}_i) = 0, i = 1, \ldots, m$, so $B(\hat{x}, \hat{w}) = 0$ for all $\hat{w} \in W$. Thus $\hat{x} \in W^\perp$.

Thus, the null space of T, $\mathcal{N}(T)$, is a subspace of $W^\perp$.

This means $\dim \mathcal{N}(T) \leq \dim W^\perp$.

Since T is a linear transformation and V is a finite dimensional vector space, we have
$$\dim V = \dim \mathcal{R}(T) + \dim \mathcal{N}(T). \text{ So}$$
$$\dim \mathcal{N}(T) = \dim V - \dim \mathcal{R}(T) \geq \dim V - m = n - m. \text{ Thus } n - m \leq \dim W^\perp.$$

(5.) Suppose that $B|_W$ is nondegenerate and $\hat{z} \in W \cap W^\perp$. Since $\hat{z} \in W^\perp$, $B(\hat{z}, \hat{w}) = 0$ for all $\hat{z} \in W$ because B is nondegenerate on W. Thus
$$W \cap W^\perp = \{\hat{0}\}.$$
The converse is immediate.

(6.) Suppose $V = W \oplus W^\perp$. Then $W \cap W^\perp = \{\hat{0}\}$, so by (5.) $B|_W$ is nondegenerate.

Conversely, assume $B|_W$ is nondegenerate. We will show $V = W \oplus W^\perp$. By (5.), we have $W \cap W^\perp = \{\hat{0}\}$ and so we only need to show $V = W + W^\perp$. This is true if and only if $\dim W + \dim W^\perp = \dim V$.

By 4., we have if $\dim V = n$ and $\dim W = m$, then $\dim W^\perp \geq n - m$. Thus $\dim W + \dim W^\perp \geq m + n - m = n = \dim V$. Thus $V = W + W^\perp$.

Section 13.3 Quadratic Forms

A quadratic form in the variables $x_1, \ldots, x_n$ is an expression of the form

$$a_1 x_1^2 + \cdots + a_n x_n^2 + \sum_{i,j; i \neq j; i, j=1}^{n} b_{ij} x_i x_j$$

that is, a homogeneous polynomial of degree 2.

A quadratic form on V can be expressed as

$$\hat{x}^T A \hat{x}$$

where $\hat{x} \in V$ and A is a symmetric matrix.

Since a symmetric bilinear form is associated with a symmetric matrix, it should not be a surprise that symmetric bilinear forms and quadratic forms have a close association.

Let B be a symmetric bilinear form on the finite dimensional vector space V. Define $Q : V \to \mathbb{R}$ by $Q(\hat{v}) = B(\hat{v}, \hat{v})$.

We demonstrate in later examples that Q is a quadratic form.

We recover a symmetric bilinear form B from Q by

$$B(\hat{u}, \hat{v}) = \frac{1}{2} \left[Q(\hat{u} + \hat{v}) - Q(\hat{u}) - Q(\hat{v}) \right].$$

If A is a symmetric matrix, then $Q(\hat{x}) = \hat{x}^T A \hat{x}$ is a quadratic form, and the Spectral Theorem says that there is a basis of V for which a quadratic form has no cross terms.

Classifying Quadratic Forms

Definition:

When A is an $n \times n$ symmetric matrix the quadratic form $Q(\hat{x}) = \hat{x}^T A \hat{x}$

(1.) is positive definite if $Q(\hat{x}) > 0$ for all $\hat{x} \in V$. This is true if and only if all the eigenvalues of A are positive.

(2.) is positive semidefinite if $Q(\hat{x}) \geq 0$ for all $\hat{x} \in V$.

(3.) is negative definite if $Q(\hat{x}) < 0$ for all $\hat{x} \in V$. This is true if and only if all the eigenvalues of A are negative.

(4.) is negative semidefinite if $Q(\hat{x}) \le 0$ for all $\hat{x} \in V$

(5.) is indefinite if $Q(\hat{x})$ takes on both positive and negative values.

Singular Values

If A is an $m \times n$ matrix, then $A^T A$ is a symmetric matrix since

$$\left(A^T A\right)^T = A^T \left(A^T\right)^T = A^T A.$$

Furthermore, the eigenvalues of $A^T A$ are nonnegative since, if $\hat{v}$ is a normalized eigenvector of $A^T A$ with eigenvalue λ then

$$\|A\hat{v}\|^2 = \left(A\hat{v}\right)^T \left(A\hat{v}\right) = \hat{v}^T \left(A^T A\right)\hat{v} = \hat{v}^T \lambda \hat{v} = \lambda.$$

Order the eigenvalues of $A^T A$ by

$$\lambda_1 \ge \lambda_2 \ge \cdots \ge \lambda_n \ge 0.$$

The singular values of A are $\sigma_i = \sqrt{\lambda_i}$. These are the lengths of $A\hat{v}_1, \ldots, A\hat{v}_n$ where $\hat{v}_i$ are the eigenvectors of $A^T A$.

Signature of Quadratic Forms

Let V be a finite dimensional vector space over $\mathbb{R}$. We demonstrate Sylvester's law of inertia.
if B is a symmetric nondegenerate bilinear form, then there is a basis of V in which the matrix is a diagonal matrix and the diagonal entries are 1 or –1.

This gives the signature of the matrix, which is the number of positive ones in the transformed quadratic form and the number of negative ones.

Sylvester's law of inertia states that the signature of the bilinear form is an invariant.

The theory for finding the matrix that gives the signature of a quadratic form is simple. If Q is the matrix of the quadratic form, then Q is symmetric and can be diagonalized by a matrix whose columns are the eigenvectors.

Thus resulting matrix will be

$$\begin{pmatrix} \lambda_1 & 0 & \cdots & 0 \\ 0 & \lambda_2 & & \vdots \\ 0 & & \ddots & 0 \\ 0 & \cdots & 0 & \lambda_{n1} \end{pmatrix}.$$

This can be put into the desired form by multiplying on the left and right by the symmetric matrix

$$Q = \begin{pmatrix} \dfrac{1}{\sqrt{|\lambda_1|}} & 0 & & \cdots & 0 \\[2ex] 0 & \dfrac{1}{\sqrt{|\lambda_2|}} & & & \vdots \\[2ex] & & \ddots & & 0 \\[1ex] 0 & & & & \dfrac{1}{\sqrt{|\lambda_n|}} \\ 0 & \cdots & & 0 & \end{pmatrix}$$

for the non-zero eigenvalues.

Example:

Let

$$Q(\hat{x}) = 4x^2 + 2xy + 2xz + 4y^2 + 2yz + 4z^2.$$

The matrix for Q is

$$\begin{pmatrix} 4 & 1 & 1 \\ 1 & 4 & 1 \\ 1 & 1 & 4 \end{pmatrix}.$$

The eigenvectors for the matrix are

$$\hat{v}_1 = \begin{pmatrix} -1 \\ 1 \\ 0 \end{pmatrix}, \ \hat{v}_2 = \begin{pmatrix} -1 \\ 0 \\ 1 \end{pmatrix}, \ \hat{v}_3 = \begin{pmatrix} -1 \\ 1 \\ 0 \end{pmatrix}.$$

An orthonormal basis of eigenvectors is

$$\hat{u}_1 = \begin{pmatrix} -\dfrac{1}{\sqrt{2}} \\[2ex] \dfrac{1}{\sqrt{2}} \\[2ex] 0 \end{pmatrix}, \ \hat{u}_2 = \begin{pmatrix} -\dfrac{\sqrt{6}}{6} \\[2ex] -\dfrac{\sqrt{6}}{6} \\[2ex] \dfrac{\sqrt{6}}{3} \end{pmatrix}, \ \hat{u}_3 = \begin{pmatrix} \dfrac{1}{\sqrt{3}} \\[2ex] \dfrac{1}{\sqrt{3}} \\[2ex] \dfrac{1}{\sqrt{3}} \end{pmatrix}.$$

Let

$$P = \begin{pmatrix} -\dfrac{1}{\sqrt{2}} & -\dfrac{\sqrt{6}}{6} & \dfrac{1}{\sqrt{3}} \\[2ex] \dfrac{1}{\sqrt{2}} & -\dfrac{\sqrt{6}}{6} & \dfrac{1}{\sqrt{3}} \\[2ex] 0 & \dfrac{\sqrt{6}}{3} & \dfrac{1}{\sqrt{3}} \end{pmatrix}.$$

We have

$$P^T Q P = \begin{pmatrix} 3 & 0 & 0 \\ 0 & 3 & 0 \\ 0 & 0 & 6 \end{pmatrix}.$$

Now

$$\begin{pmatrix} \frac{1}{\sqrt{3}} & 0 & 0 \\ 0 & \frac{1}{\sqrt{3}} & 0 \\ 0 & 0 & \frac{1}{\sqrt{6}} \end{pmatrix} \begin{pmatrix} 3 & 0 & 0 \\ 0 & 3 & 0 \\ 0 & 0 & 6 \end{pmatrix} \begin{pmatrix} \frac{1}{\sqrt{3}} & 0 & 0 \\ 0 & \frac{1}{\sqrt{3}} & 0 \\ 0 & 0 & \frac{1}{\sqrt{6}} \end{pmatrix} = \begin{pmatrix} 1 & 0 & 0 \\ 0 & 1 & 0 \\ 0 & 0 & 1 \end{pmatrix}.$$

Thus the signature of the quadratic form is 3,0.

This example was computationally simple because the eigenvalues and eigenvectors were easily expressible. We next present an example where the computations are not as simple and another method is used.

Example

Let

$$Q(x) = x^2 + 2xy - 4xz + 2yz - 6z^2.$$

The matrix of this quadratic form is

$$Q = \begin{pmatrix} 1 & 1 & -2 \\ 1 & 0 & 1 \\ -2 & 1 & -6 \end{pmatrix}.$$

The eigenvalues of Q are $-1.16098\ldots$, $1.40174\ldots$, $6.75923\ldots$, and computations as in the previous example seem impractical. Instead, we present an alternate procedure.

We want to express the quadratic form as the sum of three squares. We have

$$(x + y - 2z)^2 = x^2 + 2xy - 4xz + y^2 - 4yz + 4z^2$$

so

$$x^2 + 2xy - 4xz = (x + y - 2z)^2 - y^2 + 4yz - 4z^2.$$

Then

$$\begin{aligned} Q(x) &= (x^2 + 2xy - 4xz) + 2yz - 6z^2 \\ &= \left[(x + y - 2z)^2 - y^2 + 4yz - 4z^2\right] + 2yz - 6z^2 \\ &= (x + y - 2z)^2 - y^2 + 6yz - 10z^2 \\ &= (x + y - 2z)^2 - (y^2 - 6yz + 9z^2) + 9z^2 - 10z^2 \\ &= (x + y - 2z)^2 - (y - 3z)^2 - z^2. \end{aligned}$$

Next, we find the change of basis matrix. We seek $\hat{v}_1, \hat{v}_2$ and $\hat{v}_3$ for which

$$\begin{pmatrix} x \\ y \\ z \end{pmatrix} = (x + y - 2z)\hat{v}_1 + (y - 3z)\hat{v}_2 + z\hat{v}_3.$$

Set

$$\hat{v}_1 = \begin{pmatrix} a \\ b \\ c \end{pmatrix}, \hat{v}_2 = \begin{pmatrix} d \\ e \\ f \end{pmatrix}, \hat{v}_3 = \begin{pmatrix} g \\ h \\ i \end{pmatrix}.$$

Then

$$\begin{pmatrix} x \\ y \\ z \end{pmatrix} = (x + y - 2z)\begin{pmatrix} a \\ b \\ c \end{pmatrix} + (y - 3z)\begin{pmatrix} d \\ e \\ f \end{pmatrix} + z\begin{pmatrix} g \\ h \\ i \end{pmatrix}.$$

So

$$x = (x + y - 2z)a + (y - 3z)d + zg. \qquad \text{Thus, } a = 1, d = -1, g = -1.$$

$$y = (x + y - 2z)b + (y - 3z)e + hz. \qquad \text{Thus, } b = 0, e = 1, h = 3.$$

$$z = (x + y - 2z)c + (y - 3z)f + zi. \qquad \text{Thus, } c = 0, f = 0, i = 1.$$

Hence

$$\hat{v}_1 = \begin{pmatrix} 1 \\ 0 \\ 0 \end{pmatrix}, \hat{v}_2 = \begin{pmatrix} -1 \\ 1 \\ 0 \end{pmatrix}, \hat{v}_3 = \begin{pmatrix} -1 \\ 3 \\ 1 \end{pmatrix}.$$

The change of basis matrix is

$$P = (\hat{v}_1, \hat{v}_2, \hat{v}_3) = \begin{pmatrix} 1 & -1 & -1 \\ 0 & 1 & 3 \\ 0 & 0 & 1 \end{pmatrix}.$$

$$P^T AP = \begin{pmatrix} 1 & 0 & 0 \\ -1 & 1 & 0 \\ -1 & 3 & 1 \end{pmatrix} \begin{pmatrix} 1 & 1 & -2 \\ 1 & 0 & 1 \\ -2 & 1 & -6 \end{pmatrix} \begin{pmatrix} 1 & -1 & -1 \\ 0 & 1 & 3 \\ 0 & 0 & 1 \end{pmatrix} = \begin{pmatrix} 1 & 0 & 0 \\ 0 & -1 & 0 \\ 0 & 0 & -1 \end{pmatrix}.$$

There are one 1 and two -1s, so the signature is 1,2. This says the number of positive ones in the transformed coordinates is one and the number of negative ones is two.

Exercises

For the quadratic forms below, find the basis that corroborates Sylvester's law of inertia and give the signature of the matrix.

1. $Q(x) = x^2 - 4xy$
2. $Q(x) = x^2 + xy - 3xz + y^2 - 2yz - z^2$
3. $Q(x) = -2x^2 + xz + yz - 3z^2$

Chapter 14

Introduction to Tensor Product

This chapter is intended to be a superficial introduction to the tensor product of two vector spaces. We discuss some of the theory of tensor products and an example of a particular tensor product.

The construction of the tensor product of two finite dimensional vector spaces U and V over the same field starts by taking the external direct sum of the vector spaces. In Chapter 3 we discussed the direct sum of subspaces, which would be more properly called the internal direct sum.

Recapping that construction, if U and V are subspaces of the vector space W with $W = U + V$ and $U \cap V = \{\hat{0}\}$ then W is the direct sum of U and V.

A second direct sum is an external direct sum. Here we have two vector spaces U and V over the same field from which we form the Cartesian product $U \times V$.

Vector addition is defined by

$$\left(\hat{u}_1, \hat{v}_1\right) + \left(\hat{u}_2, \hat{v}_2\right) = \left(\hat{u}_1 + \hat{u}_2, \hat{v}_1 + \hat{v}_2\right)$$

And scalar multiplication is defined by

$$\alpha\left(\hat{u}, \hat{v}\right) = \left(\alpha\hat{u}, \alpha\hat{v}\right).$$

With tensor products, we deal with external direct sums.

We give a very brief description of the two equivalent constructions of the tensor product.

Construction 1

The tensor product of U and V is the dual space of the set of bilinear forms on $U \oplus V$.

Explanation:

The representation of a bilinear form depends on the basis of the vector space. Suppose that $\{\hat{u}_1, \ldots, \hat{u}_m\}$ is a basis for U and $\{\hat{v}_1, \ldots, \hat{v}_n\}$ is a basis for V.

Each bilinear form with respect to given bases is represented by a matrix, and the set of these matrices is a vector space.

A basis for this vector space is the set of matrices

$$\{E_{ij}|i=1,\ldots,m,\ j=1,\ldots,n\} \text{ with } E_{ij}\left(a,b\right)=\begin{cases}1 & \text{if } i=a \text{ and } j=b \\ 0 & \text{otherwise}\end{cases}$$

So the bilinear functional E_{ij} has the property

$$\hat{e}_r^T E_{ij}\hat{e}_s = \delta_{ri}\delta_{sj}$$

where $\hat{e}_r$ is the rth standard basis vector in U and $\hat{e}_s$ is the sth standard basis vector in V. The linear functional determined by $\hat{e}_r$ and $\hat{e}_s$ is denoted $e_r \otimes e_s$. Because these linear functionals are derived from bilinear forms, we have

$$x \otimes \left(y+z\right) = x \otimes y + x \otimes z,$$

$$\left(x+y\right) \otimes z = x \otimes z + y \otimes z,$$

$$\alpha\left(x \otimes y\right) = \left(\alpha x\right) \otimes y = x \otimes \left(\alpha y\right)$$

Example

Let $U = \mathbb{R}^2$ and $V = \mathbb{R}^3$ with bases

$$\mathcal{B}_1 = \{u_1,u_2\} = \left\{\begin{pmatrix}1 \\ 0\end{pmatrix},\begin{pmatrix}0 \\ 1\end{pmatrix}\right\} \text{ and } \mathcal{B}_2 = \{v_1,v_2,v_3\} = \left\{\begin{pmatrix}1 \\ 0 \\ 0\end{pmatrix},\begin{pmatrix}0 \\ 1 \\ 0\end{pmatrix},\begin{pmatrix}0 \\ 0 \\ 1\end{pmatrix}\right\}$$

respectively. Define the bilinear forms w_{rs} on $\mathbb{R}^2 \oplus \mathbb{R}^3$ by

$$w_{rs}\left\langle u_i,v_j\right\rangle = \delta_{ri}\delta_{sj}.$$

Then a bilinear form w on $\mathbb{R}^2 \oplus \mathbb{R}^3$ can be expressed as

$$w = \sum_{i=1}^{2}\sum_{j=1}^{3}\alpha_{ij}w_{ij}.$$

Suppose that

$$u = \begin{pmatrix}a_1 \\ a_2\end{pmatrix}, \quad v = \begin{pmatrix}b_1 \\ b_2 \\ b_3\end{pmatrix}, \quad \text{and } w = \sum_{i=1}^{2}\sum_{j=1}^{3}\alpha_{ij}w_{ij}$$

so that

$$u = a_1 u_1 + a_2 u_2 \quad \text{and} \quad v = b_1 v_1 + b_2 v_2 + b_3 v_3$$

and

$$(u \otimes v)(w) = w \langle u, v \rangle$$

$$= \alpha_{11} w_{11} \langle u, v \rangle + \alpha_{12} w_{12} \langle u, v \rangle + \alpha_{13} w_{13} \langle u, v \rangle + \alpha_{21} w_{21} \langle u, v \rangle$$

$$+ \alpha_{22} w_{22} \langle u, v \rangle + \alpha_{23} w_{23} \langle u, v \rangle.$$

Now,

$$\langle u, v \rangle = \langle a_1 u_1 + a_2 u_2, b_1 v_1 + b_2 v_2 + b_3 v_3 \rangle$$

$$= a_1 b_1 \langle u_1, v_1 \rangle + a_1 b_2 \langle u_1, v_2 \rangle + a_1 b_3 \langle u_1, v_3 \rangle$$

$$+ a_2 b_1 \langle u_2, v_1 \rangle + a_2 b_2 \langle u_2, v_2 \rangle + a_2 b_3 \langle u_2, v_3 \rangle$$

so that

$$\alpha_{ij} w_{ij} \langle u, v \rangle = \alpha_{ij} a_i b_j w_{ij} \langle u_i, v_j \rangle = \alpha_{ij} a_i b_j.$$

Thus,

$$(u \otimes v)(w) = \sum_{i=1}^{2} \sum_{j=1}^{3} \alpha_{ij} a_i b_j.$$

We note that

$$(u \otimes v)(w) = \sum_{i=1}^{2} \sum_{j=1}^{3} \alpha_{ij} a_i b_j = \begin{pmatrix} a_1 b_1 \\ a_1 b_2 \\ a_1 b_3 \\ a_2 b_1 \\ a_2 b_2 \\ a_2 b_3 \end{pmatrix} \cdot \begin{pmatrix} \alpha_{11} \\ \alpha_{12} \\ \alpha_{13} \\ \alpha_{21} \\ \alpha_{22} \\ \alpha_{23} \end{pmatrix}$$

and some sources write

$$u \otimes v = \begin{pmatrix} a_1 \\ a_2 \end{pmatrix} \otimes \begin{pmatrix} b_1 \\ b_2 \\ b_3 \end{pmatrix} = \begin{pmatrix} a_1 b_1 \\ a_1 b_2 \\ a_1 b_3 \\ a_2 b_1 \\ a_2 b_2 \\ a_2 b_3 \end{pmatrix} \quad \text{and} \quad w = \begin{pmatrix} \alpha_{11} \\ \alpha_{12} \\ \alpha_{13} \\ \alpha_{21} \\ \alpha_{22} \\ \alpha_{23} \end{pmatrix}.$$

We also note that

$$v \otimes u = \begin{pmatrix} b_1 \\ b_2 \\ b_3 \end{pmatrix} \otimes \begin{pmatrix} a_1 \\ a_2 \end{pmatrix} = \begin{pmatrix} b_1 a_1 \\ b_1 a_2 \\ b_2 a_1 \\ b_2 a_2 \\ b_3 a_1 \\ b_3 a_2 \end{pmatrix}$$

so

$$u \otimes v \neq v \otimes u.$$

We note that $u \otimes v$ does not depend on the choice of basis, though the proof of this fact is beyond the scope of this book.

Construction 2

In this formulation, the tensor product of U and V is the quotient space of two vector spaces.

Suppose that $\{\hat{u}_1, \ldots, \hat{u}_m\}$ is a basis for U and $\{\hat{v}_1, \ldots, \hat{v}_n\}$ is a basis for V.

We create the symbols $u_i \otimes v_j$. Whatever these new symbols are, we define them as being linearly independent. We construct a vector space W that is the set of linear combinations

$$\sum_{i=1, j=1}^{i=m, j=n} c_{ij} u_i \otimes v_j$$

and the subspace generated by

$$\left\{ \begin{matrix} (a+b) \otimes c - a \otimes c - b \otimes c, & a \otimes (b+c) - a \otimes b - a \otimes c, \alpha (a \otimes b) - (\alpha a) \otimes b, \\ \alpha (a \otimes b) - (a \otimes \alpha b) \end{matrix} \right\}$$

we denote it by X. The tensor product of U and V is the quotient space W/X.

Example of a Particular Tensor Product

We describe the so-called covariant tensor product on $\mathbb{R}^2 \times \mathbb{R}^3$. In this example we see several characteristics of tensor products.

We denote basis column vectors by $\hat{e}_i$ and basis row vectors by $\hat{e}^i$; e.g.,

$$\hat{e}_2 = \begin{pmatrix} 0 \\ 1 \\ 0 \end{pmatrix} \text{ in } \mathbb{R}^3 \text{ and } \hat{e}^1 = (1,0) \text{ in } \mathbb{R}^2.$$

We denote a column vector by a and the companion row vector by a^T so if $a = \begin{pmatrix} a_1 \\ a_2 \\ a_3 \end{pmatrix}$ then $a^T = (a_1 \ a_2 \ a_3)$ so, for example, $(\hat{e}_2)^T = \hat{e}^2$.

$$\text{For } a = \begin{pmatrix} a_1 \\ a_2 \\ a_3 \end{pmatrix} \text{ and } b = \begin{pmatrix} b_1 \\ b_2 \end{pmatrix}$$

We define $a \otimes b$ by

$$a \otimes b = \begin{pmatrix} a_1 \\ a_2 \\ a_3 \end{pmatrix} \otimes \begin{pmatrix} b_1 \\ b_2 \end{pmatrix} = \begin{pmatrix} a_1\begin{pmatrix} b_1 \\ b_2 \end{pmatrix} \\ a_2\begin{pmatrix} b_1 \\ b_2 \end{pmatrix} \\ a_3\begin{pmatrix} b_1 \\ b_2 \end{pmatrix} \end{pmatrix} = \begin{pmatrix} a_1b_1 \\ a_1b_2 \\ a_2b_1 \\ a_2b_2 \\ a_3b_1 \\ a_3b_2 \end{pmatrix}.$$

It is easy to check that $f(a,b) = a \otimes b$ is bilinear.

Note that while $a \otimes b$ and $b \otimes a$ are both in $\mathbb{R}^5$, $a \otimes b \neq b \otimes a$.

Also

$$a^T \otimes b^T = \begin{pmatrix} a_1 & a_2 & a_3 \end{pmatrix} \otimes \begin{pmatrix} b_1 & b_2 \end{pmatrix}$$
$$= \begin{pmatrix} a_1\begin{pmatrix} b_1 & b_2 \end{pmatrix} & a_2\begin{pmatrix} b_1 & b_2 \end{pmatrix} & a_3\begin{pmatrix} b_1 & b_2 \end{pmatrix} \end{pmatrix}$$
$$= \begin{pmatrix} a_1b_1 & a_1b_2 & a_2b_1 & a_2b_2 & a_3b_1 & a_3b_2 \end{pmatrix}$$

so

$$a^T \otimes b^T = (a \otimes b)^T.$$

We also have

$$a \otimes b^T = \begin{pmatrix} a_1 \\ a_2 \\ a_3 \end{pmatrix} \otimes \begin{pmatrix} b_1 & b_2 \end{pmatrix} = \begin{pmatrix} a_1\begin{pmatrix} b_1 & b_2 \end{pmatrix} \\ a_2\begin{pmatrix} b_1 & b_2 \end{pmatrix} \\ a_3\begin{pmatrix} b_1 & b_2 \end{pmatrix} \end{pmatrix} = \begin{pmatrix} a_1b_1 & a_1b_2 \\ a_2b_1 & a_2b_2 \\ a_3b_1 & a_3b_2 \end{pmatrix}$$

and

$$b^T \otimes a = \begin{pmatrix} b_1 & b_2 \end{pmatrix} \otimes \begin{pmatrix} a_1 \\ a_2 \\ a_3 \end{pmatrix} = \begin{pmatrix} b_1\begin{pmatrix} a_1 \\ a_2 \\ a_3 \end{pmatrix} & b_2\begin{pmatrix} a_1 \\ a_2 \\ a_3 \end{pmatrix} \end{pmatrix} = \begin{pmatrix} b_1a_1 & b_2a_1 \\ b_1a_2 & b_2a_2 \\ b_1a_3 & b_2a_3 \end{pmatrix}$$

so

$$a \otimes b^T = b^T \otimes a.$$

Example

Suppose that

$$u = \begin{pmatrix} u_1 \\ u_2 \end{pmatrix}, \ v = \begin{pmatrix} v_1 \\ v_2 \\ v_3 \end{pmatrix} \text{ and } w = \begin{pmatrix} w_1 \\ w_2 \\ w_3 \end{pmatrix}.$$

Then

$$(u \otimes v) + (u \otimes w) = \begin{pmatrix} u_1 v_1 \\ u_1 v_2 \\ u_1 v_3 \\ u_2 v_1 \\ u_2 v_2 \\ u_2 v_3 \end{pmatrix} + \begin{pmatrix} u_1 w_1 \\ u_1 w_2 \\ u_1 w_3 \\ u_2 w_1 \\ u_2 w_2 \\ u_2 w_3 \end{pmatrix} = \begin{pmatrix} u_1 v_1 + u_1 w_1 \\ u_1 v_2 + u_1 w_2 \\ u_1 v_3 + u_1 w_3 \\ u_2 v_1 + u_2 w_1 \\ u_2 v_2 + u_2 w_2 \\ u_2 v_3 + u_2 w_3 \end{pmatrix}$$

$$= \begin{pmatrix} u_1 (v_1 + w_1) \\ u_1 (v_2 + w_2) \\ u_1 (v_3 + w_3) \\ u_2 (v_1 + w_1) \\ u_2 (v_2 + w_2) \\ u_2 (v_3 + w_3) \end{pmatrix} = u \otimes (v + w).$$

We leave it as an exercise to show

$$\alpha (a \otimes b) = (\alpha a) \otimes b = a \otimes (\alpha b).$$

Suppose that

$$u = \begin{pmatrix} u_1 \\ u_2 \end{pmatrix}, \; v = \begin{pmatrix} v_1 \\ v_2 \\ v_3 \end{pmatrix} \text{ and } w = \begin{pmatrix} w_1 \\ w_2 \\ w_3 \end{pmatrix}.$$

Then

$$(u \otimes v) + (u \otimes w) = \begin{pmatrix} u_1 v_1 \\ u_1 v_2 \\ u_1 v_3 \\ u_2 v_1 \\ u_2 v_2 \\ u_2 v_3 \end{pmatrix} + \begin{pmatrix} u_1 w_1 \\ u_1 w_2 \\ u_1 w_3 \\ u_2 w_1 \\ u_2 w_2 \\ u_2 w_3 \end{pmatrix} = \begin{pmatrix} u_1 v_1 + u_1 w_1 \\ u_1 v_2 + u_1 w_2 \\ u_1 v_3 + u_1 w_3 \\ u_2 v_1 + u_2 w_1 \\ u_2 v_2 + u_2 w_2 \\ u_2 v_3 + u_2 w_3 \end{pmatrix}$$

$$= \begin{pmatrix} u_1 (v_1 + w_1) \\ u_1 (v_2 + w_2) \\ u_1 (v_3 + w_3) \\ u_2 (v_1 + w_1) \\ u_2 (v_2 + w_2) \\ u_2 (v_3 + w_3) \end{pmatrix} = u \otimes (v + w).$$

The following result can be proven using the ideas above.

Theorem 14.1:

If U, V and W are vector spaces over the field F, then

$$U \otimes (V + W) = (U \otimes V) + (U \otimes W).$$

Theorem 14.2:

Suppose that $\mathcal{B}_1 = \{u_1,\ldots,u_n\}$ is a basis for U and $\mathcal{B}_2 = \{v_1,\ldots,v_m\}$ is a basis for V. Then

$$\{u_i \otimes v_j \mid i = 1,\ldots,n;\ j = 1,\ldots,m\}$$

is a basis for $U \otimes V$.

The next example shows that not every term in $V \otimes W$ can be written as $v \otimes w$.

Example

Consider $\mathbb{R}^2 \otimes \mathbb{R}^2$. We have

$$\begin{pmatrix}1\\0\end{pmatrix}\otimes\begin{pmatrix}1\\0\end{pmatrix}=\begin{pmatrix}1\\0\\0\\0\end{pmatrix},\begin{pmatrix}1\\0\end{pmatrix}\otimes\begin{pmatrix}0\\1\end{pmatrix}=\begin{pmatrix}0\\1\\0\\0\end{pmatrix},$$

$$\begin{pmatrix}0\\1\end{pmatrix}\otimes\begin{pmatrix}1\\0\end{pmatrix}=\begin{pmatrix}0\\0\\1\\0\end{pmatrix},\begin{pmatrix}0\\1\end{pmatrix}\otimes\begin{pmatrix}0\\1\end{pmatrix}=\begin{pmatrix}0\\0\\0\\1\end{pmatrix}$$

so

$$\begin{pmatrix}a\\b\end{pmatrix}\otimes\begin{pmatrix}c\\d\end{pmatrix}=\left[a\begin{pmatrix}1\\0\end{pmatrix}+b\begin{pmatrix}0\\1\end{pmatrix}\right]\otimes\left[c\begin{pmatrix}1\\0\end{pmatrix}+d\begin{pmatrix}0\\1\end{pmatrix}\right]$$

$$=a\begin{pmatrix}1\\0\end{pmatrix}\otimes\left[c\begin{pmatrix}1\\0\end{pmatrix}+d\begin{pmatrix}0\\1\end{pmatrix}\right]+b\begin{pmatrix}0\\1\end{pmatrix}\otimes\left[c\begin{pmatrix}1\\0\end{pmatrix}+d\begin{pmatrix}0\\1\end{pmatrix}\right]$$

$$=a\begin{pmatrix}1\\0\end{pmatrix}\otimes c\begin{pmatrix}1\\0\end{pmatrix}+a\begin{pmatrix}1\\0\end{pmatrix}\otimes d\begin{pmatrix}0\\1\end{pmatrix}+b\begin{pmatrix}0\\1\end{pmatrix}\otimes c\begin{pmatrix}1\\0\end{pmatrix}$$

$$+b\begin{pmatrix}0\\1\end{pmatrix}\otimes d\begin{pmatrix}0\\1\end{pmatrix}$$

$$=ac\begin{pmatrix}1\\0\end{pmatrix}\otimes\begin{pmatrix}1\\0\end{pmatrix}+ad\begin{pmatrix}1\\0\end{pmatrix}\otimes\begin{pmatrix}0\\1\end{pmatrix}+bc\begin{pmatrix}0\\1\end{pmatrix}\otimes\begin{pmatrix}1\\0\end{pmatrix}$$

$$+bd\begin{pmatrix}0\\1\end{pmatrix}\otimes\begin{pmatrix}0\\1\end{pmatrix}$$

$$=ac\begin{pmatrix}1\\0\\0\\0\end{pmatrix}+ad\begin{pmatrix}0\\1\\0\\0\end{pmatrix}+bc\begin{pmatrix}0\\0\\1\\0\end{pmatrix}+bd\begin{pmatrix}0\\0\\0\\1\end{pmatrix}$$

$$=\begin{pmatrix}ac\\0\\0\\0\end{pmatrix}+\begin{pmatrix}0\\ad\\0\\0\end{pmatrix}+\begin{pmatrix}0\\0\\bc\\0\end{pmatrix}+\begin{pmatrix}0\\0\\0\\bd\end{pmatrix}=\begin{pmatrix}ac\\ad\\bc\\bd\end{pmatrix}.$$

We note there are no values for *a*, *b*, *c*, *d* for which

$$\begin{pmatrix} ac \\ ad \\ bc \\ bd \end{pmatrix} = \begin{pmatrix} 1 \\ 0 \\ 1 \\ 1 \end{pmatrix}$$

Since $ac = 1$, $ad = 0$ forces $d = 0$.

Tensor Product of Matrices

For matrices, if

$$A = \begin{pmatrix} a_{11} & a_{12} \\ a_{21} & a_{22} \end{pmatrix} \quad B = \begin{pmatrix} b_{11} & b_{12} & b_{13} \\ b_{21} & b_{22} & b_{23} \\ b_{31} & b_{32} & b_{33} \end{pmatrix}$$

Then, following the pattern for $u \otimes v$, it is plausible that

$$A \otimes B = \begin{pmatrix} a_{11}B & a_{12}B \\ a_{21}B & a_{22}B \end{pmatrix}$$

$$= \begin{pmatrix} a_{11}b_{11} & a_{11}b_{12} & a_{11}b_{13} & a_{12}b_{11} & a_{12}b_{12} & a_{12}b_{13} \\ a_{11}b_{21} & a_{11}b_{22} & a_{11}b_{23} & a_{12}b_{21} & a_{12}b_{22} & a_{12}b_{23} \\ a_{11}b_{31} & a_{11}b_{32} & a_{11}b_{33} & a_{12}b_{31} & a_{12}b_{32} & a_{12}b_{33} \\ a_{21}b_{11} & a_{21}b_{12} & a_{21}b_{13} & a_{22}b_{11} & a_{22}b_{12} & a_{22}b_{13} \\ a_{21}b_{21} & a_{21}b_{22} & a_{21}b_{23} & a_{22}b_{21} & a_{22}b_{22} & a_{22}b_{23} \\ a_{21}b_{31} & a_{21}b_{32} & a_{21}b_{33} & a_{22}b_{31} & a_{22}b_{32} & a_{22}b_{33} \end{pmatrix}.$$

We demonstrate the fact that $(A \otimes B)(u \otimes v) = (Au) \otimes (Bv)$ in the 2×2 case. There we have

$$A = \begin{pmatrix} a_{11} & a_{12} \\ a_{21} & a_{22} \end{pmatrix}, \ B = \begin{pmatrix} b_{11} & b_{12} \\ b_{21} & b_{22} \end{pmatrix}, \ \hat{u} = \begin{pmatrix} u_1 \\ u_2 \end{pmatrix}, \ \hat{v} = \begin{pmatrix} v_1 \\ v_2 \end{pmatrix}$$

so

$$A \otimes B = \begin{pmatrix} a_{11}b_{11} & a_{11}b_{12} & a_{12}b_{11} & a_{12}b_{12} \\ a_{11}b_{21} & a_{11}b_{22} & a_{12}b_{21} & a_{12}b_{22} \\ a_{21}b_{11} & a_{21}b_{12} & a_{22}b_{11} & a_{22}b_{12} \\ a_{21}b_{21} & a_{21}b_{22} & a_{22}b_{21} & a_{22}b_{22} \end{pmatrix},$$

$$u \otimes v = \begin{pmatrix} u_1 v_1 \\ u_1 v_2 \\ u_2 v_1 \\ u_2 v_2 \end{pmatrix}$$

and

$$\left(A \otimes B\right)\left(\hat{u} \otimes \hat{v}\right) = \begin{pmatrix} a_{11}b_{11}u_1v_1 + a_{11}b_{12}u_1v_2 + a_{12}b_{11}u_2v_1 + a_{12}b_{12}u_2v_2 \\ a_{11}b_{21}u_1v_1 + a_{11}b_{22}u_1v_2 + a_{12}b_{21}u_2v_1 + a_{12}b_{22}u_2v_2 \\ a_{21}b_{11}u_1v_1 + a_{21}b_{12}u_1v_2 + a_{22}b_{11}u_2v_1 + a_{22}b_{12}u_2v_2 \\ a_{21}b_{21}u_1v_1 + a_{21}b_{22}u_1v_2 + a_{22}b_{21}u_2v_1 + a_{22}b_{22}u_2v_2 \end{pmatrix}.$$

Also

$$A\hat{u} = \begin{pmatrix} a_{11} & a_{12} \\ a_{21} & a_{22} \end{pmatrix} \begin{pmatrix} u_1 \\ u_2 \end{pmatrix} = \begin{pmatrix} a_{11}u_1 + a_{12}u_2 \\ a_{21}u_1 + a_{22}u_2 \end{pmatrix}$$

$$B\hat{v} = \begin{pmatrix} b_{11} & b_{12} \\ b_{21} & b_{22} \end{pmatrix} \begin{pmatrix} v_1 \\ v_2 \end{pmatrix} = \begin{pmatrix} b_{11}v_1 + b_{12}v_2 \\ b_{21}v_1 + b_{22}v_2 \end{pmatrix}$$

so

$$\begin{aligned} \left(A\hat{u}\right) \otimes \left(B\hat{v}\right) &= \begin{pmatrix} a_{11}u_1 + a_{12}u_2 \\ a_{21}u_1 + a_{22}u_2 \end{pmatrix} \otimes \begin{pmatrix} b_{11}v_1 + b_{12}v_2 \\ b_{21}v_1 + b_{22}v_2 \end{pmatrix} \\ &= \begin{pmatrix} \left(a_{11}u_1 + a_{12}u_2\right)\left(b_{11}v_1 + b_{12}v_2\right) \\ \left(a_{11}u_1 + a_{12}u_2\right)\left(b_{21}v_1 + b_{22}v_2\right) \\ \left(a_{21}u_1 + a_{22}u_2\right)\left(b_{11}v_1 + b_{12}v_2\right) \\ \left(a_{21}u_1 + a_{22}u_2\right)\left(b_{21}v_1 + b_{22}v_2\right) \end{pmatrix} \\ &= \begin{pmatrix} a_{11}b_{11}u_1v_1 + a_{11}b_{12}u_1v_2 + a_{12}b_{11}u_2v_1 + a_{12}b_{12}u_2v_2 \\ a_{11}b_{21}u_1v_1 + a_{11}b_{22}u_1v_2 + a_{12}b_{21}u_2v_1 + a_{12}b_{22}u_2v_2 \\ a_{21}b_{11}u_1v_1 + a_{21}b_{12}u_1v_2 + a_{22}b_{11}u_2v_1 + a_{22}b_{12}u_2v_2 \\ a_{21}b_{21}u_1v_1 + a_{21}b_{22}u_1v_2 + a_{22}b_{21}u_2v_1 + a_{22}b_{22}u_2v_2 \end{pmatrix}. \end{aligned}$$

Appendix I

A Brief Guide to MATLAB

MATLAB (MATrix LABoratory) is a numerical computation tool that is particularly good at matrix computations and linear algebra algorithms. It does many other things, and we will use only a small part of its capabilities.

It is easy to get started. Here is a view of the screen when a MATLAB session begins:

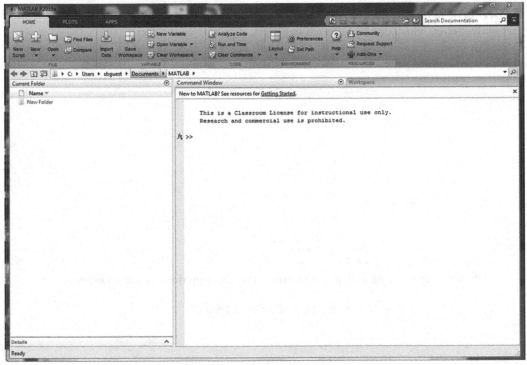

Matrices are assigned and operations are carried out in the Command Window.

CREATING MATRICES AND VECTORS

To enter the matrix $\begin{bmatrix} 3 & -2 & 9 \\ 4 & 0 & 8 \\ -1 & 5 & 4 \end{bmatrix}$ and assign it the name A, type the following after the $\gg$ prompt in the Command Window:

$$A = \begin{bmatrix} 3 & -2 & 9; & 4 & 0 & 8; & -1 & 5 & 4 \end{bmatrix}$$

and press Enter. Notice that in the assignment statement, one row is separated from the next by a semicolon. Successive entries can be separated by one or more spaces or by commas.

```
>> A = [3 -2 9; 4 0 8; -1 5 4]

A =

     3    -2     9
     4     0     8
    -1     5     4
```

MATLAB names are case-sensitive; e.g., A is a different name than a. A variable name must start with a letter, which can be followed by letters, digits, or underscores.

At this point, if we enter A, then A is displayed, but if we enter "a", MATLAB reminds us that a hasn't been defined yet.

```
>> A

A =

     3    -2     9
     4     0     8
    -1     5     4

>> a
Undefined function or variable 'a'.

Did you mean:
>> A
```

An efficient way to create a diagonal matrix is the *diag* function, as shown below.

```
>> diag([2,5,-1,4])

ans =

     2     0     0     0
     0     5     0     0
     0     0    -1     0
     0     0     0     4
```

If a matrix or other result from a computation is not assigned to a name, its default name is *ans*. We could create a 3-by-3 identity matrix using *diag*. Another way is with the MATLAB command *eye*.

```
>> I=eye(3)

I =

        1       0       0
        0       1       0
        0       0       1
```

Matrices can be combined to create larger matrices. For example, try out

$$>> D = [I, A]$$

A **vector** is a matrix consisting of either one column or one row. Not surprisingly, MATLAB distinguishes between an n-by-1 matrix and a 1-by-n matrix.

A single entry in a matrix can be assigned a value by referencing its row and column. Here is an example.

```
I =

    1   0   0
    0   1   0
    0   0   1

>> I(3,1)=6

I =

    1   0   0
    0   1   0
    6   0   1
```

The notation A(3, :) refers to all of row 3 of matrix A. Similarly, A(:, 2) refers to the second column of matrix A. This is a nice way to make separate vectors of the rows or columns of A.

```
A =
    3   -2   9
    4    0   8
   -1    5   4

>> c1=A( :, 1)

c1 =
    3
    4
   -1

>> c2=A( : , 2)

c2 =
   -2
    0
    5

>> r3=A(3, :)

r3 =
   -1   5   4
```

CORRECTING TYPING ERRORS

Pressing the up-arrow on the keyboard will scroll back through the commands in a current session. This is handy if you need to correct a typing error. Rather than re-type the full line, you can up-arrow to bring the line back, correct the error, and re-enter. This is illustrated in the next box.

```
>> B=[1 3 5 , 2,5,7 ; 4 6 8]

Error using vertcat

Dimensions of matrices being concatenated
are not consistent.

 >> B=[1 3 5 , 2,5,7 ; 4 6 8]

>> B=[1 3 5 ; 2,5,7 ; 4 6 8]

B =

    1   3   5
    2   5   7
    4   6   8
```

Use up-arrow to re-call the last command.
Correct the error and press enter.

OPERATIONS

Matrix addition in MATLAB is straightforward. The sum of matrices A and B is denoted A+B.

In MATLAB, the symbol for matrix multiplication is *. To try it out, create the matrix

$$B = \begin{bmatrix} 3 & 1 & 4 \\ 2 & -1 & 1 \end{bmatrix}$$

MATLAB calculates the product B*A, but not the product A*B.

```
>> B

B =

       3     1     4
       2    -1     1

>> C = B*A

C =

       9    14    51
       1     1    14

>> A*B
Error using  *
Inner matrix dimensions must agree.
```

The * symbol also serves for scalar multiplication; for example, 3 times matrix A is 3*A.

The transpose of a matrix A is denoted in MATLAB by A' where ' is the apostrophe on the keyboard.

```
C =

       2     3     4
       1    -1     6
       9     1     2

>> C'

ans =

       2     1     9
       3    -1     1
       4     6     2
```

There are at least two ways to calculate the dot product of two vectors in MATLAB. One way is to use the transpose operator, and another way is to use the function dot().

Let x = [3, 4, 1, 6] and y = [-1, 0, 2, 4], both row vectors. Then their dot product is x*y' or dot(x,y).

```
>> x*y'

ans =

        23

>> dot (x, y)

ans =

        23
```

One way to find the inverse of a square matrix is the function inv() as shown below.

```
>> F=inv(A)
F =
  -0.3704   0.4907  -0.1481
  -0.2222   0.1944   0.1111
   0.1852  -0.1204   0.0741
>> A*F
ans =
   1.0000      0       0
      0    1.0000      0
  -0.0000   0.0000   1.0000
```

The minus sign in the last line of output results from round-off in the computations. You can change the precision of calculations in MATLAB—see *digits* in the documentation.

Another way to get a matrix inverse is with the exponentiation operator, ^.

```
>> A^(-1)

ans =

  -0.3704   0.4907  -0.1481
  -0.2222   0.1944   0.1111
   0.1852  -0.1204   0.0741
```

The determinant of a square matrix is calculated by the function det(). What is the relationship between the matrices K and L below?

```
K =
   2   1
   3  -1

>> L=(1/det(K))*[-1,-1; -3 2]

L =
   0.2000   0.2000
   0.6000  -0.4000
```

HELP, WORKSPACE, SAVING

MATLAB has thorough documentation that a user can search via the Help button in the upper right corner of the screen, or via the F1 key.

Clicking the Workspace tab on the upper right, opens a window that shows what objects are currently in working memory. You can SAVE the workspace contents to have them available later. To do this and to see other Workspace actions, click on the symbol for the drop-down menu on the Workspace tab.

CREATING ELEMENTARY MATRICES

We can reference particular rows of a matrix by specifying the row numbers in square brackets and using a colon to include all columns, as illustrated below.

```
G =
    1   2   3   4
    5   6   7   8
    9  10  11  12
   13  14  15  16
>> G([1 3],:)

ans =

    1   2   3   4
    9  10  11  12
```

We can use this to switch the first and third rows of G , as shown.

```
>> G([1 3],:)=G([3 1],:)

G =

    9  10  11  12
    5   6   7   8
    1   2   3   4
   13  14  15  16
```

To multiply a row by a constant, replace the row with the multiple.

```
>> G(4,:)= (1/13)*G(4,:)

G =

   9.0000  10.0000  11.0000  12.0000
   5.0000   6.0000   7.0000   8.0000
   1.0000   2.0000   3.0000   4.0000
   1.0000   1.0769   1.1538   1.2308
```

Similarly, we can add to one row, a scalar multiple of another row. For example:

```
>> G(2,:)=G(2,:) + (-5)*G(3,:)

G =

   9.0000  10.0000  11.0000
12.0000
        0   -4.0000  -8.0000 -
12.0000
   1.0000   2.0000   3.0000   4.0000
   1.0000   1.0769   1.1538   1.2308
```

Now any desired elementary matrix can be created by starting with the right size identity matrix and applying the row operations.

ROW REDUCED FORM

The MATLAB function for producing row reduced form of matrix A is rref(A).

```
>> A = [3 4 -2 5; 1 -2 1 0; 7 4 -3 3]
A =
   3    4   -2    5
   1   -2    1    0
   7    4   -3    3
>> B = rref(A)
B =
   1.0000        0        0   1.0000
        0   1.0000        0   3.5000
        0        0   1.0000   6.0000
```

COMPLEX NUMBERS

MATLAB uses the lower-case letter *i* for the square root of -1, unless *i* has been specifically defined otherwise (which is not recommended.) MATLAB carries out Complex number calculations and will return complex number results of calculations. Examples:

```
>> (2+3i)*(-6+i)
ans =
 -15.0000 -16.0000i

>> (-4)^.5
ans =
   0.0000 + 2.0000i
```

ORTHONORMAL BASIS

The function orth() takes a matrix as input and returns a matrix whose columns are an orthonormal basis for the range of A.

```
M =
   1   2   1
   0  -5   1
   1   4   1

>> N=orth(M)
N =
  -0.3170  -0.5788  -0.7514
   0.7267  -0.6573   0.1997
  -0.6094  -0.4827   0.6290

>>N'*N
ans =
   1.0000  -0.0000  -0.0000
  -0.0000   1.0000  -0.0000
  -0.0000  -0.0000   1.0000
```

CHARACTERISTIC POLYNOMIAL, EIGENVALUES, AND EIGENVECTORS

To find the characteristic polynomial of a matrix, first declare the variable x as symbolic:

>> syms x

Now you can have MATLAB evaluate det(A − xI) using the eye() function for the identity.

```
A =
   2   0   0
  -3   0   1
   0   1   0

>> det(A-x*eye(3))
ans =
- x^3 + 2*x^2 + x - 2
```

To calculate eigenvalues of A, MATLAB provides the function eig(A) which returns a column vector containing the eigenvalues of A.

A more informative option is described in MATLAB Help as follows:

[V,D] = eig(A) returns diagonal matrix D of eigenvalues and matrix V whose columns are the corresponding right eigenvectors, so that A*V = V*D.

```
A =
   2   0   0
  -3   0   1
   0   1   0

>> eig(A)
ans =
   1
  -1
   2

>> [V,D]=eig(A)
V =
        0        0       0.4082
   0.7071   0.7071  -0.8165
   0.7071  -0.7071  -0.4082

D =
   1   0   0
   0  -1   0
   0   0   2
>> A*V
ans =
        0        0   0.8165
   0.7071  -0.7071  -1.6330
   0.7071   0.7071  -0.8165

>> V*D
ans =
        0        0   0.8165
   0.7071  -0.7071  -1.6330
   0.7071   0.7071  -0.8165
```

The MATLAB function for finding the length of a vector, i.e., $\sqrt{x \cdot x'}$, is norm().

```
r3 =
  -1   5   4
>> norm(r3)
ans =
   6.4807
>> norm(r3)^2
ans =
   42
>> x=[2+i, 3, 4-i]
x =
   2.0000 + 1.0000i   3.0000 + 0.0000i
   4.0000 - 1.0000i
>> norm(x)
ans =
   5.5678
```

MATRIX DECOMPOSITIONS AND FACTORIZATIONS

The LU factorization gives a square matrix A as the product LU where L is a permutation of a lower triangular matrix with ones on its diagonal, and U is an upper triangular matrix. The MATLAB function is lu().

```
A =
   3  -2   9
   4   0   8
  -1   5   4

>> [L,U]=lu(A)

L =
   0.7500  -0.4000   1.0000
   1.0000        0        0
  -0.2500   1.0000        0

U =
   4.0000        0   8.0000
        0   5.0000   6.0000
        0        0   5.4000

>> L*U

ans =
   3  -2   9
   4   0   8
  -1   5   4
```

The Cholesky factorization expresses a positive definite, symmetric matrix as the product of a triangular matrix and its transpose

$$A = R'R,$$

where R is an upper triangular matrix. The MATLAB function for this factorization is chol().

```
B =
  10   2   3
   2  16   5
   3   5   6

>> R=chol(B)

R =
   3.1623   0.6325   0.9487
        0   3.9497   1.1140
        0        0   1.9644

>> R'*R

ans =
   10.0000    2.0000    3.0000
    2.0000   16.0000    5.0000
    3.0000    5.0000    6.0000
```

Here is the MATLAB documentation for the singular value decomposition:

svd

Singular value decomposition

Syntax

```
s = svd(X)
    [U,S,V] = svd(X)
    [U,S,V] = svd(X,0)
    [U,S,V] = svd(X,'econ')
```

Description

The svd command computes the matrix singular value decomposition.

s = svd(X) returns a vector of singular values.

[U,S,V] = svd(X) produces a diagonal matrix S of the same dimension as X, with nonnegative diagonal elements in decreasing order, and unitary matrices U and V so that X = U*S*V'.

[U,S,V] = svd(X,0) produces the "economy size" decomposition. If X is m-by-n with m > n, then svd computes only the first n columns of U and S is n-by-n.

[U,S,V] = svd(X,'econ') also produces the "economy size" decomposition. If X is m-by-n with m >= n, it is equivalent to svd(X,0). For m < n, only the first m columns of V are computed and S is m-by-m.

Examples

For the matrix

```
X =
    1  2
    3  4
    5  6
    7  8
```

the statement

```
[U,S,V] = svd(X)
```

produces

```
U =
    -0.1525  -0.8226  -0.3945  -0.3800
    -0.3499  -0.4214  0.2428  0.8007
    -0.5474  -0.0201  0.6979  -0.4614
    -0.7448  0.3812  -0.5462  0.0407
S =
    14.2691    0
    0  0.6268
    0    0
    0    0
V =
    -0.6414   0.7672
    -0.7672  -0.6414
```

The economy size decomposition generated by

```
[U,S,V] = svd(X,0)
```

produces

```
U =
   -0.1525   -0.8226
   -0.3499   -0.4214
   -0.5474   -0.0201
   -0.7448    0.3812
S =
   14.2691        0
   0    0.6268
V =
   -0.6414    0.7672
   -0.7672   -0.6414
```

Finally, there is the QR factorization. MATLAB documentation says:

"The orthogonal, or QR, factorization expresses any rectangular matrix as the product of an orthogonal or unitary matrix and an upper triangular matrix. A column permutation might also be involved:

$$A = QR$$

or

$$AP = QR,$$

where Q is orthogonal or unitary, R is upper triangular, and P is a permutation."

The MATLAB function is qr().

Example:

```
D =

   1   2   3   4
   5   2   4   3
  -2   4   1   0

>> [Q,R]= qr(D)

Q =
  -0.1826   0.3853  -0.9045
  -0.9129   0.2752   0.3015
   0.3651   0.8808   0.3015

R =
  -5.4772  -0.7303  -3.8341  -
3.4689
       0   4.8442   3.1377   2.3671
       0        0  -1.2060  -2.7136

>> Q*R

ans =
   1.0000   2.0000   3.0000   4.0000
   5.0000   2.0000   4.0000   3.0000
  -2.0000   4.0000   1.0000   0.0000


>> Q'*Q
ans =   1.0000   0.0000  -0.0000
        0.0000   1.0000  -0.0000
       -0.0000  -0.0000   1.0000
```

Appendix II

R for Linear Algebra

Downloading R to Your Computer

The main page for the R Project is http://www.r-project.org/

Locate the box titled "Getting Started" and click on the link to CRAN Mirrors. In general, the download will be faster if you choose a location closer to you.

From the box titled "Download and Install R", click on the version (Linux, MacOS X, or Windows) that you need. If you are using Windows, the next step is to click on the link to the "base" subdirectories. "Base" is the essential part of the R software; many additional packages are available for specialized computations and functions. One or two more clicks will download the software. Once the download is done, install R by double-clicking on the icon for the downloaded file.

Getting Started with R

R is a computer language for statistical computing that is widely used by statisticians and statistics students. Not only is it flexible and powerful, it is free! Because matrices and matrix operations and decompositions are useful in statistics, R is also a powerful tool for doing linear algebra computations. This little manual is a starting point for learning the language; it is intended to give a minimal yet sufficient set of R commands for an introductory linear algebra course. If you are interested in more details, there are many textbooks and references for R, as well as online help.

This manual is designed to accompany a linear algebra textbook; it only describes *how* to do a computation, not *why* or under what conditions.

You will probably want to save your inputs and results in a personal folder. Set up a folder now on your flash drive.

Now double click on the icon for R on the desktop.

The first thing you should do at every session with R is to change into the directory where you are keeping your work. Click on the File menu on the upper left corner of the screen, and select "Change dir...". Select your folder.

Calculations in R

We communicate commands to R by typing them at the ">" prompt and pressing the Enter key. Try these out:

```
>2/6*12
>3^2
```

You may add comments if they are preceded by a # symbol. This is a good way to make notes about what you are doing. Try out:

```
>100/10^2    # exponentiation takes precedence over division
```

Try:

```
>3 + 7 *
```

Instead of the usual prompt, R returns a + prompt to indicate that the command is incomplete; R waits for the rest of the command:

```
> 3 + 7 *
 + 5
 [1] 38
```

Note: in this guide, R commands and output will be indented, to make it easier to distinguish them from the rest of the text.

Assigning Values to Variables

In R, acceptable variable names use letters, numbers, a dot, or an underscore. A variable name must start with a letter or a dot. R is case-sensitive, so for example, A and a are not the same variables. Some examples of acceptable variable names are x3, yr12, yr.birth, birth_month, R2D2.

Variables are assigned values using an equal sign on the right of the variable. Try out the following commands in R.

```
>x=12
> x^2
>year = 2011
>year=year+1
>year
>e      # the letter e is not a constant in R.
>e=exp(1)    # this assigns the "usual" value to the variable e
>pi     # "pi" cannot be used as a variable name, since it
represents a defined constant in R
>3=x    # this will generate an error message
>wt<-12.63    # the combination <- is another assignment operator
>wt
```

Vectors and Matrices in R

We can create a vector using the c() function, as follows.

```
>b = c(9, 8, 5, 12, 4, 17)
>b
[1] 9 8 5 12 4 17
```

The index [1] denotes the first entry of the vector. If the vector were long enough to continue onto the next line, there would be an index at the beginning of the next line to indicate the position in the vector for the next data value. We can inspect any position in a vector using its index. To demonstrate, enter

```
>b[4]
```

You can add components to a vector in a variety of ways. Try these:

```
>b [7]=9
>b=c(b,c(5,8,11))
>b
```

Note 1: Since the character c is a defined function in R, it cannot be used as a variable name.

Note 2: Changing the position of the cursor on a line *cannot* be accomplished in R with a mouse click; the left and right arrow keys must be used. If you want to move left to correct a typing error, you must either backspace or use the left arrow key; move back to the end of the line using the right arrow key or the "End" key.

We can apply mathematical operations and functions to vectors. Investigate the following. Inspect the variable after each operation.

```
>b -9 # This operation does not alter the entries of the object b.
>1/b
>sort(b)
>x=c(1,5,3,5)
>b + x
>sum(b)
>length(b)
```

The sequence of integers from n to m is denoted by n:m in R. For example,

```
> 2:9
[1] 2 3 4 5 6 7 8 9
> s=1:12
> s
[1] 1 2 3 4 5 6 7 8 9 10 11 12
```

An R workspace is a portion of computer memory where objects such as matrices and vectors are stored while R is running. It is frequently helpful to remind yourself what objects you have currently available in your workspace. To see the contents of your workspace, use ls() or objects(). To delete an object, use the "remove" operator, rm(). For example, try

```
>rm(x)
>ls()
```

If you discover that you have mistyped a command, of course you can re-type it and re-enter it. But it may be more efficient to backtrack using arrow keys and then edit. The up arrow and down arrow keys let you scroll through previous commands. Try this out.

You can edit the current line and re-execute it. For example, scroll up to

```
>sort(b).
```

Use the left arrow to move to the > prompt and change the line to

```
>y=sort(b)
```

and enter.

Now the variables y and b contain the same entries but in different orders.

To enter the matrix $\begin{bmatrix} 3 & -2 & 9 \\ 4 & 0 & 8 \\ -1 & 5 & 4 \end{bmatrix}$ and assign it the name A, type the following after the > prompt:

```
> A=matrix(c(3,4,-1,-2,0,5,9,8,4),ncol=3)
> A
     [,1] [,2] [,3]
[1,]   3   -2   9
[2,]   4    0   8
[3,]  -1    5   4
>
```

Notice that the entries are listed column-by-column in the assignment statement. If you would prefer to type the entries row-by-row, use

```
> B=matrix(c(3,-2,9,4,0,8,-1,5,4), ncol=3, byrow = T)
> B
     [,1] [,2] [,3]
[1,]   3   -2   9
[2,]   4    0   8
[3,]  -1    5   4
>
```

There is a shortcut for creating a diagonal matrix:

```
> D=diag(c(4,6,3,2))
> D
     [,1] [,2] [,3] [,4]
[1,]   4    0    0   0
[2,]   0    6    0   0
[3,]   0    0    3   0
[4,]   0    0    0   2
>
```

Try out

```
>diag(1,4)
```

A single entry of a matrix can be edited by referencing its row and column. For example,

```
> D[2,1]=7
> D
```

```
       [,1]  [,2]  [,3]  [,4]
 [1,]    2    0    0    0
 [2,]    7    5    0    0
 [3,]    0    0   -1    0
 [4,]    0    0    0    4
```

Parts of a matrix can be extracted by specifying rows and columns in square brackets. Study the following examples.

```
> G=matrix(1:28,ncol=7,byrow=T)
> G
       [,1]  [,2]  [,3]  [,4]  [,5]  [,6]  [,7]
 [1,]    1    2    3    4    5    6    7
 [2,]    8    9   10   11   12   13   14
 [3,]   15   16   17   18   19   20   21
 [4,]   22   23   24   25   26   27   28
> G[3,]                   # the third row of G
[1] 15 16 17 18 19 20 21
> G[,6]                     # the sixth column of G
[1] 6 13 20 27
> G[c(1,4),]            # rows 1 and 4 of G
       [,1]  [,2]  [,3]  [,4]  [,5]  [,6]  [,7]
 [1,]    1    2    3    4    5    6    7
 [2,]   22   23   24   25   26   27   28
>
```

Matrices can be "pasted together" using the functions cbind() and rbind(). Test out the following examples:

```
> A
     [,1]  [,2]  [,3]
 [1,]   3   -2    9
 [2,]   4    0    8
 [3,]  -1    5    4
> B=matrix(c(5,7,2,0,1,-3),ncol=2,byrow=T)
> B
     [,1]  [,2]
 [1,]   5    7
 [2,]   2    0
 [3,]   1   -3
> C=cbind(A,B)
> C
     [,1]  [,2]  [,3]  [,4]  [,5]
 [1,]   3   -2    9    5    7
 [2,]   4    0    8    2    0
 [3,]  -1    5    4    1   -3
D=rbind(A,t(B))  # t(B) is the transpose of B
> D
     [,1]  [,2]  [,3]
 [1,]   3   -2    9
 [2,]   4    0    8
 [3,]  -1    5    4
 [4,]   5    2    1
 [5,]   7    0   -3
```

Matrix Operations

The R notation for matrix multiplication is %*%. For example:

```
> M=matrix(c(1,2,1,3,1,3,2,1,2), ncol=3)
> M
     [,1] [,2] [,3]
[1,]    1    3    2
[2,]    2    1    1
[3,]    1    3    2
> M%*%A
     [,1] [,2] [,3]
[1,]   13    8   41
[2,]    9    1   30
[3,]   13    8   41
```

R declines to multiply matrices if their dimensions are not conformable:

```
> M%*%D
Error in M %*% D : non-conformable arguments
```

However, R will treat a vector as either a row or a column, in order to make it conform to multiplication. For example:

```
> V
[1] 1 5 4
> A%*%V
     [,1]
[1,]   29
[2,]   36
[3,]   40
> V%*%A
     [,1] [,2] [,3]
[1,]   19   18   65
```

Observe the crucial distinction between the two operations, * and %*%, when applied to vectors:

```
> V1=c(1,4,-1,2)
> V2=c(3,0,-2,5)
> V1*V2
[1] 3 0 2 10
> V1%*%V2
     [,1]
[1,]   15
```

A straightforward way to find the length of a vector is to use sqrt() and %*%.

```
> V1
[1] 1 4 -1 2
> sqrt(V1%*%V1)
     [,1]
[1,] 4.690416
```

The symbols for matrix addition and subtraction are the usual + and – signs.

```
> A+M
  [,1] [,2] [,3]
[1,]   4   1  11
[2,]   6   1   9
[3,]   0   8   6
> A-M
  [,1] [,2] [,3]
[1,]   2  -5   7
[2,]   2  -1   7
[3,]  -2   2   2
```

The inverse of a matrix is produced by the solve() function in R, as illustrated:

```
> B=solve(A)
> B
     [,1]       [,2]        [,3]
[1,] -0.3703704 0.4907407 -0.14814815
[2,] -0.2222222 0.1944444 0.11111111
[3,] 0.1851852 -0.1203704 0.07407407
> A%*%B
     [,1] [,2] [,3]
[1,] 1.000000e+00 2.081668e-16   0
[2,] 0.000000e+00 1.000000e+00   0
[3,] -1.110223e-16 5.551115e-17  1
```

Because of inevitable tiny round-off errors, the product of A with its inverse is very nearly but not exactly the identity matrix.

We can round off the last product by applying the round() function:

```
> round(A%*%B, digits = 5)
  [,1] [,2] [,3]
[1,]   1   0   0
[2,]   0   1   0
[3,]   0   0   1
```

The solve function will produce the solution of the matrix equation Ay = x, as in the following example.

```
> A
  [,1] [,2] [,3]
[1,]   3  -2   9
[2,]   4   0   8
[3,]  -1   5   4
> x = c(4, -2,5)
> y = solve(A,x)
> y
[1] -3.2037037 -0.7222222 1.3518519
> A%*%y
  [,1]
[1,]   4
[2,]  -2
[3,]   5
```

The determinant function is denoted by det().

```
> det(A)
[1] 108
> det(B)
[1] 0.009259259
> 1/108              # as expected, det(A⁻¹) = 1/(det A)
[1] 0.009259259
```

The symbol for scalar multiplication is *.

```
> 5*A
     [,1]  [,2]  [,3]
[1,]  15  -10   45
[2,]  20    0   40
[3,]  -5   25   20
```

The transpose of matrix A is t(A).

```
> t(A)
     [,1]  [,2]  [,3]
[1,]   3    4   -1
[2,]  -2    0    5
[3,]   9    8    4
```

Elementary Row Operations

To multiply a row by a constant, replace the row with the multiple. In the following, the second row of matrix G is replaced by 1/5 times the second row.

```
> G
     [,1]  [,2]  [,3]  [,4]
[1,]   1    2    3    4
[2,]   5    6    7    8
[3,]   9   10   11   12
[4,]  13   14   15   16
> G[2,]=(1/5)*G[2,]
> G
     [,1]  [,2]  [,3]  [,4]
[1,]   1  2.0  3.0  4.0
[2,]   1  1.2  1.4  1.6
[3,]   9 10.0 11.0 12.0
[4,]  13 14.0 15.0 16.0
```

To interchange two rows, follow this example:

```
> G
     [,1]  [,2]  [,3]  [,4]  [,5]  [,6]  [,7]
[1,]   1    2    3    4    5    6    7
[2,]   8    9   10   11   12   13   14
[3,]  15   16   17   18   19   20   21
[4,]  22   23   24   25   26   27   28
```

```
> G[c(1,3),]=G[c(3,1),]    # interchange the first and third rows
> G
   [,1] [,2] [,3] [,4] [,5] [,6] [,7]
[1,]  15  16  17  18  19  20  21
[2,]   8   9  10  11  12  13  14
[3,]   1   2   3   4   5   6   7
[4,]  22  23  24  25  26  27  28
>
```

To add to a row a scalar multiple of another row:

```
> G[4,]=G[4,]+(-22)*G[3,]    # add to row 4, -22 times row 3.
> G
  [,1] [,2] [,3] [,4] [,5] [,6] [,7]
[1,]  15  16  17  18  19  20  21
[2,]   8   9  10  11  12  13  14
[3,]   1   2   3   4   5   6   7
[4,]   0 -21 -42 -63 -84 -105 -126
```

Row Reduced Form

To obtain the row-reduced form of a matrix, we need to install a package called **pracma**.
The command for this is

```
> install.packages("pracma")
```

You will be prompted to choose a CRAN mirror. Installation will be faster from a mirror that is nearer to your geographic location. During the installation, respond "yes" to allow the code to be stored in a library on your computer.

Once the download is complete, you need to add the package to the working library for the current session. Use the command

```
> library("pracma")
```

This command must be used again whenever you start up a new session in R. Once the package is accessible, the row echelon form of a matrix is produced using the rref() function.

```
> K
   [,1] [,2] [,3] [,4] [,5] [,6]
[1,]   1   2   3   4   5   6
[2,]   7   8   9  10  11  12
[3,]  13  14  15  16  17  18
[4,]  19  20  21  22  23  24
> rref(K)
   [,1] [,2] [,3] [,4] [,5] [,6]
[1,]   1   0  -1  -2  -3  -4
[2,]   0   1   2   3   4   5
[3,]   0   0   0   0   0   0
[4,]   0   0   0   0   0   0
```

Complex Numbers in R

The lower-case i is not automatically recognized as the complex i, but it can be added to your workspace by defining it, as follows.

```
> i=complex(real=0,im=1)
> i
[1] 0+1i
```

R does not automatically give complex numbers as output when inputs are real, but does give complex number results when the inputs are complex. For example:

```
> sqrt(-1)
[1] NaN                    # NaN means Not a Number
Warning message:
In sqrt(-1) : NaNs produced
```

However,

```
> sqrt(-1+0i)
[1] 0+1i
```

You can convert a real number to the real part of a complex number using as.complex().

```
> as.complex(15)
[1] 15+0i
> sqrt(-3)
[1] NaN
Warning message:
In sqrt(-3) : NaNs produced
> sqrt(as.complex(-3))
[1] 0+1.732051i
```

Finding Eigenvalues and Eigenvectors

The eigen() function takes a matrix as input and produces the eigenvalues and eigenvectors in an object called a list. This particular list consists of two components, $values and $vectors. Consider the next example.

```
> K=matrix(1:16,ncol=4,byrow = T)
> K
     [,1] [,2] [,3] [,4]
[1,]  1    2    3    4
[2,]  5    6    7    8
[3,]  9   10   11   12
[4,]  13  14   15   16
> KE=eigen(K)
> KE                 # the whole list is displayed.
$values
[1] 3.620937e+01 -2.209373e+00 -1.941536e-15 -3.467987e-16
$vectors
```

```
          [,1]      [,2]      [,3]      [,4]
[1,]  -0.1511543 0.7270500 -0.3051507 0.05761073
[2,]  -0.3492373 0.2832088 0.7458883 0.32916941
[3,]  -0.5473203 -0.1606324 -0.5763245 -0.83117101
[4,]  -0.7454033 -0.6044736 0.1355869 0.44439087
```

The vector of eigenvalues and the matrix of eigenvectors can be referenced separately:

```
> KE$values
[1] 3.620937e+01 -2.209373e+00 -1.941536e-15 -3.467987e-16
> KE$vectors
          [,1]      [,2]      [,3]      [,4]
[1,]  -0.1511543 0.7270500 -0.3051507 0.05761073
[2,]  -0.3492373 0.2832088 0.7458883 0.32916941
[3,]  -0.5473203 -0.1606324 -0.5763245 -0.83117101
[4,]  -0.7454033 -0.6044736 0.1355869 0.44439087
```

Orthonormal Basis

Suppose we want to find an orthonormal basis for the column space of the matrix $M = \begin{bmatrix} 1 & 2 & 1 & 5 \\ 0 & -5 & 1 & 1 \\ 1 & 4 & 1 & 0 \end{bmatrix}$.

The R function svd(M) will return the singular value decomposition of M as a list. The list consists of matrices u and v and a vector d with the properties:

- The columns of u are an orthonormal basis for the column space of M.
- The columns of v are an orthonormal basis of the column space of the transpose of M (i.e., the row space of M.)
- If D is a diagonal matrix with diagonal d, then uDv = M.
- The non-zero entries of d are the singular values of M.

All of this is demonstrated in the following script:

```
> M
   [,1] [,2] [,3] [,4]
[1,]  1   2    1    5
[2,]  0   -5   1    1
[3,]  1   4    1    0
> SVDM=svd(M)
> SVDM          #display the list produced by the svd function
$d
[1] 6.893554 5.154879 1.380627
$u
          [,1]      [,2]      [,3]
[1,]  0.5061532 -0.8400709 -0.1951661
[2,]  -0.6369880 -0.5166982 0.5720745
[3,]  0.5814251 0.1652389 0.7966435
$v
```

```
        [,1]     [,2]     [,3]
  [1,]  0.1577674 -0.1309113 0.43565527
  [2,]  0.9462386 0.3034609 -0.04645035
  [3,]  0.0653640 -0.2311461 0.85001375
  [4,]  0.2747172 -0.9150657 -0.29244398
  > U=SVDM$u        # for convenience, assign the elements of the list
to shorter names
  > V=SVDM$v
  > D=diag(SVDM$d)
  > U%*%t(U)          # check that the columns of U are orthonormal
        [,1]     [,2]       [,3]
  [1,]  1.000000e+00 -6.289728e-17 2.032879e-19
  [2,]  -6.289728e-17 1.000000e+00 6.732895e-17
  [3,]  2.032879e-19 6.732895e-17 1.000000e+00
  > t(V)%*%V          # check that the rows of V are orthonormal
        [,1]     [,2]       [,3]
  [1,]  1.000000e+00 1.196146e-16 1.980702e-17
  [2,]  1.196146e-16 1.000000e+00 1.246832e-17
  [3,]  1.980702e-17 1.246832e-17 1.000000e+00
  > U%*%D%*%t(V)     #check that M = UDVt
        [,1] [,2] [,3]     [,4]
  [1,]  1.000000e+00  2  1 5.000000e+00
  [2,]  -5.811324e-17  -5  1 1.000000e+00
  [3,]  1.000000e+00  4  1 3.760555e-16
```

Characteristic Polynomial

The "pracma" package includes a function charpoly() to calculate the coefficients of the characteristic function. Here is an example.

```
  > A
     [,1] [,2] [,3]
  [1,]  3  -2  9
  [2,]  4  0  8
  [3,]  -1  5  4
  > charpoly(A)
  [1]  1  -7 -11 -108      # The characteristic polynomial of A is x³
  - 7x² - 11x - 108
```

Matrix Decompositions and Factorizations

LU Factorization

The "pracma" package provides a function lu() which accepts a positive definite, square matrix A as input. The output is a pair of matrices L and U where L is lower triangular, U is upper triangular, and A=LU.

```
  > A
     [,1] [,2] [,3]
  [1,]  3  -2  9
```

```
[2,]   4   0   8
[3,]  -1   5   4
> LUA=lu(A)
> LUA
$L
       [,1]      [,2]  [,3]
[1,]  1.0000000 0.000   0
[2,]  1.3333333 1.000   0
[3,] -0.3333333 1.625   1
$U
     [,1]     [,2]    [,3]
[1,]  3 -2.000000 9.0
[2,]  0 2.666667 -4.0
[3,]  0 0.000000 13.5
> LUA$L%*%LUA$U   # Check that LU = A
     [,1] [,2] [,3]
[1,]   3   -2   9
[2,]   4    0   8
[3,]  -1    5   4
```

The **Cholesky factorization** expresses a positive definite, symmetric matrix as the product of an upper triangular matrix and its transpose. The function chol() returns the upper triangular matrix, as shown in the following example. Note that this function is part of base R; no added package is needed to obtain it.

```
> B
     [,1] [,2] [,3]
[1,]  10   2   3
[2,]   2  16   5
[3,]   3   5   6
> R=chol(B)
>R
       [,1]      [,2]      [,3]
[1,] 3.162278 0.6324555 0.9486833
[2,] 0.000000 3.9496835 1.1140133
[3,] 0.000000 0.0000000 1.9644272
> t(R)%*%R        # check that B is the product of R by its transpose
     [,1] [,2] [,3]
[1,]  10   2   3
[2,]   2  16   5
[3,]   3   5   6
```

The **singular value decomposition** of a matrix A consists of a diagonal matrix S of the same dimension as A, with nonnegative diagonal elements in decreasing order, and unitary matrices U and V such that $A = U*S*V^T$. The R function svd() takes A as input and returns a list containing U, V, and the diagonal of S.

Example:

```
> A
     [,1] [,2] [,3]
[1,]   3   -2   9
[2,]   4    0   8
[3,]  -1    5   4
```

```
> K=svd(A)
> K
$d
[1] 13.438450 5.786093 1.388958
$u
     [,1]       [,2]        [,3]
[1,] -0.7075215 -0.28897604 -0.6449079
[2,] -0.6606568 -0.05351371 0.7487783
[3,] -0.2508904 0.95583949 -0.1530519
$v
     [,1]       [,2]       [,3]
[1,] -0.33592425 -0.3520204 0.8736341
[2,] 0.01195011 0.9258665 0.3776618
[3,] -0.94181319 0.1373058 -0.3068143
> KU=K$u              #rename the components of the list K, for
convenience
> KV=K$v
> D=diag(K$d)
> KU %*%t(KU)              #check that KU is unitary
     [,1]       [,2]        [,3]
[1,] 1.000000e+00 -2.298509e-17 9.147278e-17
[2,] -2.298509e-17 1.000000e+00 -2.254802e-16
[3,] 9.147278e-17 -2.254802e-16 1.000000e+00
> KV%*%t(KV)              #Check that KV is unitary
     [,1]       [,2]       [,3]
[1,] 1.000000e+00 -1.661540e-17 -4.816568e-17
[2,] -1.661540e-17 1.000000e+00 9.845911e-17
[3,] -4.816568e-17 9.845911e-17 1.000000e+00
> KU%*%D%*%t(KV)              #Check that A is the product.
     [,1]       [,2] [,3]
[1,]  3 -2.000000e+00  9
[2,]  4 -8.637839e-16  8
[3,]  -1 5.000000e+00  4
```

Saving and Retrieving the Workspace

Before you close an R session, you should explicitly save the workspace so that you can retrieve it later. Click on the floppy disk icon in the toolbar above the R console. Locate the folder where you want to save the workspace and assign a filename to the workspace. Use the extension .RData. For example, save the workspace as "Chapter1Data.RData".

It is a good practice to save the workspace periodically even if you aren't exiting the program, since in the event of a system crash or power outage, what hasn't been saved is lost.

Using separate workspaces for different projects can help reduce clutter and improve organization.

When you're ready to end your session with R, enter the quit command: q()

When you want to access the workspace later, open R, change to the appropriate directory, click on the "load workspace" icon, and click on the workspace file. You may have to wait while the workspace loads and it may seem slow. If it takes very long, click on the Stop sign in the toolbar.

You may want to bookmark or print off this resource:

Quick reference list for R syntax and commands
http://cran.r-project.org/doc/contrib/Short-refcard.pdf

Reference

1. R Development Core Team (2010). *R: A language and environment for statistical computing*. R Foundation for Statistical Computing, Vienna, Austria. ISBN 3-900051-07-0, URL http://www.R-project.org/.

Answers to Selected Exercises

Section 1.1

1. (a.) $\begin{pmatrix} 4 & -14 \\ 1 & 9 \end{pmatrix}$

   (b.) $\begin{pmatrix} -7 & -5 & 10 \\ -9 & -15 & 6 \\ -11 & 5 & 26 \end{pmatrix}$

   (c.) undefined as a single matrix

   (d.) undefined as a single matrix

   (e.) $\begin{pmatrix} 4 & 2 \\ 0 & 2 \\ 14 & 4 \end{pmatrix}$

3. (a.) $\begin{pmatrix} 4 & -2 & 0 \\ 0 & 0 & 3 \\ 0 & 0 & 0 \end{pmatrix}$

   (b.) $\begin{pmatrix} 0 & 0 & 1 \\ 0 & 1 & 0 \\ 1 & 0 & 0 \end{pmatrix}$

   (c.) $\begin{pmatrix} 1 & 1 & 1 \\ 2 & 2 & 2 \\ -1 & -1 & -1 \end{pmatrix}$

11. $\begin{pmatrix} 1/2 & -9 & 7/2 \\ 5/2 & -5/2 & 45/2 \end{pmatrix}$

Section 1.2

3. $\begin{pmatrix} a & 0 \\ c & a \end{pmatrix}$

5. $A - 3B = \begin{pmatrix} -5 & -3 \\ 7 & -15 \end{pmatrix}$

$BA^T = \begin{pmatrix} 2 & 8 \\ -16 & -4 \end{pmatrix}$

$AB^T = \begin{pmatrix} 2 & -16 \\ 8 & -4 \end{pmatrix}$

$\left(AB^T\right)^T = \begin{pmatrix} 2 & 8 \\ -16 & -4 \end{pmatrix}$

7. A is a square matrix

9. (b.) The second column of AB is all zeroes

11. (a.) $(A+B)^2 = \begin{pmatrix} 116 & 144 \\ 180 & 224 \end{pmatrix}$ $A^2 + 2AB + B^2 = \begin{pmatrix} 112 & 132 \\ 192 & 228 \end{pmatrix}$

15. (a.) $AB = \begin{pmatrix} a_{11}b_{11} & 0 & 0 \\ * & a_{22}b_{22} & 0 \\ ** & *** & a_{33}b_{33} \end{pmatrix}$

   (b.) $(AB)_{23} = a_{21}b_{13} + a_{22}b_{23} + a_{23}b_{33} = a_{21}0 + a_{22}0 + 0b_{33} = 0$

19. (a.) $\det \begin{pmatrix} a & b & c \\ a & b & c \\ d & e & f \end{pmatrix} = 0$

23. (a.) $a = -2, b = -1$

   (b.) $a = -n, b = -1$

25. (b.) A 2×2 matrix whose trace is 0 is of the form $\begin{pmatrix} a & b \\ c & -a \end{pmatrix}$

27. (a.) $\begin{pmatrix} 1 & 0 & 0 \\ 0 & 1/3 & 0 \\ 0 & 0 & 1 \end{pmatrix}$

   (c.) $\begin{pmatrix} 0 & 1 & 0 \\ 1 & 0 & 0 \\ 0 & 0 & 1 \end{pmatrix}$

29. (a.) $\begin{pmatrix} 1 & 0 & -2 \\ 3 & 4 & -1 \\ 1 & 1 & 2 \end{pmatrix}$

(b.) The action done on the rows when multiplied by the elementary matrix on the left is instead action done on the columns when multiplied by the elementary on the right.

30. In this problem, while the inverse is unique, it can be achieved by different orderings of the elementary matrices

(a.)
$$\begin{pmatrix} 1 & -2 \\ 0 & 1 \end{pmatrix} \begin{pmatrix} 1 & 0 \\ -1 & 1 \end{pmatrix} \begin{pmatrix} 1/2 & 0 \\ 0 & 1 \end{pmatrix} \begin{pmatrix} 2 & 4 \\ 1 & 3 \end{pmatrix} = \begin{pmatrix} 1 & 0 \\ 0 & 1 \end{pmatrix}$$

$$\begin{pmatrix} 2 & 4 \\ 1 & 3 \end{pmatrix}^{-1} = \begin{pmatrix} 3/2 & -2 \\ -1/2 & 1 \end{pmatrix}$$

(c.)
$$\begin{pmatrix} 1 & -4 \\ 0 & 1 \end{pmatrix} \begin{pmatrix} 1 & 0 \\ 0 & -\dfrac{1}{3} \end{pmatrix} \begin{pmatrix} 1 & 0 \\ -1 & 1 \end{pmatrix} \begin{pmatrix} 1 & 0 \\ 0 & \dfrac{1}{5} \end{pmatrix} \begin{pmatrix} 1 & 4 \\ 5 & 5 \end{pmatrix} = \begin{pmatrix} 1 & 0 \\ 0 & 1 \end{pmatrix}$$

$$\begin{pmatrix} 1 & 4 \\ 5 & 5 \end{pmatrix}^{-1} = \begin{pmatrix} -1/3 & 4/15 \\ 1/3 & -1/15 \end{pmatrix}$$

(e.)
$$\begin{pmatrix} 1 & 0 & -4 \\ 0 & 1 & 0 \\ 0 & 0 & 1 \end{pmatrix} \begin{pmatrix} 1 & 0 & 0 \\ 0 & 1 & -11 \\ 0 & 0 & 1 \end{pmatrix} \begin{pmatrix} 1 & 0 & 0 \\ 0 & 1 & 0 \\ 0 & 0 & -1/21 \end{pmatrix} \begin{pmatrix} 1 & 0 & 0 \\ 0 & 1 & 0 \\ 0 & -2 & 1 \end{pmatrix}$$

$$\begin{pmatrix} 1 & 0 & 0 \\ 0 & 1 & 0 \\ -1 & 0 & 1 \end{pmatrix} \begin{pmatrix} 1 & 0 & 0 \\ 2 & 1 & 0 \\ 0 & 0 & 1 \end{pmatrix} \begin{pmatrix} 1 & 0 & 4 \\ -2 & 1 & 3 \\ 1 & 2 & 5 \end{pmatrix} = \begin{pmatrix} 1 & 0 & 0 \\ 0 & 1 & 0 \\ 0 & 0 & 1 \end{pmatrix}$$

$$\begin{pmatrix} 1 & 0 & 4 \\ -2 & 1 & 3 \\ 1 & 2 & 5 \end{pmatrix}^{-1} = \begin{pmatrix} 1/21 & -8/21 & 4/21 \\ -13/21 & -1/21 & 11/21 \\ 5/21 & 2/21 & -1/21 \end{pmatrix}$$

Section 1.3

1. $L = \begin{pmatrix} 1 & 0 & 0 \\ 2 & 1 & 0 \\ 3 & -1 & 1 \end{pmatrix}$ $U = \begin{pmatrix} 1 & 3 & 6 \\ 0 & 3 & -14 \\ 0 & 0 & -27 \end{pmatrix}$

3. $L = \begin{pmatrix} 1 & 0 & 0 \\ 2 & 1 & 0 \\ 0 & -8 & 1 \end{pmatrix}$ $U = \begin{pmatrix} 3 & 1 & 4 \\ 0 & -1 & 0 \\ 0 & 0 & 5 \end{pmatrix}$

5. $L = \begin{pmatrix} 1 & 0 & 0 \\ 2 & 1 & 0 \\ 4 & 2/9 & 1 \end{pmatrix}$ $U = \begin{pmatrix} 1 & 2 & 4 \\ 0 & -9 & 16 \\ 0 & 0 & 193/9 \end{pmatrix}$

$$7.\ L = \begin{pmatrix} 2 & 0 & 0 & 0 \\ 4 & 1 & 0 & 0 \\ 1 & -1/2 & 1 & 0 \\ -3 & -2 & 2/3 & 1 \end{pmatrix} \quad U = \begin{pmatrix} 1 & 1 & 0 & 1/2 \\ 0 & -4 & -3 & 3 \\ 0 & 0 & 3/2 & 7 \\ 0 & 0 & 0 & 17/6 \end{pmatrix}$$

Section 2.1

1. Solution is valid.
3. Solution is not valid.

Section 2.2

$$1.\ \begin{pmatrix} 3 & 5 & 9 \\ 1 & -2 & -7 \\ 0 & 1 & 5 \end{pmatrix}$$

$$3.\ \begin{pmatrix} 1 & 1 & 1 & 0 \\ 1 & 1 & 1 & 3 \\ 1 & 2 & 1 & 5 \end{pmatrix}$$

5. $8x_1 + 3x_2 = 6$

$$-4x_1 + 2x_2 + x_3 = 2$$

Section 2.4

1. Free variables $x_2 = r,\ x_4 = s,\ x_5 = t$

   $x_1 = -2r + s - 3t + 4$, leading variables x_1, x_3

   $x_2 = -2x_2 + x_3 - 3x_2 + 4,;\ x_3 = -2s - 2$

2. $x_2 = r, x_4 = s, x_5 = t;$

   $x_1 = -2r + s - 3t + 4,\ x_3 = -2s - 2$

$$5.\ \begin{pmatrix} 2 & -3 & 6 \\ 1 & 5 & 9 \end{pmatrix} \rightarrow \begin{pmatrix} 1 & 0 & \dfrac{57}{13} \\ 0 & 1 & \dfrac{12}{13} \end{pmatrix} \quad x = \frac{57}{13},\ y = \frac{12}{13}$$

$$7.\ \begin{pmatrix} 3 & -5 & 1 & -2 & 0 \\ 6 & -10 & 2 & -4 & -6 \end{pmatrix} \rightarrow \begin{pmatrix} 1 & -\dfrac{5}{3} & \dfrac{1}{3} & -\dfrac{2}{3} & 0 \\ 0 & 0 & 0 & 0 & 1 \end{pmatrix} \text{ no solution}$$

9. $\begin{pmatrix} 4 & 3 & -2 & 14 \\ -2 & 1 & -4 & -2 \\ 2 & 4 & -6 & 12 \end{pmatrix} \rightarrow \begin{pmatrix} 1 & 0 & 1 & 2 \\ 0 & 1 & -2 & 2 \\ 0 & 0 & 0 & 0 \end{pmatrix}$ z is the free variable;

x and y are leading variables; $x = -z + 2$, $y = 2z + 2$
$z = t$; $x = -t + 2$, $y = 2t + 2$

13. There is a unique solution unless $h = -8/3$.

15. $\left(-\dfrac{8}{11}t + \dfrac{43}{11}, -\dfrac{31}{22}t - \dfrac{3}{22}, t \right)$

17. (a.) $49x + 15y + 52z = 116$

   (c.) $41x - 6y - 5z = 80$

19. (a.) $\left(\dfrac{170}{23}, \dfrac{-31}{23}, \dfrac{-55}{23} \right)$

Section 2.5

1. (a.) (i) $x_1 \begin{pmatrix} 2 \\ -4 \\ 1 \end{pmatrix} + x_1 \begin{pmatrix} -5 \\ 0 \\ 6 \end{pmatrix} + x_1 \begin{pmatrix} 1 \\ 1 \\ -4 \end{pmatrix} = \begin{pmatrix} -2 \\ 7 \\ 0 \end{pmatrix}$

   (ii) $\begin{pmatrix} 2 & -5 & 1 \\ -4 & 0 & 1 \\ 1 & 6 & -4 \end{pmatrix} \begin{pmatrix} x_1 \\ x_2 \\ x_3 \end{pmatrix} = \begin{pmatrix} -2 \\ 7 \\ 0 \end{pmatrix}$

   (b.) (i) $x_1 \begin{pmatrix} 1 \\ 0 \end{pmatrix} + x_2 \begin{pmatrix} -1 \\ 0 \end{pmatrix} + x_3 \begin{pmatrix} 1 \\ 1 \end{pmatrix} + x_4 \begin{pmatrix} 1 \\ 0 \end{pmatrix} = \begin{pmatrix} 0 \\ 9 \end{pmatrix}$

   (ii) $\begin{pmatrix} 1 & -1 & 1 & 1 \\ 0 & 0 & 1 & 0 \end{pmatrix} \begin{pmatrix} x_1 \\ x_2 \\ x_3 \\ x_4 \end{pmatrix} = \begin{pmatrix} 0 \\ 9 \end{pmatrix}$

3. (a.) (i) $-3x + 5y + 2z = -5$

   $2w + 4x - 2y + 3z = 7$

   $w + x + 2y + 6z = 0$

   (ii) $w \begin{pmatrix} 0 \\ 2 \\ 1 \end{pmatrix} + x \begin{pmatrix} -3 \\ 4 \\ 1 \end{pmatrix} + y \begin{pmatrix} 5 \\ -2 \\ 2 \end{pmatrix} + z \begin{pmatrix} 2 \\ 3 \\ -6 \end{pmatrix} = \begin{pmatrix} -5 \\ 7 \\ 0 \end{pmatrix}$

5. No solution
9. (a.) yes
   (b.) yes

Section 3.1

1. $\hat{u}\cdot\hat{v}=6, \|\hat{u}\|=\sqrt{13}, \|\hat{v}\|=\sqrt{40}, \|\hat{u}+\hat{v}\|=\sqrt{65}$

3. (a.) $\|\hat{u}\|=\sqrt{17}$

   (b.) $\dfrac{\hat{u}}{\|\hat{u}\|}=\left(\dfrac{1}{\sqrt{17}},0,\dfrac{-4}{\sqrt{17}}\right)$

   (c.) $\left\|\dfrac{\hat{u}}{\|\hat{u}\|}\right\|=1$

5. 12.03

7. $\left(\dfrac{-a}{\sqrt{a^2+b^2}},\dfrac{b}{\sqrt{a^2+b^2}}\right),\left(\dfrac{a}{\sqrt{a^2+b^2}},\dfrac{-b}{\sqrt{a^2+b^2}}\right)$

9. Analytically, this is because the equations

$2x+4y=6$ and $3x+6y=6$ are inconsistent. Geometrically, this is because the vectors $\begin{pmatrix}2\\4\end{pmatrix}$ and $\begin{pmatrix}3\\6\end{pmatrix}$ are parallel and the vector $\begin{pmatrix}6\\6\end{pmatrix}$ is not parallel to them.

13. (b.) $\left(1,x^3\right)=\displaystyle\int_{-1}^{1}1\cdot x^3\,dx=\left.\dfrac{x^4}{4}\right|_{-1}^{1}=\dfrac{1^4}{4}-\dfrac{(-1)^4}{4}=0$

Section 3.2

1. Is a vector space.
3. Is not a vector space. Not closed under addition or scalar multiplication.
5. Not a vector space. Not closed under vector addition or scalar multiplication. Does not have the zero vector.
7. Is a vector space.
9. Not a vector space. $\begin{pmatrix}-2\\-5\end{pmatrix}+\begin{pmatrix}5\\2\end{pmatrix}=\begin{pmatrix}3\\-3\end{pmatrix}$.

   Not closed under addition or scalar multiplication.
11. Is a vector space.

Section 3.3

1. (a.) subspace (b.) subspace (c.) not a subspace.
3. (a.) subspace (b.) subspace (c.) not a subspace.

7. (a.) It is linear combinations of $\begin{pmatrix}1/2\\1\\0\end{pmatrix}$ and $\begin{pmatrix}1/2\\0\\1\end{pmatrix}$

Section 3.4

1. Independent.
3. dependent
5. Independent.
11 (a.) not in span.
   (c.) not in span.
13. (a.) $c = 0$, a and b any values.
   (c.) a any value, $b = 2a$, $c = 3a$.

19. $\left\{ \begin{pmatrix} 1 \\ 3 \\ 6 \end{pmatrix}, \begin{pmatrix} 2 \\ 4 \\ 0 \end{pmatrix} \right\} \begin{pmatrix} 1 \\ 0 \\ 2 \end{pmatrix}$

Section 3.5

1. $\left\{ \begin{pmatrix} 1 \\ -2 \\ 4 \end{pmatrix}, \begin{pmatrix} 11 \\ 14 \\ 8 \end{pmatrix} \right\}$ is a maximal independent set but not a basis.

$\left\{ \begin{pmatrix} 1 \\ -2 \\ 4 \end{pmatrix}, \begin{pmatrix} 11 \\ 14 \\ 8 \end{pmatrix}, \begin{pmatrix} 1 \\ 0 \\ 0 \end{pmatrix} \right\}$ is a basis.

3. $\left\{ \begin{pmatrix} 5 \\ 6 \end{pmatrix} \right\}$ is not a basis.

$\left\{ \begin{pmatrix} 5 \\ 6 \end{pmatrix}, \begin{pmatrix} 1 \\ 0 \end{pmatrix} \right\}$ is a basis.

5. The set given is a basis.

7. $\left\{ \begin{pmatrix} 1 \\ 2 \\ 0 \\ 5 \end{pmatrix}, \begin{pmatrix} 1 \\ 0 \\ 0 \\ 0 \end{pmatrix}, \begin{pmatrix} 0 \\ 1 \\ 0 \\ 0 \end{pmatrix}, \begin{pmatrix} 0 \\ 0 \\ 1 \\ 0 \end{pmatrix} \right\}$

9. $\left\{ \begin{pmatrix} 1 \\ 2 \\ 1 \\ 0 \end{pmatrix} \right\}$ is a basis for $U \cap W$;

$\left\{ \begin{pmatrix} 1 \\ 0 \\ 2 \\ 1 \end{pmatrix}, \begin{pmatrix} 1 \\ 1 \\ 0 \\ 0 \end{pmatrix}, \begin{pmatrix} 6 \\ -1 \\ 2 \\ 3 \end{pmatrix} \right\}$. Is a basis for $U + W$.

11. $h = 38/3$

Section 3.7

1. $[\hat{u}]_B = \begin{pmatrix} 6 \\ -\dfrac{22}{3} \end{pmatrix}$

3. $[\hat{u}]_B = \begin{pmatrix} \dfrac{3}{13} \\ \dfrac{11}{13} \\ \dfrac{6}{13} \end{pmatrix}$

5. $[\hat{u}]_B = \begin{pmatrix} 71/2 \\ -15/2 \\ -11/2 \\ 59/2 \end{pmatrix}$

7. $[\hat{u}]_B = \begin{pmatrix} -\dfrac{12}{5} \\ \dfrac{4}{5} \\ \dfrac{9}{5} \end{pmatrix} = \left(-\dfrac{12}{5}\right)(1+2x) + \left(\dfrac{4}{5}\right)(x-x^2) + \left(\dfrac{9}{5}\right)(3+x^2)$

9. $[\hat{u}]_B = \begin{pmatrix} 5 \\ -4 \\ 2 \\ 1 \end{pmatrix} = 5\begin{pmatrix} 1 & 1 \\ 1 & 1 \end{pmatrix} - 4\begin{pmatrix} 1 & 1 \\ 1 & 0 \end{pmatrix} + 2\begin{pmatrix} 1 & 0 \\ 0 & 1 \end{pmatrix} + \begin{pmatrix} 0 & 2 \\ 1 & 1 \end{pmatrix} = \begin{pmatrix} 3 & 3 \\ 2 & 8 \end{pmatrix}$

13. $P_{B_1}[\hat{x}]_{B_1} = P_{B_2}[\hat{x}]_{B_2} \quad [\hat{v}]_{B_1} = \begin{pmatrix} 2 \\ 5 \end{pmatrix} [\hat{v}]_{B_2} = \begin{pmatrix} 6 \\ -2 \end{pmatrix}$

$P_{B_1} = \begin{pmatrix} 0 & 1 \\ 3 & 2 \end{pmatrix} \quad P_{B_2} = \begin{pmatrix} 4 & 1 \\ -1 & 1 \end{pmatrix}$

$[\hat{v}]_{B_2} = P_{B_2}{}^{-1}P_{B_1}[\hat{v}]_{B_1} = \begin{pmatrix} \dfrac{-11}{5} \\ \dfrac{69}{5} \end{pmatrix}$

$$[\hat{v}]_{B_1} = P_{B_1}^{-1}P_{B_2}[\hat{v}]_{B_2} = \begin{pmatrix} -52 \\ 3 \\ 22 \end{pmatrix}$$

15. $P_{B_1} = \begin{pmatrix} 1 & -3 & 4 \\ 5 & 0 & 1 \\ -2 & 2 & 4 \end{pmatrix}$ $P_{B_2} = \begin{pmatrix} 2 & 1 & 0 \\ 2 & 6 & 1 \\ -2 & 3 & 0 \end{pmatrix}$

$$[\hat{v}]_{B_1} = \begin{pmatrix} 2 \\ -3 \\ 1 \end{pmatrix} \quad [\hat{v}]_{B_2} = \begin{pmatrix} 2 \\ 0 \\ 5 \end{pmatrix}$$

$$[\hat{v}]_{B_2} = P_{B_2}^{-1}P_{B_1}[\hat{v}]_{B_1} = \begin{pmatrix} \dfrac{51}{8} \\[6pt] \dfrac{9}{4} \\[6pt] -\dfrac{61}{4} \end{pmatrix} \quad [\hat{v}]_{B_1} = P_{B_1}^{-1}P_{B_2}[\hat{v}]_{B_2} = \begin{pmatrix} \dfrac{23}{13} \\[6pt] -\dfrac{7}{13} \\[6pt] \dfrac{2}{13} \end{pmatrix}$$

Section 3.8

3. (a.) Row reduced form of the matrix $\begin{pmatrix} 1 & 3 \\ 0 & 0 \end{pmatrix}$ basis for the row space $\{(1 \ \ 3)\}$ basis for the

column space $\left\{ \begin{pmatrix} 1 \\ 2 \end{pmatrix} \right\}$; basis for the null space $\left\{ \begin{pmatrix} -3 \\ 1 \end{pmatrix} \right\}$

(b.) Row reduced form of the matrix $\begin{pmatrix} 1 & 0 & -1/3 \\ 0 & 1 & 2 \end{pmatrix}$ basis for the row space

$\{(1 \ \ 0 \ \ -1/3), (0 \ \ 1 \ \ 2)\}$ basis for the column space $\left\{ \begin{pmatrix} 3 \\ 3 \end{pmatrix}, \begin{pmatrix} 0 \\ 1 \end{pmatrix} \right\}$; basis for the null

space $\left\{ \begin{pmatrix} 1/3 \\ -2 \\ 1 \end{pmatrix} \right\}$

Section 3.9

1. $\left\{ \begin{pmatrix} 0 \\ 0 \\ 1 \\ 0 \end{pmatrix} \begin{pmatrix} 0 \\ 0 \\ 0 \\ 1 \end{pmatrix} \right\}$

3. $\begin{pmatrix} a & c \\ c & b \end{pmatrix} + \begin{pmatrix} d & e \\ e & f \end{pmatrix} = \begin{pmatrix} a+d & c+e \\ c+e & b+f \end{pmatrix}$

so the sum of symmetric matrices is symmetric

$$\alpha \begin{pmatrix} a & c \\ c & b \end{pmatrix} = \begin{pmatrix} \alpha a & \alpha c \\ \alpha c & \alpha b \end{pmatrix}$$

so a scalar multiple of a symmetric matrix is symmetric.

Thus, S is a subspace of $M_{2 \times 2}(\mathbb{R})$. Likewise, T is a subspace of $M_{2 \times 2}(\mathbb{R})$.

$\left\{ \begin{pmatrix} 1 & 0 \\ 0 & 0 \end{pmatrix}, \begin{pmatrix} 0 & 0 \\ 0 & 1 \end{pmatrix}, \begin{pmatrix} 0 & 1 \\ 1 & 0 \end{pmatrix} \right\}$ is a basis for S so the dimension of S is 3 and $\left\{ \begin{pmatrix} 0 & 1 \\ -1 & 0 \end{pmatrix} \right\}$ is a basis

for T so the dimension of T is 1.

The problem can be finished by showing $S \cap T = \emptyset$.

Section 4.1

1. (a.) linear.
   (b.) not linear.
   (c.) not linear.
3. (a.) not linear.
   (b.) linear.

Section 4.2

1. Matrix of $T = \begin{pmatrix} 4 & -3 \\ 1 & 0 \\ 0 & 5 \end{pmatrix}$;

3. (a.) Matrix of $T = (3)$.

5. $T \begin{pmatrix} x \\ y \end{pmatrix} = \begin{pmatrix} x - 3y \\ 7x \\ 5x + 9y \end{pmatrix}$

7. $T(x) = (2x)$

9. (a.) $T \begin{pmatrix} 1 \\ 0 \end{pmatrix} = \begin{pmatrix} -3/4 \\ -1/2 \\ -1 \end{pmatrix}, T \begin{pmatrix} 0 \\ 1 \end{pmatrix} = \begin{pmatrix} -5/4 \\ -5/2 \\ -1 \end{pmatrix}$

(b.) $T \begin{pmatrix} 5 \\ -6 \end{pmatrix} = \begin{pmatrix} 15/4 \\ 25/2 \\ 1 \end{pmatrix}$

(c.) $T\begin{pmatrix} x \\ y \end{pmatrix} = \begin{pmatrix} \dfrac{-3}{4}x - \dfrac{5}{4}y \\ \dfrac{-1}{2}x - \dfrac{5}{2}y \\ -x - y \end{pmatrix}$

11. $m = 2/3$

15. (a.) $\begin{pmatrix} 0 & 1 & 0 & 0 & \cdots & 0 \\ 0 & 0 & 2 & 0 & & 0 \\ \vdots & 0 & 0 & 3 & & \vdots \\ & \vdots & 0 & 0 & & \\ & \vdots & \vdots & & 0 \\ 0 & 0 & 0 & 0 & & k-1 \end{pmatrix}$ (b). $\{1\}$ (c). range $= \mathbb{R}^3$

17. (a). $\begin{pmatrix} 0 & 1 & 0 \\ 1 & 0 & 2 \\ 1 & 1 & 0 \end{pmatrix}$ (b). $\{1\}$ (c). $\left\{1, x, x^2, \ldots, x^{n-1}\right\}$

19. (c.) $\left(\dfrac{7}{53}\right)e^{3x}\sin x - \left(\dfrac{2}{53}\right)e^{3x}\cos x$

Section 4.3

1. (a.)   Matrix of $T = \begin{pmatrix} \dfrac{76}{7} & \dfrac{240}{7} \\ -\dfrac{18}{7} & -\dfrac{62}{7} \end{pmatrix}$.

(b.)   null space of $T = \hat{0}$

(c.)   Basis for the range of $T = \left\{ \begin{pmatrix} \dfrac{76}{7} \\ -\dfrac{18}{7} \end{pmatrix}, \begin{pmatrix} \dfrac{240}{7} \\ -\dfrac{62}{7} \end{pmatrix} \right\}$

2. $P_C = \begin{pmatrix} 1 & 0 & 0 \\ 1 & 1 & 1 \\ 0 & 0 & 1 \end{pmatrix}$ $P_B = \begin{pmatrix} 1 & -1 & 3 \\ 0 & 1 & 2 \\ 0 & 2 & 0 \end{pmatrix}$ $T = \begin{pmatrix} 1 & 0 & 4 \\ 4 & 0 & 0 \\ 1 & 1 & 1 \end{pmatrix}$

$P_C^{-1}TP_B = \begin{pmatrix} 1 & 7 & 3 \\ 2 & -13 & 4 \\ 1 & 2 & 5 \end{pmatrix}$

4. $P_C = \begin{pmatrix} 1 & 0 & 0 \\ 1 & 1 & 1 \\ 0 & 0 & 1 \end{pmatrix}$ $P_B = \begin{pmatrix} 1 & -1 & 1 \\ 0 & 1 & 2 \\ 1 & 0 & 0 \end{pmatrix}$ $T = \begin{pmatrix} 3 & 0 & -4 \\ 4 & 0 & 0 \\ 1 & 1 & 1 \end{pmatrix}$

$P_C^{-1}TP_B = \begin{pmatrix} -1 & -3 & 3 \\ 3 & -1 & -2 \\ 2 & 0 & 3 \end{pmatrix}$

Section 4.4

5. (a.) $T\begin{pmatrix} 1 \\ 0 \end{pmatrix} = \begin{pmatrix} 1 \\ 2 \end{pmatrix}, T\begin{pmatrix} 0 \\ 1 \end{pmatrix} = \begin{pmatrix} -1 \\ 3 \end{pmatrix}, S\left(T\begin{pmatrix} 1 \\ 0 \end{pmatrix}\right) = \begin{pmatrix} 4 \\ 6 \end{pmatrix},$

$$S\left(T\begin{pmatrix} 0 \\ 1 \end{pmatrix}\right) = \begin{pmatrix} 6 \\ 19 \end{pmatrix}$$

Section 5.1

1. Characteristic polynomial $(\lambda - 3)(\lambda - 2)^2$ eigenvalues $\lambda = 2,3$ eigenvectors:
$\lambda = 2, v = \begin{pmatrix} 0 \\ 1 \\ 0 \end{pmatrix}$; $\lambda = 3, v = \begin{pmatrix} -1 \\ 1 \\ 1 \end{pmatrix}$

3. Characteristic polynomial $(\lambda - 2)^3$ eigenvalues $\lambda = 2$ eigenvectors: $\lambda = 2, v = \begin{pmatrix} 1 \\ 0 \\ 0 \end{pmatrix}$

5. Characteristic polynomial $(\lambda - 1)(\lambda - 5)(\lambda + 5)$ eigenvalues $\lambda = 1,5,-5$ eigenvectors:

$\lambda = 1, v = \begin{pmatrix} \frac{1}{7} \\ \frac{4}{7} \\ 1 \end{pmatrix}$; $\lambda = 5, v = \begin{pmatrix} -1 \\ 0 \\ 1 \end{pmatrix}$; $\lambda = -5, v = \begin{pmatrix} 1 \\ -2 \\ 1 \end{pmatrix}$

7. Characteristic polynomial $(\lambda)(\lambda - 2)(\lambda - 3)$ eigenvalues $\lambda = 0,2,3$ eigenvectors:

$$\lambda = 0, v = \begin{pmatrix} 2 \\ -1 \\ 0 \end{pmatrix}; \lambda = 2, v = \begin{pmatrix} 0 \\ 1 \\ 0 \end{pmatrix}, \lambda = 3, v = \begin{pmatrix} -2 \\ 1 \\ 3 \end{pmatrix}$$

9. Characteristic polynomial $(\lambda + 3)(\lambda - 4)(\lambda - 3)^2$ eigenvalues $\lambda = -3,4,3$ eigenvectors:

$$\lambda = 3, v = \begin{pmatrix} 0 \\ 0 \\ 1 \\ 0 \end{pmatrix}; \lambda = 4, v = \begin{pmatrix} 6 \\ 1 \\ 0 \\ 0 \end{pmatrix}; \lambda = -3, v = \begin{pmatrix} -1 \\ 1 \\ 0 \\ 0 \end{pmatrix}$$

13. (a.) No, (b.) Yes.

Section 5.2

1. Cannot be diagonalized.
3. Cannot be diagonalized.

5. $P = \begin{pmatrix} \dfrac{1}{7} & -1 & 1 \\ \dfrac{4}{7} & 0 & -2 \\ 1 & 11 & 1 \end{pmatrix}$ $D = \begin{pmatrix} 1 & 0 & 0 \\ 0 & 5 & 0 \\ 0 & 0 & -5 \end{pmatrix}$ is one correct answer.

7. $P = \begin{pmatrix} -2 & -1 & 0 \\ 1 & 0 & 1 \\ 1 & 1 & 0 \end{pmatrix}$ $D = \begin{pmatrix} 1 & 0 & 0 \\ 0 & 2 & 0 \\ 0 & 0 & 2 \end{pmatrix}$ is one correct answer.

9. Cannot be diagonalized.

Section 5.3

$$P_\mathcal{B} = \begin{pmatrix} 2 & 1 \\ 1 & 1 \end{pmatrix}, \; T = \begin{pmatrix} 2 & -1 \\ 5 & 3 \end{pmatrix}, \; [T]_\mathcal{B} = P_\mathcal{B}^{-1} T P_\mathcal{B} = \begin{pmatrix} -10 & -7 \\ 23 & 15 \end{pmatrix}$$

3. $P_\mathcal{B} = \begin{pmatrix} 2 & 0 & 3 \\ 0 & 4 & 3 \\ 1 & -1 & 0 \end{pmatrix}, \; T = \begin{pmatrix} 1 & 0 & 1 \\ 1 & 2 & 1 \\ 2 & 3 & 0 \end{pmatrix}, \; [T]_\mathcal{B} = \begin{pmatrix} 8 & 28 & 33 \\ 4 & 16 & 18 \\ -13/3 & -19 & -21 \end{pmatrix}$

Section 5.4

1. Eigenvalues are 1,5, –5 with eigenvectors $\begin{pmatrix} 1 \\ 4 \\ 7 \end{pmatrix}, \begin{pmatrix} -1 \\ 0 \\ 1 \end{pmatrix}, \begin{pmatrix} 1 \\ -2 \\ 1 \end{pmatrix}$ respectively.

$$x_1(t) = c_1 e^t - c_2 e^{5t} + c_3 e^{-5t}$$

$$x_2(t) = 4c_1 e^t - 2c_3 e^{-5t}$$

$$x_3(t) = 7c_1 e^t + c_2 e^{5t} + c_3 e^{-5t} \quad c_1 = \frac{1}{4}, c_2 = -\frac{5}{4}, c_3 = \frac{1}{2}$$

3. $a = -2$

Section 6.2

1. (b.) $f^2 = \int_0^1 (x^2 + x)^2 \, dx = \dfrac{31}{30}$

(c.) $f - g^2 = \int_0^1 \left[(x^2 + x) - (x-1) \right]^2 \, dx = \dfrac{28}{15}$

3. $\begin{pmatrix} x_1 \\ y_1 \end{pmatrix} + \begin{pmatrix} x_2 \\ y_2 \end{pmatrix}, \begin{pmatrix} x_3 \\ y_3 \end{pmatrix} = \begin{pmatrix} x_1 + x_2 \\ y_1 + y_2 \end{pmatrix}, \begin{pmatrix} x_3 \\ y_3 \end{pmatrix} = (x_1 + x_2)^2 x_3 + (y_1 + y_2) y_3$

$\begin{pmatrix} x_1 \\ y_1 \end{pmatrix}, \begin{pmatrix} x_3 \\ y_3 \end{pmatrix} + \begin{pmatrix} x_2 \\ y_2 \end{pmatrix}, \begin{pmatrix} x_3 \\ y_3 \end{pmatrix} = (x_1)^2 x_3 + (y_1 + y_2) y_3 + (x_2)^2 x_3 + (y_1 + y_2) y_3$

Section 6.3

1. (a.) orthogonal $\begin{pmatrix} 7 \\ -3 \end{pmatrix} = \dfrac{19}{17} \begin{pmatrix} 1 \\ -4 \end{pmatrix} + \dfrac{25}{17} \begin{pmatrix} 4 \\ 1 \end{pmatrix}$

(c.) orthonormal $\begin{pmatrix} 0 \\ 3 \end{pmatrix} = \dfrac{-6}{\sqrt{5}} \begin{pmatrix} \dfrac{1}{\sqrt{5}} \\ -\dfrac{2}{\sqrt{5}} \end{pmatrix} - \dfrac{3}{\sqrt{5}} \begin{pmatrix} -\dfrac{2}{\sqrt{5}} \\ -\dfrac{1}{\sqrt{5}} \end{pmatrix}$

(e.) neither orthogonal nor orthogonal.

Section 6.4

1. The sets of vectors in these answers are orthogonal but not orthonormal.

(a.) $\hat{v}_1 = \begin{pmatrix} 1 \\ 0 \end{pmatrix}, \hat{v}_2 = \begin{pmatrix} 0 \\ 7 \end{pmatrix}$

(c.) $\hat{v}_1 = \begin{pmatrix} 2 \\ 7 \end{pmatrix}, \hat{v}_2 = \begin{pmatrix} -119/53 \\ 34/53 \end{pmatrix}$

(e.) $\hat{v}_1 = \begin{pmatrix} -2 \\ 5 \\ -5 \end{pmatrix}, \hat{v}_2 = \begin{pmatrix} 35/27 \\ 61/27 \\ 47/27 \end{pmatrix}, \hat{v}_3 = \begin{pmatrix} -90/53 \\ 27/106 \\ 99/106 \end{pmatrix}$

(g.) $\hat{v}_1 = \begin{pmatrix} 1 \\ 1 \\ 1 \end{pmatrix}, \hat{v}_2 = \begin{pmatrix} 7/3 \\ 13/3 \\ -20/3 \end{pmatrix}, \hat{v}_3 = \begin{pmatrix} 33/103 \\ -27/103 \\ -6/103 \end{pmatrix}$

(i.) $\hat{v}_1 = \begin{pmatrix} 2 \\ 3 \\ 0 \\ 0 \end{pmatrix}, \hat{v}_2 = \begin{pmatrix} 33/13 \\ -22/13 \\ 6 \\ 3 \end{pmatrix}, \hat{v}_3 = \begin{pmatrix} 1359/706 \\ -453/353 \\ -544/353 \\ 515/706 \end{pmatrix}, \hat{v}_4 = \begin{pmatrix} 729/5831 \\ -486/5831 \\ 27/343 \\ -1809/5831 \end{pmatrix}$

3. An orthogonal set of vectors is

$$\hat{v}_1 = 2 + x, \hat{v}_2 = \frac{13x^2 - 8x - 3}{13}, \hat{v}_3 = \frac{5x^2 + 2x - 2}{13}$$

The norms are

$$\hat{v}_1^2 = \frac{26}{3}, \hat{v}_2^2 = \frac{88}{195}, \hat{v}_3^2 = \frac{22}{507}$$

so an orthonormal set is

$$\sqrt{\frac{3}{26}}(2 + x), \sqrt{\frac{195}{88}}\left(\frac{13x^2 - 8x - 3}{13}\right), \sqrt{\frac{507}{22}}\left(\frac{5x^2 + 2x - 2}{13}\right)$$

5. An orthogonal set is $\hat{v}_1 = \begin{pmatrix} -1 \\ 3 \end{pmatrix}, \hat{v}_2 = \begin{pmatrix} 9/2 \\ 1/2 \end{pmatrix}$

$$\hat{v}_1^2 = 28 \ \hat{v}_2^2 = 90/4$$

Section 6.5

$$\begin{pmatrix} -2\sqrt{2} & 4\sqrt{2} \\ 0 & 6\sqrt{2} \end{pmatrix}$$

Section 6.6

1. (a.) $W^\perp = \text{span}\left\{ \begin{pmatrix} 4 \\ -\dfrac{9}{2} \\ 1 \end{pmatrix} \right\}$

(b.) $W^{\perp} = \text{span} \left\{ \begin{pmatrix} -\dfrac{1}{2} \\ 0 \\ 1 \end{pmatrix} \right\}$

(c.) $W^{\perp} = \text{span} \left\{ \begin{pmatrix} -2/3 \\ 0 \\ 0 \\ 1 \end{pmatrix}, \begin{pmatrix} 1 \\ -4 \\ 1 \\ 0 \end{pmatrix} \right\}$

Section 6.7

$\begin{pmatrix} \dfrac{8}{41} \\ \dfrac{-10}{41} \end{pmatrix}$

3. Orthonormal basis of $W = \left\{ \begin{pmatrix} \dfrac{1}{\sqrt{3}} \\ -\dfrac{1}{\sqrt{3}} \\ \dfrac{1}{\sqrt{3}} \end{pmatrix}, \begin{pmatrix} \dfrac{1}{\sqrt{6}} \\ \dfrac{2}{\sqrt{6}} \\ \dfrac{1}{\sqrt{6}} \end{pmatrix} \right\}$ $\text{proj}_W \hat{v} = \begin{pmatrix} \dfrac{7}{2} \\ 2 \\ 2 \\ \dfrac{7}{2} \end{pmatrix}$

5. Orthonormal basis of $W = \left\{ \begin{pmatrix} \dfrac{1}{2} \\ \dfrac{1}{2} \\ \dfrac{1}{2} \\ \dfrac{1}{2} \end{pmatrix}, \begin{pmatrix} \dfrac{2}{\sqrt{18}} \\ -\dfrac{3}{\sqrt{18}} \\ \dfrac{2}{\sqrt{18}} \\ -\dfrac{1}{\sqrt{18}} \end{pmatrix} \right\}$ $\text{proj}_W \hat{v} = \begin{pmatrix} \dfrac{43}{36} \\ -\dfrac{29}{12} \\ \dfrac{43}{36} \\ -\dfrac{35}{36} \end{pmatrix}$

7. Orthonormal basis of $W = \left\{ \begin{pmatrix} \dfrac{-1}{2} \\ \dfrac{1}{2} \\ \dfrac{-1}{2} \\ \dfrac{1}{2} \end{pmatrix}, \begin{pmatrix} \dfrac{1}{\sqrt{12}} \\ \dfrac{3}{\sqrt{12}} \\ \dfrac{1}{\sqrt{12}} \\ -\dfrac{1}{\sqrt{12}} \end{pmatrix} \right\}$ $\text{proj}_W \hat{v} = \begin{pmatrix} 0 \\ 4 \\ 0 \\ 0 \end{pmatrix}$

Section 7.1

1. (a.) $\hat{w} = \begin{pmatrix} 4 \\ -2 \end{pmatrix}$

(c.) $\hat{w} = \begin{pmatrix} 1 \\ 0 \\ 0 \end{pmatrix}$

(e.) $\hat{w} = \begin{pmatrix} i \\ 3+2i \\ 5-4i \end{pmatrix}$

3. (a.) $\hat{w} = \begin{pmatrix} 5 \\ 3 \\ -4 \end{pmatrix}$

(b.) $P_B = \begin{pmatrix} 1 & 1 & 2 \\ 1 & 1 & 0 \\ 1 & 0 & 0 \end{pmatrix}$ $P_B[\hat{w}]_B = [\hat{w}]; \ [\hat{w}]_B = P_B^{-1}[\hat{w}] = \begin{pmatrix} -4 \\ 7 \\ 1 \end{pmatrix}$

Section 7.2

1. (a.) matrix of $T = \begin{pmatrix} 1 & 2 & -3 \\ 3 & 0 & 5 \end{pmatrix}$; $T^*(w) = \begin{pmatrix} a+3b \\ 2a \\ -3-5b \end{pmatrix}$;

matrix of $T^* = \begin{pmatrix} 1 & 3 \\ 2 & 0 \\ -3 & 5 \end{pmatrix}$

(c.) matrix of $T = \begin{pmatrix} 4 \\ -3 \end{pmatrix}$; $T^*(w) = (4a-3b)$;

matrix of $T^* = \begin{pmatrix} 4 & -3 \end{pmatrix}$

(e.) matrix of $T = \begin{pmatrix} 3 & 4+5i \\ 6i & -2i \\ 6-2i & 1-i \end{pmatrix}$

$T^*(w) = \begin{pmatrix} 3a - 6ib + (6+2i)c \\ (4-5i)a + 2ib + (1+i)c \end{pmatrix}$

matrix of $T^* = \begin{pmatrix} 3 & -6i & 6+2i \\ 4-5i & 2i & 1+i \end{pmatrix}$

Section 7.3

$$\hat{v}_1 = \begin{pmatrix} \dfrac{1}{\sqrt{2}} \\ -\dfrac{1}{\sqrt{2}} \end{pmatrix} \quad \hat{v}_2 = \begin{pmatrix} \dfrac{1}{\sqrt{2}} \\ \dfrac{1}{\sqrt{2}} \end{pmatrix} \quad D = \begin{pmatrix} 1 & 0 \\ 0 & 3 \end{pmatrix}$$

3. $\hat{v}_1 = \begin{pmatrix} 1/\sqrt{2} \\ -1/\sqrt{2} \end{pmatrix} \quad \hat{v}_2 = \begin{pmatrix} 1/\sqrt{2} \\ 1/\sqrt{2} \end{pmatrix} \quad D = \begin{pmatrix} 1 & 0 \\ 0 & 5 \end{pmatrix}$

5. $\hat{v}_1 = \begin{pmatrix} 1 \\ -1 \\ 0 \end{pmatrix} \quad \hat{v}_2 = \begin{pmatrix} 1 \\ 1 \\ 1 \end{pmatrix} \quad \hat{v}_3 = \begin{pmatrix} 1 \\ 1 \\ -2 \end{pmatrix} \quad D = \begin{pmatrix} 2 & 0 & 0 \\ 0 & 3 & 0 \\ 0 & 0 & 6 \end{pmatrix}$

Index

Printed in the United States
By Bookmasters